# NEW DEVELOPMENTS IN LASERS AND ELECTRO-OPTICS RESEARCH

# NEW DEVELOPMENTS IN LASERS AND ELECTRO-OPTICS RESEARCH

WILLIAM T. ARKIN
EDITOR

**Nova Science Publishers, Inc.**
*New York*

For permission to use material from this book please contact us:
Telephone 631-231-7269; Fax 631-231-8175
Web Site: http://www.novapublishers.com



LIBRARY OF CONGRESS CATALOGING-IN-PUBLICATION DATA
New developments in lasers and electro-optics research / William T. Arkin (editor).
p. cm.
Includes index.
ISBN 1-59454-771-8
1. Electrooptics. 2. Lasers. I. Arkin, William T.
TA1750.N49 2007
621.36'6--dc22 2005027405

*Published by Nova Science Publishers, Inc. ✢ New York*

# CONTENTS

# PREFACE

It is expected that ongoing advances in optics will revolutionize the 21st century as they began doing in the last quarter of the 20th. Such fields as communications, materials science, computing and medicine are leaping forward based on developments in optics. This new series presents leading edge research on optics and lasers from researchers spanning the globe.

For many years, the development of effective ablation techniques for the fabrication of micro-optical components has been the key factor in the birth of new photonic devices and systems. Excimer lasers have recently shown their great potential in micromachining transparent materials for optoelectronic applications. In chapter 1, the ablation characteristics of two of the most important transparent materials, polymers and glasses, are studied. Simple shaped microcavities are first machined for studying the fundamental ablation parameters, including threshold fluence, effective absorption coefficient, and ablation (material removal) rate as well as the associated morphologies of ablated surfaces. In the ablation experiments on polymers, five standard grades and nine commercially proprietary polymeric compounds are selected. Ablation techniques using these polymers for making arrayed ferrules and curved microlenses are presented. Applications of these ablated microstructures for optical fiber connectors, optical fiber coupling and alignment, and transparent chip encapsulants, are introduced and examined with emphasis on the quality of the ablated areas to satisfy the required performance. In glass ablation, several glasses are considered and their associated ablation behaviors are studied. The procedures to ablate different types of glass-based arrayed microstructures with flat and curved surfaces are described. The utilizations of these arrayed microstructures for optical waveguide, wave absorber, and beam guider are specifically discussed. Finally, recommendations for future efforts and trends are presented.

The UV $Cu^+$ laser excited in a nanosecond pulsed longitudinal Ne-CuBr discharge with a resource above 700 hours is developed and investigated in chapter 2. A comparative study on the temporal behavior of the average output power of the laser with inserts made of a quartz and a ceramic tube is reported. Spectral investigations of the discharge are also made. The use of a ceramic tube results in an absence of the tube metalization, and allows to maintain a constant gas mixture composition. The influence of different additives – air, synthetic air, $N_2$, $O_2$, $CO_2$, He, Ar, Xe, and $H_2$, to the neon buffer gas on the average output power is investigated. It is obtained that addition of a small amount of hydrogen 0.02-0.04 Torr leads to an increase of more than twice of the average output power at a simultaneous laser oscillation of the five $Cu^+$ lines - 248.6, 252.9, 259.7, 260.0, and 270.3 nm. The use of the IC excitation scheme leads to a 30 % increase of the average output power in compare with the use of the Blumlein scheme. The laser is operated at a periodic refreshment of the buffer gas mixture - 16.7 Torr Ne and 0.03 Torr $H_2$. In this regime for a 700-hour laser action the average output power is maintained about 600 mW. That make this laser usable for a number of scientific applications in a laboratory conditions. The influence of UV 248.6 nm laser treatment on the characteristics of the conducting polymer PEDOT is investigated. The refracting index of liquid stilbene films is measure in this UV region. It is shown the possibility to make micro holes in different materials which are transparences for visible and IR light.

Chapter 3 reviews works concerning improvement of reconstructed images from computer- generated holograms for 2D, 3D Fourier-transform and Fresnel types of holograms. The authors show experiments of reconstruction of 2 level phase quantized 2D Fourier transformed holograms produced by using iterative method, error diffusion method and a combination method of the iterative and the error diffusion methods. In particular, use of a combination method was effective for image improvement of 2D objects, but in 3D objects, single use of the iterative method was more effective in improving the reconstructed images. Since 3D computer-generated hologram that is created from the iterative operations requires much computing time, the 3D hologram using 1D Fourier transform operations was proposed and performed. As a result, similar reconstructed images as 3D holograms were obtained from the holograms made with 1D Fourier transform operations. In the case of the off-axis Fresnel hologram, to improve the reconstructed images, holograms with high construct diffraction patterns were formed by using histogram method. From reconstruction experiments of the holograms, clear images were obtained and 3D images were seen directly with eyes.

Three-Dimensional Imaging and Modeling (3DIM) has been an active and interdisciplinary research field, which covers a wide range of spectrum from range image acquisition, multi-view range images registration, multi-view range images integration, geometric representation, and manipulation of 3D image. The technology of 3DIM has been widely used for industry, medicine, culture, education and so on. In chapter 4, the authors focus on the technology of multi-resolution 3DIM. The state-of-the art of techniques for multi-resolution 3D imaging will be introduced as updated methods of range imaging acquisition, which involves 3D imaging based on the strategies of a point-array encoding (incoherent illumination). In addition, a new method for generating variable frequency sinusoidal fringe patterns is presented. This method is based on acoustic-optics technique for 3D imaging (coherent illumination). The decoding and the calibration methods for 3D imaging systems with regard to different encoding schemes are described. For illustration, two typical surfaces, the free form and a step-like surface, are analyzed to demonstrate the approaches. It will be shown that the quality of reconstructed surface can be gradually improved as level-of-details, resulting in a multi-resolution acquisition. The methods of registration and integration of partial range images are also introduced in this Chapter. They will discuss some approaches for high-level range data processing, including multiple view range images registration, multiple view range images integration, and efficient geometric description of 3D graphical like subdivision. The manipulation of 3D image are also presented, which involves data reduction of 3D image and representation of 3D image with different level-of-details.

Digital holography is a quickly developing technology for high speed imaging. In-line holography has been extensively investigated due to it can effectively utilize the space-bandwidth product of digital recording instruments. However, since the reference wave and the object wave are overlapped during the hologram is recorded; the directly inverse Fresnel transform can not present the original object well. The reconstructed image is blurred by its conjugated image and the zero-order diffraction of the hologram. This shortcoming limits the application of the in-line holography. Therefore, an effective reconstruction algorithm is important for generalizing its application. In chapter 5, some approaches based on the phase retrieval algorithm for in-line hologram reconstruction are reported. Firstly, the Yang-Gu algorithm and the GS algorithm are used to restructure pure absorption and pure phase object from their in-line holograms, respectively. The differences between these two algorithms are analyzed. Then the GS algorithm is extended to reconstruct whole optical field from double or multiple holograms. At last, a new approach for reconstructing object from a hologram series is presented. Experimental results show that all these methods can reconstruct original object well.

The self-interference of a large number of random coherent waves scattered from a diffuse object surface or propagated through a medium of random refractive index fluctuations results in a granular structure known as speckle pattern. Chapter 6 illustrates a concise development on various speckle techniques with latest developments and refinements in terms of optical systems, fast and reliable image processing systems and modern fringe analysis methods for optical metrology. Particular emphasis is provided for the measurement of mechanical behavior of MEMS structures and non-destructive evaluation.

Due to the rapid technical advances of semiconductor laser diodes as a light source, the ultra-high-density optical information storage systems have been accelerated to commercialize in various formats, such as CD (Compact Disk), CD-ROM (Read-Only Memory), CD-R (-Recordable), CD-RW (-ReWriteable), DVD (Digital Versatile/Video Disc), DVD-R, and DVD-RW, for many applications, i.e., audio, music, image, movie, and computer data storage. In this complex situation, the compatibility of diverse storage media discs in a playback/writing system is one of the most important issues for system development. This problem has caused complications in the optical components as well as the non-compact system size. A semiconductor laser is an integral part among many optical components because the wavelength and output power of laser light source is a key factor of the media disc specifications. In order to realize signal compact optical pick-up module that can respond to various media discs, a multiple wavelength laser light source is indispensable. There have been many attempts to develop multi-wavelength laser diodes. The typical approaches are to use two individually packaged laser diodes or hybrid-integrated two laser chips in the same package. However, those techniques require sophisticated assembly processing resulting in low reliability and high cost. In chapter 7, the authors reviewed the recent progress of multi-wavelength laser diodes and how it can effectively realize dual or multiple wavelength laser diodes in the future. They proposed, and successfully demonstrated, new concept device structures for monolithically-integrated two laser diodes with different wavelengths on a GaAs wafer in parallel and vertical direction.

Chapter 8 presents a review of our computer algorithms developed for the shape detection technique. The study in this chapter involves the laser line projection technique, image processing and approximation neural networks. To extract topographic information of the shape of an object, it is scanned by means of a laser line. From the scanning procedure, a set of images is captured by a CCD camera. By processing these images, the object shape is recovered. The shape information is extracted from an image by detecting the behaviour of the position of the laser line in the image plane. To determine the mathematical model of the relationship between the laser line behaviour and the object shape, approximation neural networks are applied. Approximation networks such as Radial Basis Function neural networks, General Regression neural networks and Bezier neural networks are used to create a model of the behaviour of the laser line. The architecture of these approximation networks is built using data of images of a laser line projected on objects, whose dimensions are known. The data correspond to the object dimensions of the region where the laser line is projected. In this manner, the approximation network is constructed. The approach of the approximation networks in this technique is the calibration procedure of shape detection by computer algorithms. In this form, the accuracy of the shape measurements is improved. It is because errors of parameters measured on the set-up are not introduced to the system. The results of this examination are presented by computer simulation and in experimental form.

In: New Developments in Lasers and Electrc-Optics Research ISBN 1-59454-771-8
Editor: William T. Arkin, pp. 1-40 

*Chapter 1*

# USING EXCIMER LASERS FOR MICROABLATION OF TRANSPARENT MATERIALS FOR OPTOELECTRONIC APPLICATIONS

***Ampere A. Tseng****

Department of Mechanical and Aerospace Engineering
Arizona State University, Tempe, Arizona 85287-6106 USA
E-mail: ampere.tseng@asu.edu

***Kris Naessens and Roel Baets***

Department of Information Technology
Ghent University, Gent 9000 Belgium

## ABSTRACT

For many years, the development of effective ablation techniques for the fabrication of micro-optical components has been the key factor in the birth of new photonic devices and systems. Excimer lasers have recently shown their great potential in micromachining transparent materials for optoelectronic applications. In this article, the ablation characteristics of two of the most important transparent materials, polymers and glasses, are studied. Simple shaped microcavities are first machined for studying the fundamental ablation parameters, including threshold fluence, effective absorption coefficient, and ablation (material removal) rate as well as the associated morphologies of ablated surfaces. In the ablation experiments on polymers, five standard grades and nine commercially proprietary polymeric compounds are selected. Ablation techniques using these polymers for making arrayed ferrules and curved microlenses are presented. Applications of these ablated microstructures for optical fiber connectors, optical fiber coupling and alignment, and transparent chip encapsulants, are introduced and examined with emphasis on the quality of the ablated areas to satisfy the required performance. In glass ablation, several glasses are considered and their associated ablation behaviors are studied. The procedures to ablate different types of glass-based arrayed

* Corresponding author: Department of Mechanical and Aerospace Engineering, Arizona State University, 1711 South Rural Road, Tempe, Arizona 85287-6106 USA; E-mail: ampere.tseng@asu.edu; Fax: (480) 965-1384.

microstructures with flat and curved surfaces are described. The utilizations of these arrayed microstructures for optical waveguide, wave absorber, and beam guider are specifically discussed. Finally, recommendations for future efforts and trends are presented.

**Keywords:** ablation, contour scanning, excimer laser, glass, microlens, micromachining, optical fiber connector, optical waveguides, and polymer.

## INTRODUCTION

Currently there is a strong demand for a variety of optical or photonic components and devices by a variety of industries, including telecommunication, medicine, entertainment, instrumentation, optical data memory/storage, and remote sensing and imaging. Frequently, these types of optical components are made from different types of transparent materials. Polymeric or glass-based materials are often selected because of their high availability, low cost, and exceptional optical properties. Most transparent materials, especially the glasses, are very difficult to be machined, by conventional cutting or grinding processes, to a feature size at micrometer levels. In this chapter, a non-conventional method using excimer lasers for micromachining optical materials is presented.

Excimers are gas-based lasers that offer the most efficient access to the UV (ultraviolet) or deep UV region with short pulse rates and durations, leading to high pulse intensity. The short wavelengths allow for high resolution and high absorption in materials, including polymers and glasses. While the ultimate resolution of an optical system is limited by diffraction effects, which is proportional to the wavelength of the radiation involved, the high absorption with the short pulse (~10 ns) allows a large amount of energy to be absorbed in a vey thin surface layer at very short time for the efficient removal of material from a target area. As a result, excimer lasers are capable of machining material from the target directly and do not require photoresists for pattern transfer, and clean-room facilities as demanded in conventional semiconductor processes. Excimer lasers are typically used in machining materials (such as glasses), which are hard to machine with other types of lasers, or where very high precision is required (Sugioka et al., 2003; Chen et al., 2005a). They are also very effective for ablating polymeric and biological materials where a clean cut is required without causing thermal damage to the surrounding materials. Recently, excimer lasers have been used for a broad range of applications, including ink jet nozzle drilling, via-drilling in printed circuit boards, removal of short cuts in electronic circuitry, wire stripping, fabrication of microfluidics, waveguides, microlenses, microgears, and assembly alignment (Gower, 2001; Dyer, 2003; Chen et al, 2005a).

The mechanism for machining or material removal using excimer lasers is usually referred to as ablation, in which the material explodes into vapor and is ejected in the form of species such as atoms, molecules, ions, and clusters due to the interaction of an intense laser pulse with the material. The process is often envisaged as a sequence of steps, initiated by the laser radiation interacting with the solid target, absorption of energy and localised heating of the surface, and subsequent material evaporation. The macroscopic effects of ablation include plasma, acoustic shocks, and cratering of the surface. To ablate a high-precision and high-quality structure at micrometer scales, the interaction between the laser and material has to be well understood and controlled. As a result, in this chapter, the responses of both optical polymers and glasses during excimer laser ablation are examined with a goal of making high precision and high quantity microstructures for optoelectronic applications.

Following the Introduction, the characteristics of several major excimer lasers are assessed and the basic components of an excimer laser system for ablation of optical materials are presented. By examining the ablation behavior of different polymeric and glass materials, the major material properties or parameters for governing the ablation process are identified for guiding the subsequent ablation process in machining microstructures, which can act as components, devices, or systems for different optical applications. By conducting different ablation experiments, the specific values of these governing parameters, including the threshold fluence and absorption coefficient, are obtained for a range of polymers and glasses. These values are compared with each other as well as with other published data and their differences are discussed. The ablated surface morphology is also examined and the rationale for causing the undesirable surface roughness and debris are discussed for develping a more accurate control of the ablation process to achieve the required dimensional accuracy and surface quality.

In addition to standard common polymers, several commercially proprietary polymers are selected for ablating various optical microstructures for different applications. One particular example is to illustrate the capabilities of excimer laser ablation for fast prototyping of microlenses with the contour scanning ablation technique. In glass materials, the technique to ablate different types of arrayed microstructures with flat and curved surfaces are described. The illustrations of these arrayed microstructures for optical waveguide, wave absorber, and beam guider, are specifically discussed. Finally, recommendations for future efforts and trends are included. It is to be noted that the present article is focused on direct ablation of optical microstructures. Other manufacturing approaches by excimer lasers, such as surface treatment, microlithography, laser-induced etching, and laser induced deposition, are beyond the scope of this paper.

## Excimer Laser and Basic Microablation System

Excimer lasers refer to a cluster of rare-gas halide lasers that emit powerful pulses of light whose wavelengths are in the UV or deep UV spectral region. The word “excimer” is a contraction of the term “excited dimer”. This describes a diatomic molecule that is bound in its electronically excited upper state, but is repulsive or only weakly bound in its lower ground state. The active medium in an excimer laser is a high-pressure plasma consisting of helium and/or neon and a trace amount active species, typically a mixture of a rare-gas such as argon, krypton, and xenon and a halide like fluorine, chlorine, and bromine. Since each of these binary mixtures has a characteristic emission wavelength, the excimer laser is normally designated by the name of the mixture. As a result, the wavelength of an excimer laser can be altered by changing the gas mixture with associated modifications of mirrors to obtain maximum output. Table 1 shows the most commonly used excimer laser, including XeF (351 nm), XeCl (308 nm), KrF (248 nm), ArF (193 nm), and $F_2$ (157 nm) and their associated photon energy. The practical attributes of each laser are also included. In addition to those commonly used, excimer lasers can also operate on a number of other wavelengths, such as 222 nm employing KrCl, 337 nm utilizing (molecular) $N_2$, 428 nm working with N ions, and 675-750 nm with atomic F.

**Table 1. Wavelengths and photon energy for excimer lasers**

| Laser (active gas) | Wavelength [nm] | Photon energy [eV] | Comments |
|---|---|---|---|
| Molecular Fluoride ($F_2$) | 157 | 7.90 | Ablating ceramics with the best ablated surface quality and resolution. Laser energy limited by optics degradation, even using $CaF_2$ optics. The laser beam path must be sealed from the ambient $O_2$ molecules to prevent of the laser energy absorbed by $O_2$. |
| Argon Fluoride (ArF) | 193 | 6.42 | Good combination of power and lasing efficiency for ablating ceramic and glass materials with high ablation rate and good ablated surface quality. A vacuum or inert gas purged beam path is recommended to prevent photon absorption by environmental $O_2$. Optics degraded more quickly than those with longer wavelengths. |
| Krypton Fluoride (KrF) | 248 | 5.00 | Best combination of absorption characteristics in plastics and overall average power, lasing efficiency, and operational cost for ablation of polymers. |
| Xenon chloride (XeCl) | 308 | 4.02 | Very long life gas. Longer absorption lengths in polymers can lead to higher etch rates than KrF, but lower average powers generate a slower rate of material removal overall. Often used for marking applications. |
| Xenon Fluoride (XeF) | 351 | 3.53 | Lower absorption associated with slow ablation rate makes this laser less-attractive for ablation even for polymers, but appropriate for less energy intensity tasks, such as pulse annealing and laser induced etching for various materials. Proximity to Hg lamp I-line makes it an attractive candidate for exposure of I-line photoresists. |

As indicated in Table 1, because of their relatively short wavelengths, excimer lasers radiate the high-energy photons in the range of 3.53 to 7.90 eV, which is sufficient to break apart many polymer-based materials. For example, C-H, O-H, H-H, O-O, and C-C bonds, which are the main part of all organic materials, have a bond energy of 3.5, 4.5, 4.6, 5.1, and 6.2 eV, respectively. As a result, the excimer lasers can directly dissolve the molecular bonds without generating undesirable heat, having a desirable surface morphology. In contrast to that, other lasers such as the $CO_2$ and Nd:YAG can only force thermal dissociation of the type of polymer materials, which may not be able to process heat-sensitive materials, such as glass-based materials. For this type of materials, especially fused silica ($SiO_2$), the associated

Si-O bond has a binding energy slightly higher than 100-eV (Alexander et al., 1999). Thus, ablation of glasses can be tougher and more complex than that of polymers. However, excimer lasers still have more advantages than other lasers in ablating glasses.

The excimer laser beam has a relatively low spatial coherence, which allows the use of projection optics and a relatively large beam size can be obtained to satisfy the requirement for projection printing. With a mask and demagnifying optics, laser projection printing allows parallel production of a large array of demagnified microstructures simultaneously to increase the throughput. The other major fabrication mode is called direct writing, in which the microstructure is constructed by direct bombardment of a series of small spots using a focused laser beam scanning a specific pattern over the target. Since direct writing is operated in a series mode, its throughput is much lower than that of projection printing. Consequently, excimer lasers are normally used for projection printing to machine microstructures and are fundamentally different from most other industrial lasers, such as the solid-state Nd:YAG and $CO_2$ lasers, which are generally employed in direct writing during machining.

Excimer lasers are usually operated with a pulse rate and pulse duration on the order of 1 to 100 Hz and 5 to 50 ns, respectively. Nearly all technical material, including plastics, ceramics, glass, metals, semiconductors, composites and biological/medical materials can be processed by excimer lasers. However, excimer lasers exhibit relatively poor beam quality, including high-order mode structures and high divergence. Also, the beam profile is mainly determined by the discharge conditions and the laser resonator optics. With continuous operation, discharge conditions, gas quality, and cleanness of the cavity windows are subject to natural degradation or contamination. Therefore, a homogeneous beam profile at the excimer lasers exit is difficult to achieve. For stable long-term operation, a beam homogenizer is frequently used to compensate and average these influences. A typical fly's eye homogenizer can convert a non-uniform profile of an excimer laser beam into a homogeneous beam with higher than 95% uniformity.

A basic laser micromachining workstation consists of an excimer laser source, an optical lens set, a motorized stage, a mask holder, an optical table, and an outer protective shielding, as shown in Fig. 1. A homogenizer and optical lenses are normally included to obtain a uniform intensity distribution of the laser beam. The laser beam is then passed through a projection mask and projected onto the target surface by a demagnifying objective lens. Two types of excimer lasers, KrF and ArF, are selected for the present study. As indicated in Table 1, the photon energies of the ArF laser is higher than the oxygen binding energy, 5.1 eV. As a result, the ArF beam can be absorbed by $O_2$ in the air, generating ozone ($O_3$). To prevent the generation of hazardous ozone gases and to have efficient propagation without unnecessary energy absorption, the laser beam passage should be encapsulated, and flushed by an inert gas or drawn into a vacuum. For each micromachining experiment conducted, the specific laser and the selected operating parameters, including repetition rate, pulse number, and fluence (energy density), will be specifically reported. The reduction factor by the demagnifing lenses can vary from 4 to 20, depending on the requirement. In addition to adjusting the optical design, the laser fluence projected on the specimen surface can also be varied by directly reducing the output laser energy or by activating the laser attenuator.

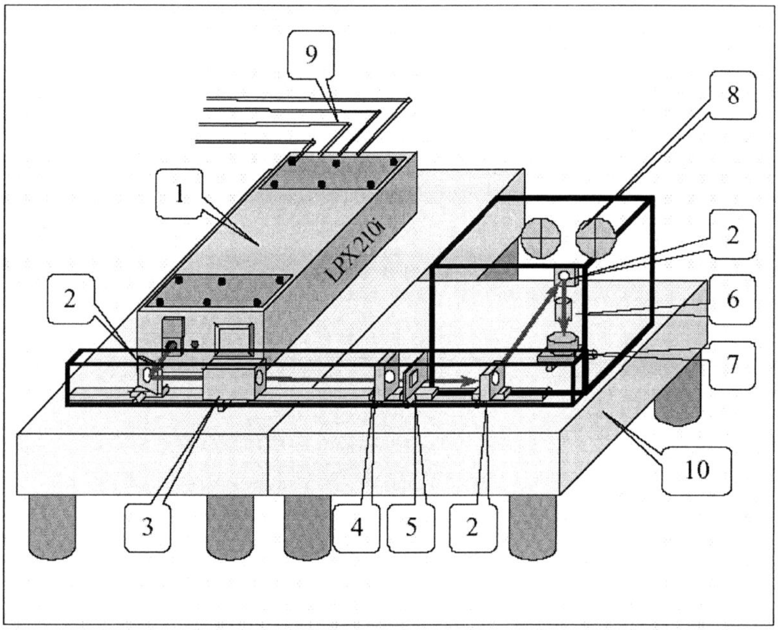


Figure 1. Basic excimer laser microablation system: (1) laser source, (2) 45° mirror, (3) homogenizer, (4) condenser lens, (5) mask, (6) optical lens, (7) moving stage, (8) air conditioner, (9) laser gas, and (10) optical table.

In a typical micromachining system, the target specimen is placed on a motorized stage that can rotate (θ) and translate in all three axes (x-y-z) through a computerized controller. The controller also triggers the output of laser pulses so that the laser pulse firing is synchronized with the movement of the stage table. Programming software can be provided with the micromachining system enabling users to program the table movement and laser firing sequence. A CCD camera is usually mounted on top of the objective lens to monitor the machining process. A beam profile viewer can be used for monitoring the homogeneity of laser intensity distribution.

In project printing, the mask is one of the most critical components. In basic excimer laser micromachining, two types of binary masks, stencil and metal-on-quartz, are normally used to define the regions to be patterned. In stencil masks, patterns are etched on metal foils. This type of mask has a long lifetime with excellent contrast because it not only exhibits high resolution or small feature sizes, but also withstands high energy laser irradiation. However, not all of the features can be designed on a single stencil mask. A circle with middle fallout is a good example. Since there is no sub-layer to hold the etched middle fallout, every feature in the etched pattern has to be connected to each other and no island or detached sections can be included in the pattern. This is also known as the "donut" problem. On the other hand, in a metal-on-quartz mask, a thin layer of Cr or other metals, such as Al, Au, or Ni, is patterned on a quartz plate, so that the quartz can hold all non-connected features in a multi-feature mask. However, the lifetime of the mask is relatively short because quartz can gradually deteriorate by absorbing some of the pass-through laser energy, especially the short-wavelength ArF laser. In general, these binary masks are well suited and optimized for ablating two-dimensional microstructures. Although gray-scale masks have also been developed for excimer laser machining, they are more appropriate for ablating 3D microstructures and are seldom used for fabricating planar structures, because of their expensive cost or less desirable resolution.

## ABSORPTION AND ABLATION

When a laser beam enters an opaque material, part of the radiation is absorbed via vibrational and/or electronic excitations. The beam intensity drops exponentially along beam propagation direction (η) following to the Beer-Lambert law:

$$I(\eta) = (1-R)I_0 e^{-\alpha\eta} \tag{1}$$

where $I_0$ the incident beam intensity, R the reflected portion of the beam at the material-ambient medium interface, α is the absorption coefficient and η is the traveling distance, also called absorption path length. Absorption takes place and absorbance of a laser beam in the target material should also drop exponentially to the path length. If the intensity of the beam is large enough and exceeds a certain threshold value, the absorbed energy can lead to ablation and liberate molecular-sized material from the surface through two different ablation mechanisms: photothermal and photochemical.

Depending on the electron-lattice coupling characteristics of the target material, some materials are more susceptible to the photothermal mechanism than photochemical, or vise versa. For example, ionic ceramics display high electron-lattice coupling strengths, which favor localized non-thermal excitations, while nonlocalized thermal relaxation mechanisms are more common in metals due to their electron-lattice scattering mechanisms. As a result, the two mechanisms can occur in varying degrees of combination in micromachining that uses high-intensity excimer lasers (Pãtzel, 2002; Tseng, Chen, and Ma, 2004). Normally, the temperature profile resulting from the photothermal mechanism can produce extensive damage in the heat-affected zone surrounding the melted or vaporized region. On the other hand, in the photochemical case, the pulse duration is far shorter than the thermal diffusion time scale. Thus, the amount of residual energy left in the substrate is typically small. Consequently, the photochemical mechanism is preferred in micromachining.

If absorption and ablation can be considered as two sequential processes that do not overlap in time, by neglecting reflections, Eq. (1) can be rewritten in terms of the intensity density or fluence for a pulsed laser:

$$F(\eta) = F_0 e^{-\alpha\eta} \tag{2}$$

where F (η) is the pulse fluence, and $F_0$ is the incident fluence. If the pulse fluence exceeds the threshold value ($F_t$), the ablation depth per pulse (Δd) can be derived directly from the above equation:

$$\Delta d = \frac{1}{\alpha}\ln\left(\frac{F_0}{F_t}\right) \qquad \text{for } F > F_t \tag{3}$$

This formulation is widely adopted for the correlation of experimental ablation data, in which only two parameters fully characterize the material behavior: $F_t$ and $\alpha$.

The above Beer-Lambert relationship may oversimplify the ablation process and frequently the predicted ablation rate does not correspond with the measurement, especially for the ablation process governed by photothermal mechanism, where the photon energy is converted to vibrational or thermal energy before the onset of ablation. An Arrhenius expression is suggested by Srinivasan, Smrtic and Babu (1986) for representing the thermal effect and the Beer-Lambert relationship can be extended to

$$\Delta d = \frac{1}{\alpha}\ln(\frac{F_0}{F_t}) + A_f \exp[-(\frac{E_a/\alpha}{F_0 - F_t})\ln(\frac{F_0}{F_t})] \quad \text{for } F_0 > F_t \tag{4}$$

where $A_f$ is the frequency factor and $E_a$ is the activation energy. Many more complex relationships or models have also been suggested, but usually are a further extension of the foregoing Beer-Lambert Arrhenius relationship and take a few more aspects of the process into account. For example, a photophysical model developed by Schmidt et al. (1998) includes not only the photothermal and photochemical dissociation, but also the plume absorption, thermal diffusion within the material, absorption saturation, incubation effects and a moving material interface during ablation. Solving the set of equations in this model requires knowledge of some fifteen parameters! In the present study, since the Beer-Lambert relationship (Eq. 3) is not used for interpreting the physical behavior and is reasonably accurate for correlating the ablation experimental data for the materials considered, only this relationship is selected for correlating the experimental data and guiding the fabrication processes. As will be demonstrated in the later sections, this simple relationship is adequate for this purpose.

## Ablation Rates of Optical Polymers

The key to laser precision ablation is the ability to operate a laser beam with a proper beam shape, intensity and wavelength to remove a required amount of material from a pre-defined location in a controllable manner. Thus, the dimensional accuracy has to be controlled within a required precision at every "ablation pulse" or "machining step". Precise knowledge of the ablation rate (Δd), especially the effects of major operating parameters, including fluence, pulse duration, and pulse number on the ablation rate, is essential to achieve the precision ablation of complicated optical structures.

A wide range of transparent polymers have been used for optical applications. Most of them are either common polymers or polymeric compounds with proprietary compositions developed for specific optical applications. In this section, five common transparent polymers which are not only popular for optical applications, but are also commercially important for other general usages. Nine proprietary polymers are selected for ablation study to provide necessary information for making optical components. The former includes polycarbonate (PC), polyethylene terephthalate (PET), polyimide (PI), polymethyl methacrylate (PMMA), and polystyrene (PS), while the latter involves three cyclic olefin copolymers (COC) and six epoxy- or acrylic-based polymers.

## Common Optical Polymers

Most polymer materials are not transparent. The five most common polymers selected offer light transmission rates greater than 80% with good impact resistance. For example, PMMA or pure acrylic has more than ten times higher impact resistance than the normal borosilicate glasses, and is easily fabricated. PET and PC have an impact resistance which is even an order of magnitude higher. Also, the light weights of this group of polymers, about one-half that of a comparable size of glass, allow their use in applications instead of glass. Certainly, most polymers have relatively low abrasion resistance, although a proper coating can be applied to increase their abrasion resistances.

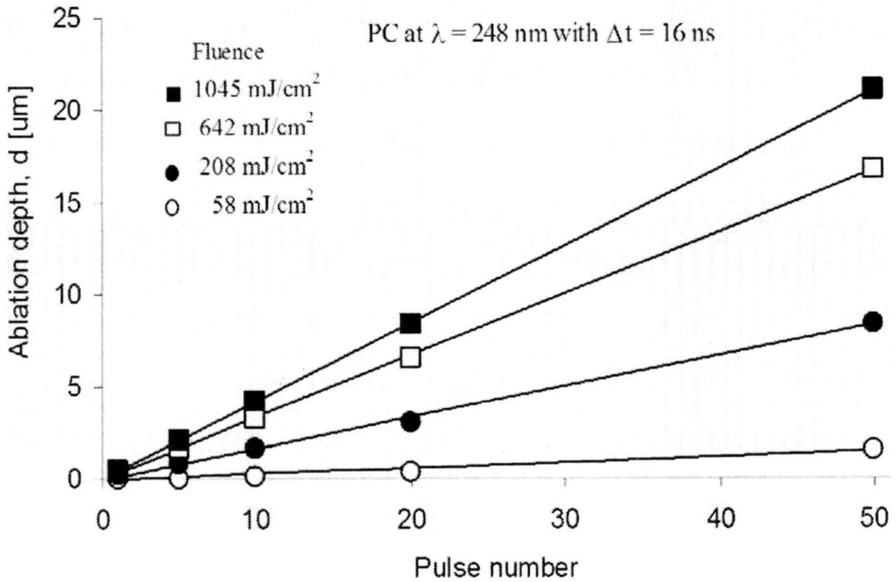


Figure 2. Correlations of ablation depth with pulse number at various fluences for PC at λ = 248-nm and Δt = 16 ns.

The ablation-rate experiments are performed with a PulseMaster 848 from Lumonics of Canada, an excimer laser operating at either 193-nm (ArF mixture) or 248-nm (KrF mixture) equipped with a projection lens providing variable demagnification from 4× to 11×. In the ablation rate evaluation, a square- or circular- shape stencil mask is used with a pulse duration (Δt) of 16 ns (Full-Width at Half-Maximum or FWHM). Although no homogenizer is used, the laser beam is carefully calibrated. Characterization of ablated cavities in terms of depth and surface roughness is performed by white light interferometry. The effects of operating parameters, including fluence, pulse number, and pulse repetition rate, on the ablation rate (Δd) are discussed for the selected polymers.

Figure 2 shows the experimental results of the relationship between the ablation depth (d) and pulse number (N) at four fluences of 58, 208, 642, and 1045 mJ/cm$^2$. The experiment was conducted by ablating a PC using a KrF (248-nm) mixer with a stencil mask to produce 250-μm × 250-μm cavity on the target material at a pulse duration of 16 ns and a repetition rate ($R_p$) of 20 Hz. As shown in Fig. 2, the ablation depth (d) for all four fluences considered increases quite linearly with the pulse number, which means that each pulse at a specific fluence ($F_0$) contributes equally to the ablation depth. Mathematically, d can be represented by

$$d = N\ \Delta d(F_0) \tag{5}$$

where N is the number of pulses. Based on the data shown in Fig. 2, it can be analyzed that the ablation rate ($\Delta d$) is independent from the pulse number and is a function of fluence. If $\Delta d$ is plotted against the incident fluence ($F_0$) in a semi-logarithm scale as shown in Fig. 3, the corresponding fluence at a zero ablation rate is $F_t$ and the inverse of the slope of the correlated curves is $\alpha$. The experimental ablation data using both 193-nm and 248-nm excimer lasers are also plotted in Fig. 3 for PET, PMMA, PI, and PS. The correlated parameter, $F_t$ and $\alpha_e$ are summarized in Table 2. The corresponding coefficient of determination ($R^2$) for each curve fitting are also reported in Table 2. Since the values of the coefficient $R^2$ are all higher than 0.98, this implies that the ablation rate correlates with the fluence extremely good and follows the Beer-Lambert relationship almost perfectly. $R^2$ is a measure of the accuracy of correlation and always lies between 0 and 1. A value of zero occurs when the two variables are totally independent of each other, while it reaches 1 when the two variables correlate perfectly, i.e., no deviation from the curve (Holman, 2001). It is to be noted that the effective absorption coefficient ($\alpha_e$) instead of $\alpha$ is used in Table 2 and Eq. (2), which implies that the value of $\alpha_e$ is directly obtained from data correlation and may not be consistent with the measured absorption coefficient by spectroscopy.

**Table 2. Correlated ablation parameters for standard optical polymers**

| Polymer | ArF (193 nm) $\Delta t = 16$ ns | | | KrF (248 nm) $\Delta t = 16$ ns | | |
|---|---|---|---|---|---|---|
| | $F_t$ [mJ/cm$^2$] | $\alpha_e$ [μm$^{-1}$] | $R^2$ | $F_t$ [mJ/cm$^2$] | $\alpha_e$ [μm$^{-1}$] | $R^2$ |
| PC | 21.5 | 18.8 | 0.98 | 57.5 | 7.15 | 1.00 |
| PET | 18.4 | 15.6 | 0.98 | 32.1 | 6.92 | 0.99 |
| PI | 25.1 | 18.6 | 0.99 | 46.7 | 9.65 | 0.98 |
| PMMA | 33.8 | 6.33 | 0.99 | 273. | 1.00 | 0.99 |
| PS | 15.3 | 24.1 | 0.99 | 136. | 4.43 | 0.98 |

To show the reliability of the present results, some of the values reported in Table 2 are also compared with the data published earlier. The present results for PI at both 193-nm and 248-nm are consistent with those obtained by Gorodetsky et al. (1985), with identical values of $F_t$ with a less than 10% difference in $\alpha_e$. For PET, at 248-nm radiation, the present values of the threshold fluence and absorption coefficient are, respectively, equal to 32.1 mJ/cm$^2$ and 6.92 μm$^{-1}$, which are 35% lower than those based on the thermocouple technique obtained by Dyer and Sidhu (1985), while the present results at 193-nm ablation are 40% higher in threshold fluencs with a 40% lower absorption coefficient, as compared with those by Dyer and Sidhu (1985). Since the threshold energy density is $F_t$ x $\alpha_e$, the corresponding density values become approximately equal and the differences in $F_t$ and $\alpha_e$ may be attributed to the fact that the different pulse durations (16 ns versus 9 ns) are used. For PC, the present values of the threshold fluence and absorption coefficient are at least 30% higher and 40% lower, respectively, than those based on the quartz-crystal microbalance technique obtained by Lazare and Granier (1988). Again, their corresponding density values become comparable and the differences in the ablation parameters may be caused by the effect of pulse duration.

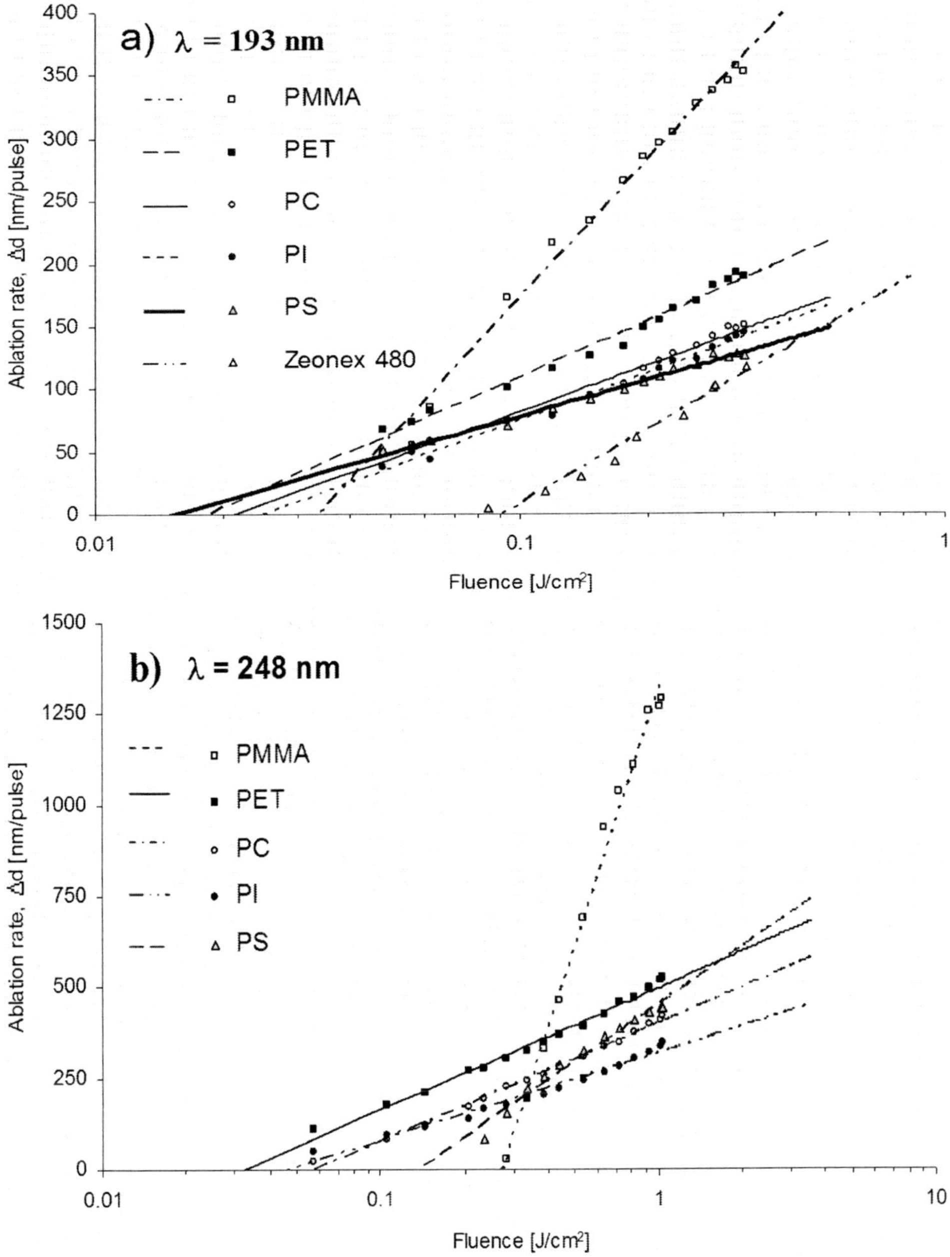


Figure 3. Correlation of ablation rate with incident fluence for various polymers with $\Delta t = 16$ ns at: a) $\lambda$ =193 nm, and b) $\lambda$ =248-nm.

The PS results obtained by Lazare and Granier (1988) are also compared to the present values. Again, the present values are much higher (more than 30%) in $F_t$ and much lower (more than 30%) in $\alpha_e$. One of the reasons for causing such big differences in $F_t$ and $\alpha_e$ could be that the polymers evaluated have different additives, because the results being compared are based on different brands and different material suppliers frequently put different small amount of additives into their brands for different purposes. Further studies on the ablation behavior, especially on PS, should be encouraged. The material supplier for the PC, PI, PMMA, and PS currently studied is Goodfellow Corp. of Devon, PA (www.goodfellow.com).

As indicated in Table 2, the absorption coefficient of PMMA, especially at 248-nm ablation, is much smaller than that of the others. The corresponding cavities produced by 50 pulses at 642 mJ/cm$^2$ fluence for PC and PMMA are shown in an SEM (Scanning Electronic Microscopy) image Fig. 4. The depth of PMMA cavity is 43 μm and is much deeper than that of PC, which is consistent with Eqs. (2) and (3). This implies that a polymer with a lower absorption coefficient (or larger penetration depth), such as PMMA, has a faster ablation rate at a given fluence above the threshold, i.e., the deeper the penetration, the more material to be ablated. However, since the pulse penetration is deeper, the corresponding absorbed energy density is lower than that of polymers with higher absorption coefficients (or smaller penetration depths). As a result, lower energy density is available for dissociation and ablation, which can result in a less desirable surface quality. As indicated in Fig. 4, the cavity definition in PMMA is much worse than that of PC. The sidewalls of the PMMA cavity have an irregular structure and the bottom surface reveals the presence of small craters normally caused by the explosions of microscopic gas bubbles during ablation. Since polymer is a dielectric material, a thin gold film (20-nm thick) is coated on the ablated polymer sample by vacuum deposition to make it conductive for SEM observation.

polycarbonate polymethylmethacrylate

Figure 4. SEM images of 200×200μm$^2$ cavities ablated by KrF laser with 50 pulses at 20 Hz, 642-mJ/cm$^2$ fluence, and 16-ns duration time: PC (left) and PMMA (right).

Ablation times for micromachining can be reduced by increasing the repetition rate or pulse frequency. Although this consideration might be of little use to prototyping applications, it is of major importance for determining throughput in a production environment. For this purpose the impact of repetition rate on ablation rate is also studied. It has been found that, for polymers with relatively high absorption coefficients, including PC, PET, PI, and PS, the ablation rates are almost independent from the repetition rate for the range considered, i.e. 20-150Hz. On the other hand, for PMMA, the ablation rate rises about

40% as the repetition rate increases from 20 to 150 Hz at a fluence of 1 J/cm$^2$. This may be due to the fact that heat is accumulated more as the repetition rate increases, and less time is available for the heat to be diffused or transferred. Consequently, the PMMA temperature rises during ablation at a higher repetition rate and the removal rate increases due to thermal softening of the polymer. Thermal coupling between pulses has previously been discussed by Burn and Cain (1996) and Liu, Feng, and Yi (2000).

## Proprietary Optical Polymers

In this section, nine proprietary polymers, including three cyclic olefin copolymers (COC) and six either epoxy- or acrylic-based polymers are studied for their ablation behavior. Three COC are Zeonex® 330R, Zeonex® 480, and Zeonor® 1600, while the other six are Emerson & Cuming Stycast-1200J, Delo Katiobond 4563, Norland NOA 61, Norland NOA 72, Norland NOA 76, and Norland NOA 83H, in which the first two are epoxy based adhesives and the latter four are acrylic based polymeric compounds.

Microlenses have been ablated by an ArF laser in thees COCs, supplied by Zeon Corp of Japan (www.zeon.co.jp). The COC grades have excellent optical properties, such as 92% light transmittance, low impurity property, and low water absorbency. They are frequently used for lenses in laser printers and optical elements in cameras and medical equipments. Stycast-1200J, provided by Emerson & Cuming of Billerica, MA (www.emersoncuming.com), is a transparent thermal curing epoxy, has low viscosity and two compounds, and is particularly suitable for encapsulants in packaging applications. Katiobond 4563, supplied by Delo of Landsberg, German (www.delo.de), is a clear, ridge, single compound, and UV curing epoxy and is also suitable for encapsulation.

Encapsulants are used for protecting IC (integrated circuits) chips from moisture and mechanical damage. In optoelectronic chip packaging, they need to fulfill the additional requirement of transparency for the light emitted from or checked by the encapsulated IC. In the present article, both Stycast-1200J and Katiobond 4563 will be assessed by laser ablation of micro encapsulating components for optoelectronic packaging applications. The four Norland NOA grades, made by Norland Products of Cranbury, NJ, are high transmitting UV curing acrylics and their suitability for ablating refractive microlenses will be studied for packaging applications.

The ablation-rate experiments for the proprietary polymers are conducted with the Lumonics PulseMaster 848 system using the ArF mixture ($\lambda$ = 193-nm) only. Circular cavities of 200 μm in diameter ablated with 2500 pulses and 290 mJ/cm$^2$ fluence for Zeonex 480 (left) and Zeonor 1600 (right) are shown in Fig. 5. The cavities depths at specific pulse numbers and incident fluence can be measured by white light interferometry. The corresponding ablation rates (depth ablated per pulse) at different incident fluences for Zeonex 480 are also plotted in Fig. 3a. Again, the extrapolated fluence at zero ablation rates and the inverse of the slops of the correlated curves are $F_t$ and $\alpha_e$, respectively. As shown in Fig. 3a, the threshold fluence of Zeonex 480 is much higher than that of the common standard polymers, e.g., 2.7 times higher than that of PMMA and 6 times higher than that of PS. Similar plots based on ablation data with 30 ArF laser pulses at 20 Hz repetition rate are also obtained for other eight proprietary compounds. The extrapolated ablation parameters, $F_t$ and $\alpha_e$, are all summarized in Table 3. For all of these compounds, no suitable published data have been found for comparing the ablation values reported in Table 3. Again, the ablation parameters obtained will be used to guide the ablation of different microstructures for optical applications later.

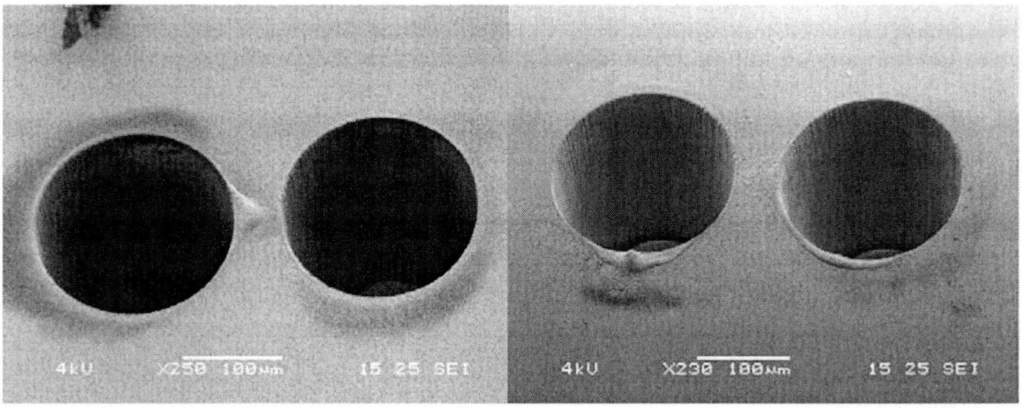


Figure 5. SEM images of circular cavities of 200 μm in diameter ablated by ArF laser with 2500 pulses at 290 mJ/cm$^2$ fluence, 16 ns duration time, and 50 Hz: Zeonex 480 (left) Zeonor 1600 (right) COC.

**Table 3. Correlated ablation parameters for commercially proprietary polymers**

| Polymer (at $\lambda$ = 193 nm, 20 Hz with 30 pulses, $\Delta t$ = 16 ns) | $F_t$ [mJ/cm$^2$] | $\alpha_e$ [μm$^{-1}$] |
|---|---|---|
| Zeonex® 330R | 93.0 | 4.2 |
| Zeonex® 480 | 90.7 | 11.8 |
| Zeonor® 1600 | 57.5 | 12.4 |
| Norland NOA 61 acrylic | 24.4 | 10.7 |
| Norland NOA 72 acrylic | 23.4 | 10.5 |
| Norland NOA 76 acrylic | 15.3 | 10.7 |
| Norland NOA 83H acrylic | 24.7 | 10.7 |

## Polymer-Based Ablated Optical Microstructures

The essence of ablating optical-related microstructures is to satisfy the required dimensional accuracy. In addition, surface quality and debris are also important for the control of optical related applications. In this section, ablated surface roughness, debris control, and other related ablation characteristics will be discussed. Then, four ablated microstructures for optical related applications are presented.

## Surface Morphology of Ablated Polymers

When a flat substrate is ablated by an excimer beam with a homogeneous intensity distribution, the ablated cavity should feature a smooth surface. In the surface morphology experiments, cavities of 100 μm x 100 μm in cross-section are ablated by ArF and KrF lasers. The corresponding roughness and debris at the bottom surface of the ablated cavity are then studied. The typical morphologies of the ablated surfaces of the five common polymers considered are shown in Fig. 6. They are all ablated by 20 ArF-laser pulses at a repetition rate of 20-Hz and a fluence of 48 mJ/cm$^2$.

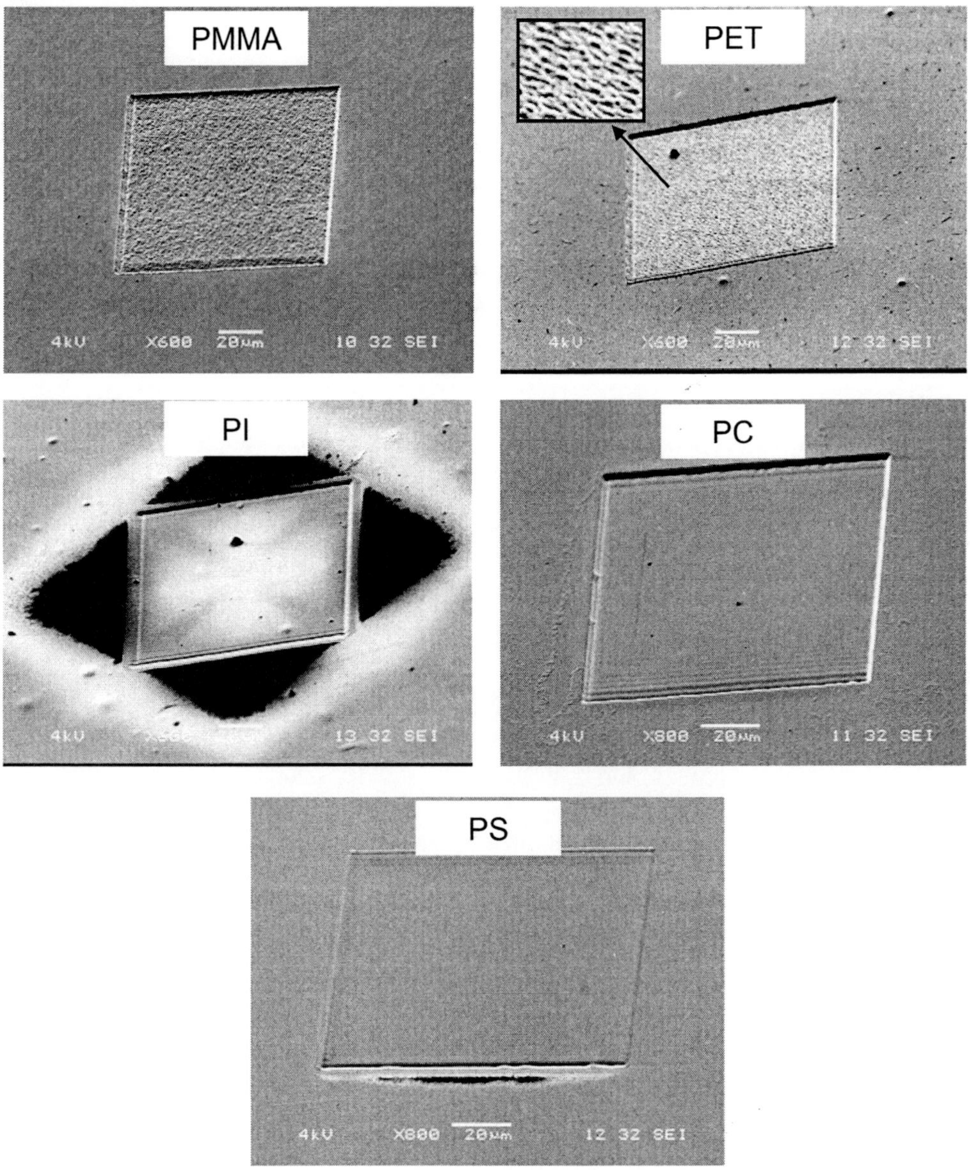


Figure 6. SEM images of cavities (100 μm x 100 μm) morphologies ablated by 20 ArF laser pulses at $\lambda$ = 48 mJ/cm$^2$, $\Delta t$ = 16 ns, and $R_p$ = 20 Hz for five commons polymers.

Cavities ablated in PMMA at 193 nm exhibit a melted surface layer with the evidence of burst gas bubbles, which is mainly induced by its low $\alpha_e$ and high $F_t$. The low $\alpha_e$ causes the pulse energy to be absorbed within a relative thick layer, ~ $1/\alpha_e$ or 160-nm, instead of 40-60 nm for the other four standard polymers, which results in less local energy density. The low absorbed energy density and the high $F_t$ cause less energy to be devoted to photochemical ablation. As a result, more remaining pulse energy is photothermally converted into heat. As indicated in Fig. 6, the heat can be accumulated until it is large enough to melt the surface, which is also known as recast (Chen et al., 2005b). The same argument can be applied for KrF ablation, in which PMMA exhibits an even lower absorption at $\lambda$ = 248 nm. PET also features a regular corrugated surface at both wavelengths but it is due to different reasons. One of the reasons is the semi-crystalline nature of PET, which favors the formation of a thin low-viscosity layer in the ablated area, leading to a regular ripple, as seen on the figure. Also, it may be due to the fact that the heat generated and/or partial crystallization during ablation locally relaxes the internal mechanical stresses, which are originally created during forming (or rolling) the PET films. For PS, it is somewhat between: considerably lower absorption than PC (or PI), but better than PMMA. Thus, the surface quality of PS is better than that of PMMA, but not as smooth as PC. The slightly black line on the lower side of the cavity must be some debris. Note that debris particles are very irregular. The Au layer on the samples, which is necessary for making SEM images, does not attach very well to these surfaces. For this reason, debris is colored black on the SEM image.

***Roughness***

To study the impact of pulse fluence and pulse count (scaling liner with the total ablation depth) on the sample roughness, cavities have been ablated in the five standard polymers at fluences ranging from 48 to 340 mJ/cm$^2$ by ArF lasers and from 58 to 1045 mJ/cm$^2$ using KrF lasers. The length of the pulse train is varied from a single pulse to 100 pulses. It has been found from these tests that the relative roughness decreases asymptotically with the number of pulses or with the accumulated ablation depth to a value of 0.5% for PC, PI and PS, within the experimental parameter range. Since the ablation workstation used is not equipped with a homogenizer, it is possible that a better relative surface roughness (lower than 0.5%) can be obtained. It can also be observed in Fig. 6 that the relative roughnesses of PMMA and PET are much higher than the other three. In general, the corresponding relative roughnesses of PMMA and PET are one-order magnitude higher than that of PI and PC. Note that the relative roughness is defined as the root-mean-square roughness ($R_q$) normalized by the total cavity depth.

***Debris***

The ablated materials, which make up the plume or ionized plasma, are not always fully decomposed into gaseous constituents. The plume or plasma can also contain micro- and macro-scale particles with high molecular content. These particles will usually drop back onto the substrate in or close to the ablated area shortly after ejection and are commonly referred to as 'debris'. The exact spread and amount of debris is determined by plume expansion dynamics and depends on ablated geometry, fluence, beam size, and ambient atmosphere. Figure 7 shows optical images of debris patterns around the cavities ablated in PI at four different fluences. The cavities are all ablated by 100 KrF laser pulses and have a $100\times100\mu m^2$ cross-section. As indicated in the figure, higher fluences (and same pulse count) give more debris as more material is ablated. More material removal means a deeper cavity. Thus more plume or plasma can be generated in and confined by the deeper cavity and by the subsequent laser pulses, where the highly trapped plume can result in more debris to be built up near the ablated area. Certainly, a comprehensive modeling effort can provide a more detailed description on the debris formation, as indicated by Bityurin et al. (2003) and Zhigilei et al. (2003). Furthermore, as indicated in Fig. 6, the cavity in PC is typically clean. However, it is expected that noticeable debris can be accumulated, if longer pulse trains or higher fluence with more pulses are applied.

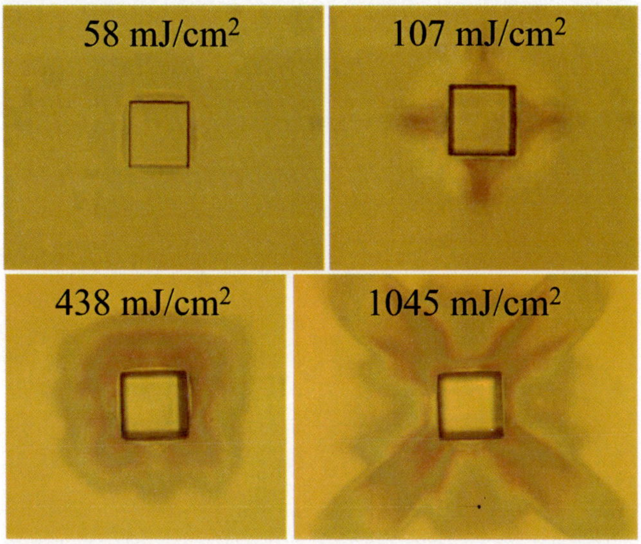


Figure 7. Optical images of four debris pattern around $100\times100\mu m^2$ cavity in PI using 100 KrF pulses.

The different material composition induced on the ablated surface during ablation can also affect the quality of ablated structures. Figure 8 shows SEM images of ablated cavities ($100\times100\mu m^2$) of PI by 100 KrF laser pulses at two different fluences. As shown, at low fluences, e.g., 48 mJ/cm$^2$, a carbon debris (cones shown in the figure) layer can be formed. The thickness of this carbon debris layer increases with each pulse. Since the carbon layer or debris is much harder to be ablated than PI, it can shield the PI below from further ablation. The corresponding threshold fluence to ablate this carbon layer is around 260 mJ/cm$^2$, according to Ball et al. (1996). At fluences higher than this threshold value (e.g., 312 mJ/cm$^2$ in the figure) the carbon debris does not develop and a smooth surface is created.

Debris can be reduced by ablation in a He or $H_2$ atmosphere. For debris removal after processing a multitude of techniques are at our disposal: laser cleaning at subthreshold fluences, ultrasonic bath, brushing and liquid wash. For critical applications, a very thin sacrificial layer can be applied on top of the surface. After the ablation, the layer (covered with debris) is removed by rinsing or stripping, revealing a top surface free of contamination.

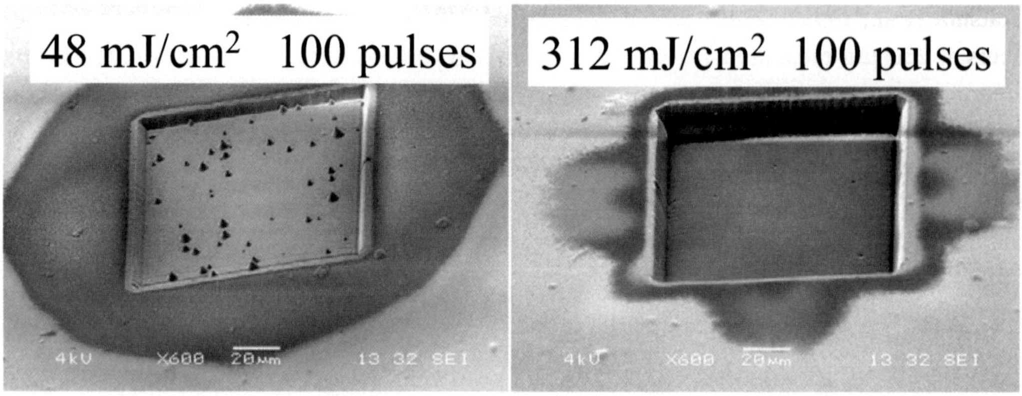


Figure 8. SEM images of ablated cavities ($100\times100\mu m^2$) of PI by KrF laser at two different fluences.

***Incubation***

The threshold fluences obtained from Fig. 3 are essentially independent of the pulse count. In some polymers, it has been observed that low UV absorption induces the so-called incubation phenomena. This means it can be ablated below the single pulse ablation threshold of the virgin target material after repetitive exposure to subthreshold fluences, as indicated by Kuper and Stuke (1989) and Blanchet, Cotts, and Fincher (2000). Among the five polymers (PC, PET, PI, PMMA, and PS) studied, PMMA has been observed to exhibit a pronounced 'incubation' effect at the KrF wavelength. The single-pulse ablation threshold can be much higher than that of a train of 10 pulses. The corresponding ablation rate does also not follow Eq. (2) and can vary from a high positive to even negative values (swelling of material). The origin of this effect is a considerable increase of absorption in PMMA that has been exposed to sub-threshold pulses. Chemical modifications take place in the irradiated area and irreversibly change the absorption coefficient. Incubation effects create an ambiguity between pulse count and ablation depth and are problematic when shallow features need to be ablated with high accuracy.

## Ablation of Microarray Ferrule for Multimode Optical Fiber Alignment

An ever increasing demand for higher data throughput drives the development of optical fiber communication with smaller system size and higher efficiency. Plastic Optical Fibers (POF) can provide very high speed data transmission over distances in the range from a few to 200 meters, which is for typical Local Area Networks (LAN) connections, for intra-vehicle links, and for Very Short Reach (VSR) interconnections (Boom et al., 2001). One of the key components in the POF communication system is the fiber connector or so-called ferrule. There are various types of ferrules depending on applications for which they are developed. Basically, the fiber is mounted in the ferrule and the end of the fiber is located at the end of the ferrule, which also provides a fiber alignment mechanism. Ferrules can be made of a variety of materials and can be constructed out of plastics (Shuto et al., 2004). One of the technologies that is able to increase the throughput of optical fiber communication is parallel optical interconnect, employing large 2D arrays of fiber for optically interconnecting chips

(Katsura et al., 1999). Multifiber communication systems with connectors featuring 8 or 12 fibers at 250 μm pitch are offered by several suppliers but the needed ferrules for 2D (two-dimensional) arrays are expensive and, sometimes, hardly commercially available.

As a demonstration, a prototype of a 4×8 fiber ferrule for integration in a plastic connector housing is fabricated by laser ablation. The ferrule consists of four PMMA plates stacked on top of each other. Each plate has a width slightly smaller than 250 μm and contains an array of 8 grooves at 250 μm pitch. Fibers are fixed in the grooves with a UV-curable adhesive and the four fiber-embedded PMMA plates are then glued together at a pitch of 250 μm. As shown in Fig. 9.a, the groove is ablated into a U-shape, which is 126-μm wide in the middle with a mouth width of 155-μm. An ArF laser operating at 50 Hz and a fluence of 270 mJ/cm$^2$ is used for the ablation. With the PMMA data of $F_t$ and $\alpha_e$ provided by Table 2, the ablation rate or the ablated depth per pulse predicted by Eq. (3) is 328 nm and thus, as indicated in Eq. (5) by regulating the pulse number, the depth of the ablated grooves can be controlled within micrometer accuracy. Also, at the ArF wavelength, PMMA can be ablated faster than other polymers due to lower absorption. Because of the size limitation of the projection lens of the laser workstation, the mask is designed to allow 4 grooves to be ablated at a time. As a result, a scanning ablation technique is used, in which the PMMA plate is moved at constant velocity, allowing simultaneous ablation of 4 grooves.

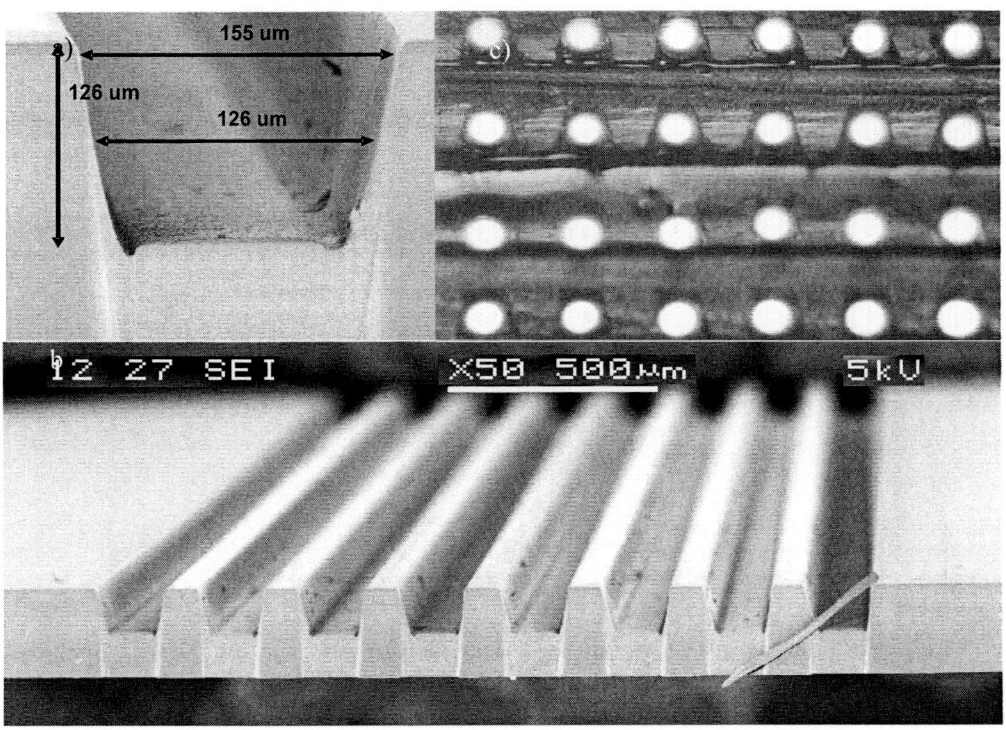


Figure 9. Ablation of 2D plastic optical fiber ferrule: a) SEM image of single groove cross-section ablated by ArF laser, b) SEM image of 8-groove array in PMMA plate, c) optical image of assembled 4×8 fiber ferrule.

The ablated PMMA plate with 8 grooves is shown in Fig. 9.b, while the ferrule, i.e. four grooved plates mounted with fibers and glued together with UV curing adhesive, is shown in

Fig. 9.c. Two types of POFs, which are clad with dielectric structures with 0.5 stepindex profiles, are selected to be inserted into the grooves of the ablated plate. One is supplied by Toray Industries of Tokyo, Japan (www.toray.com) having a fiber core diameter of 116 μm with an outer (cladding) diameter of 125-μm and the other is provided by Asahi Glass of Tokyo, Japan (www.agc.co.jp) with a 63-μm core and a 125-μm cladding diameter. The diameter tolerance of Toray POFs is controlled within ±2μm, while Asahi POFs have size variation of ±7μm. The pitches between the mounted fibers have been measured. The results indicate that the standard deviation of both horizontal and vertical pitches for the assembled ferrules using Toray POFs is approximately 5-μm, while the corresponding variation for ferrules using Asahi POFs is twice as larger at 10-μm. The higher standard deviation is likely caused by the higher diameter tolerance (or variation) of the Asahi POFs. It implies that the fiber position accuracy in the ferrule matrix can be deteriorated by large tolerances on the fiber diameter. As a result, the POF with better size accuracy, i.e., Toray POFs, should be used. It is to be noted that the fiber position or pitch accuracy in the ferrule matrix directly dictates the efficiency and quality of data communication.

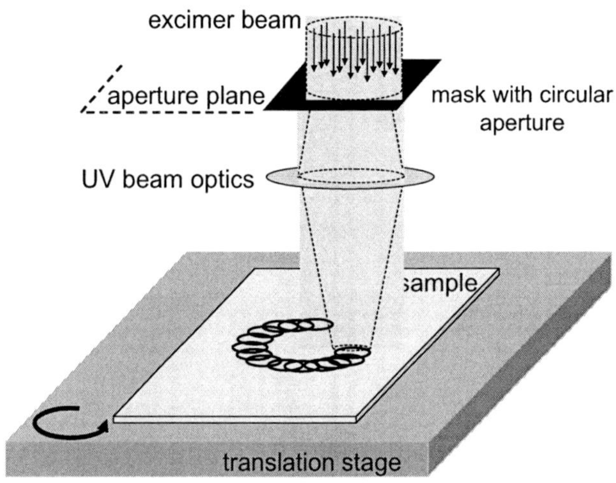


Figure 10. Schematic of contour scanning ablation technique.

## Ablation of Microlenses for Singlemode Optical Fiber Coupling

Microlenses are one of the most important optical components in the photonics industry. A variety of techniques, from the melt reflow method (He et al., 2003) to the multilevel etching technique (Liau, Nam, and Waarts, 1994), have been used to fabricate microlenses. Recently, excimer lasers have been popular for making high quality optical lenses. Techniques using excimer lasers vary from the straightforward discretization of the curved surfaces with multiple masks and exposures (Ihlemann and Wolff-Rottke, 1996) to a technique using a grayscale-mask (Hayden, 2003) and to a contour scanning approach (Naessens, 2003a; Lee, Chen, and Wu, 2005).

In this section, a special contour scanning technique is introduced for ablating microlenses, in which a lens is gradually approximated by ablating a set of ring-shaped trenches on top of one other. Each trench is ablated with the same circular beam aperture scanning the substrate along a circular contour with a certain diameter and at a certain velocity, as shown in Fig. 10. The required diameters and velocities are calculated with an algorithm based on simulated annealing, as indicated by Naessens et al. (2003a). After ablation of these trenches, the surface is covered with debris and corrugated due to the discrete nature of the pulsed ablation process. Smoothening the surface is critical for high quality lenses and can be achieved by ablating the lens deeper into the substrate with a large aperture covering the entire lens area. Fig. 11 illustrates the effect of laser smoothing, in which Fig. 11.a is the surface of a PC microlens with a lens diameter of 200-μm ablated by the scanning contour technique while the corresponding surface after laser smoothing is shown in Fig. 11.b. It clearly shows a smooth surface surrounded by a steep wall after smoothing. The laser smoothing is performed by an ArF laser with 180 shots at a fluence of 140 mJ/cm$^2$, taking away a layer of approximately 18 μm, which is also consistent with the prediction based on Eqs (3) and (4).

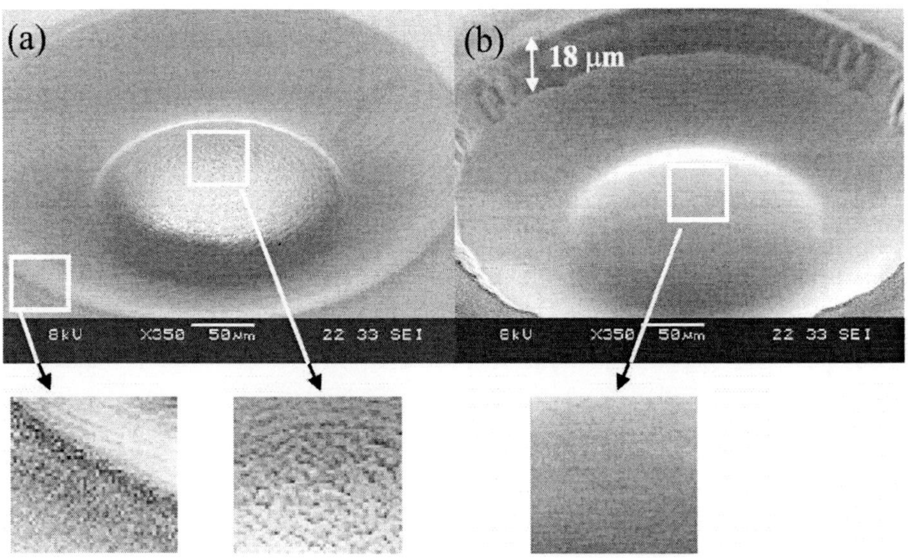


Figure 11. SEM images of PC microlens by contour scanning ablation using ArF laser with 140 mJ/cm$^2$ fluence: a) before laser smoothing, b) after laser smoothing by ablating 18-μm layer.

Various types of microlenses have been fabricated by this contour scanning technique. Many spherical microlenses made in PC with a radius of curvature (RoC) ranging from 125 to 1500 μm (f/# from 0.8 to 14) have been ablated by an ArF laser under different pulse fluences between 60 and 180 mJ/cm$^2$ and pulse repetition rates ($R_p$) from 20 Hz to 150 Hz. The contour speed can be chosen and serves as input parameter. The typical range of the contour speed is between 60 and 220 μm/s for a beam fluence of 177 mJ/cm$^2$. A stencil mask with a circular aperture of 700-μm in diameter is used and with a demagnification of 7x, a laser beam size of approximately 100-μm in diameter can be projected on the sample or target plane as illustrated in Fig. 10.

As shown in Fig. 11, although the final lens diameter is 200 μm, the actual ablation area is much larger than the lens size and is approximately twice the diameter of the lens. As a result, the deepest ablation can be located at the rim of the lens and a convex lens with a parabolic or spherical profile can be ablated. In some other cases, depending on their RoC, the ablated areas are even larger and can be three or four times larger than the lens diameter (e.g., those shown in Fig. 13). Certainly, smaller laser beam sizes on the target plans can be used, but require a much longer ablation time. In fact, all the microlenses presented in this article are made by the same mask using the same contour scanning technique. Using a WYKO white light interferometer, the geometrical differences between the ablated and theoretical lens shapes are measured. The typical deviation in RoC is less than 3%, which is a measure for the fabrication accuracy. The surface roughness ($R_q$) is nearly 20 nm for f/# > 4.0. For these lenses, the diffraction-limited performance of the lenses is obtained.

With the same contour scanning technique, a parabolic microlens is suitable for relax the alignment tolerances in fiber-to-fiber coupling in SM (single-mode) optical fiber communication. Due to the small core size of an SM fiber, the propagating light is confined to a very small region with a diameter of 9-10 μm. When light is coupled from one fiber to another, a very strict alignment is required to obtain sufficient coupling efficiency. For example, in telecom applications, 0.2 dB coupling loss in a fiber connector is considered acceptable and corresponds to a lateral misalignment of a mere ± 1.1 μm. The microlens should be right in front of the fiber facet. Before ablating the required microlens, the target plastic substrate is first adhered onto the fiber facet with UV curing adhesive. Then the lens can be formed in the plastic substrate with the contour scanning ablation technique above the fiber core by visual alignment (2 μm resolution). The lens has a parabolic profile and is designed to obtain an intermediate 28 μm waist diameter between two fibers equipped with such a microlens. By measuring the fiber-to-fiber coupling efficiency (CE) with and without microlenses, the corresponding effect on lateral alignment tolerance has been experimentally studied. The average coupling loss in case of microlens equipped fibers is -1.0 dB, which can be caused by many factors, including Fresnel reflection, surface imperfections, and the limited alignment accuracy of the microlens and the underlying fiber core. The tolerance for fiber-to-fiber misalignment however, is increased from ±1.1μm to ± 3.1 μm!

## Ablation of Microlenses in Transparent Chip Encapsulants

In IC packaging, encapsulants are normally used for protecting chips from moisture and mechanical damage. For optoelectronic chips, transparency is the additional requirement for the encapsulating substance to allow light emission from or light detection by the ICs. In many optoelectronic assemblies the divergence, direction, position and size of light beams leaving or entering the package need to be controlled accurately to obtain satisfactory operation of the chip. For that reason refractive microlenses are inserted in the light path and alignment features (e.g. pinholes for fiber connector alignment) are introduced in the package. But this approach requires accurate alignment of the lenses and microfeatures with the underlying chip, which is an expensive step in an assembly process. An alternative, more cost-effective approach, uses laser ablation for micromachining the optics and microalignment features directly into the encapsulant. Due to the direct-write and contactless nature of the process, excimer laser ablation can be employed in a final stage of an opto-electronic assembly process to significantly reduce assembly and packaging cost. The ablated microlenses developed for this application are presented in this section.

As discussed earlier in the section of Ablation of Optical Polymers, a promising class of transparent polymers is the Cyclic Olefin Copolymers (COC), including: Zeonex® 330R, Zeonex® 480, and Zeonor® 1600. These polymeric compounds have a fair heat resistance (or higher glass transition temperature) and very low water absorption compared to standard common polymers. These properties make them ideal encapsulant materials for optoelectronic chip packaging. Figure 12 shows the morphology of three ablated circular cavities of 200-μm in diameter for Zeonex 330R, 480, and Zeonor1600, respectively. The cavities are ablated by an ArF Laser with 30 pulses at a fluence of 290 mJ/cm$^2$ and a repetition rate of 20 Hz. As shown, Zeonex 480 (middle figure) and Zernor 1600 (bottom figure) provide a surface of good optical quality for ablated microlenses, while the surface finish of Zeonex 330R is not appropriate for optical applications. The main reason for the corrugated surface is that Zeonex 330R has the lowest absorption, about one third of the other two, as indicated in Table 3. Figure 13 depicts the corresponding surface quality of the microlenses ablated respectively in Zeonex 480 and Zeonor1600 by an ArF laser. The lens surfaces are smooth and well-defined. Little debris can be detected in the vicinity of the ablated areas. The microlenses are made by the contour scanning process discussed earlier with a fluence of 290 mJ/cm$^2$ at 20 Hz, which takes about 2 minutes to make one lens.

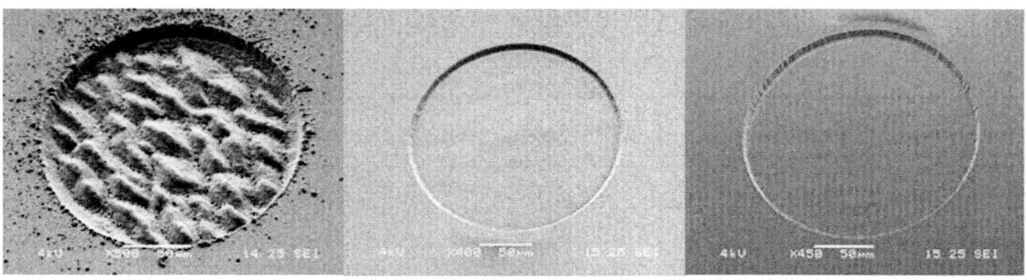


Figure 12. SEM images of circular cavities of 200-μm in diameter ablated by 30 ArF laser pulses at a fluence of 290 mJ/cm$^2$ and 20 Hz in Zeonex 330R (left), Zeonex 480 (middle), and Zeonor 1600 (right) COC.

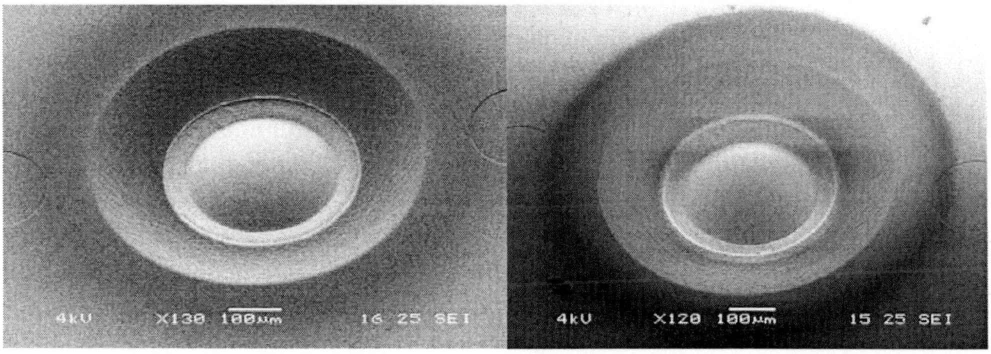


Figure 13. SEM images of microlenses ablated by ArF laser using contour scanning process a fluence of 290 mJ/cm$^2$ at 20 Hz in Zeonex 480 (left) and Zeonor 1600 (right).

A group of epoxy- and acrylate-based transparent polymers, which have been used for regular IC packing for many years, are also selected for the present ablation study to assess their feasibility for making good quality microlenses for optoelectronic chips encapsulants.

There are all commercially available, including Emerson & Cuming Stycast-1200J, Delo Katiobond 4563, Norland NOA 61, Norland NOA 72, Norland NOA 76, and Norland NOA 83H. As mentioned earlier, Stycast-1200J and Delo Katiobond 4563 are epoxy based thermosets, while the other four are acrylic based thermoplastics. The ablation parameters of all these polymers have been studied and are reported in Table 3.

Figure 14 shows two circular cavities of 200-μm in diameter in Stycast 1200J (left) and Delo Katiobond 4563 (right), respectively. They are ablated by an ArF laser using 30 pulses at a fluence of 290 mJ/cm$^2$ and a repetition rate of 20 Hz. It is clear that the surface quality of Stycast 1200J is much better than that of Delo Katiobond 4563. Again, the less desireable surface quality of Delo epoxy is caused by its low absorption coefficient, which is about one-half of Stycast's, as indicated in Table 3. The quality of microlenses made in the corresponding materials is shown in Fig. 15, indicating that the microlens in Katiobond has rougher surface quality as well as large surface cracks generated at the edge of the ablated area. It is to be noted that a rougher surface is less desirable, but a cracked surface is unacceptable. The microlenses with a lens diameter of 200 μm shown in Fig. 15 are ablated, again, using the contour scanning technique at a wavelength of 193 nm, a fluence of 345 mJ/cm$^2$, and a pulse repetition rate of 20-Hz.

Micolenses are also made in the four Norland acrylates. Figure 16 depicts the micolenses ablated in the four acrylates under the same ablation conditions, as those in Fig. 15. All the four Norland polymers exhibit similar behavior under ablation. Although quantitative measurements have not yet been performed, visual analysis of surface quality on SEM images reveal that all four Norland acrylates are suitable for making microlenses with the contour scanning technique. For the nine proprietary polymers considered (also indicated in Table 3), all, except Zeonex 330R and Delo Katiobond 4563, are appropriate for making microlenses with satisfactory quality by excimer lasers using the contour scanning technique. More comments on related experiments can be found in Naessens et al. (2003b).

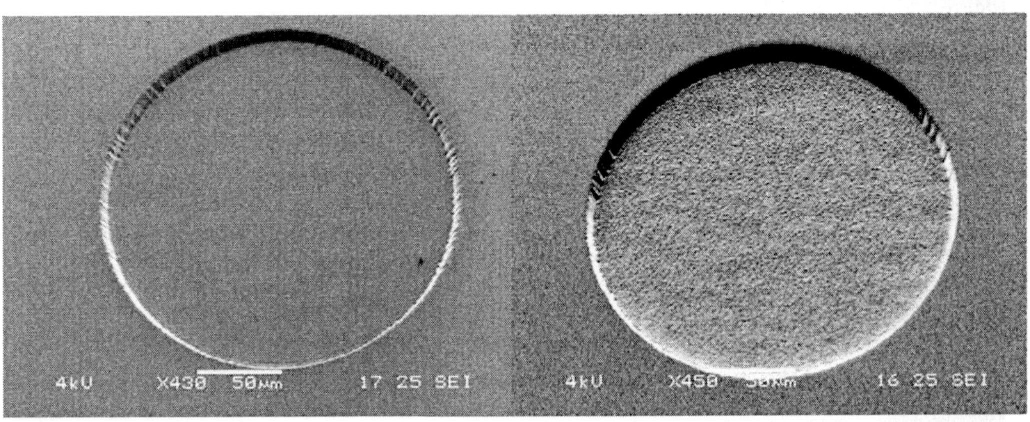


Figure 14. SEM images of laser ablated cavities with a diameter of 200 μm in Stycast 1200J (left) and Delo Katiobond (right) by 30 ArF pulses at fluence of 290 mJ/cm$^2$ and 20 Hz.

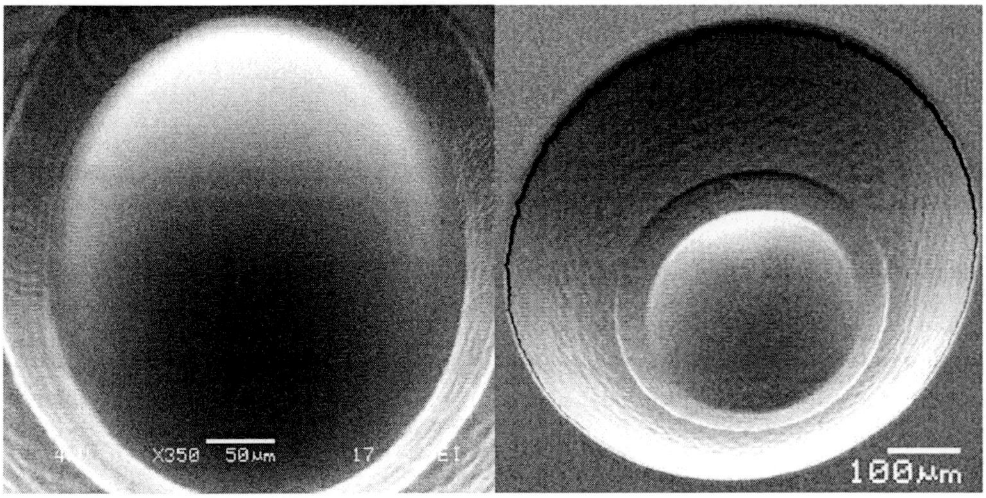


Figure 15. SEM images of microlenses with a lens diameter of 200 µm in Stycast 1200J (left) and Delo Katiobond (right showing large surface cracks at edge of ablated area) by ArF laser using contour scanning at fluence of 345 mJ/cm$^2$ and 20 Hz.

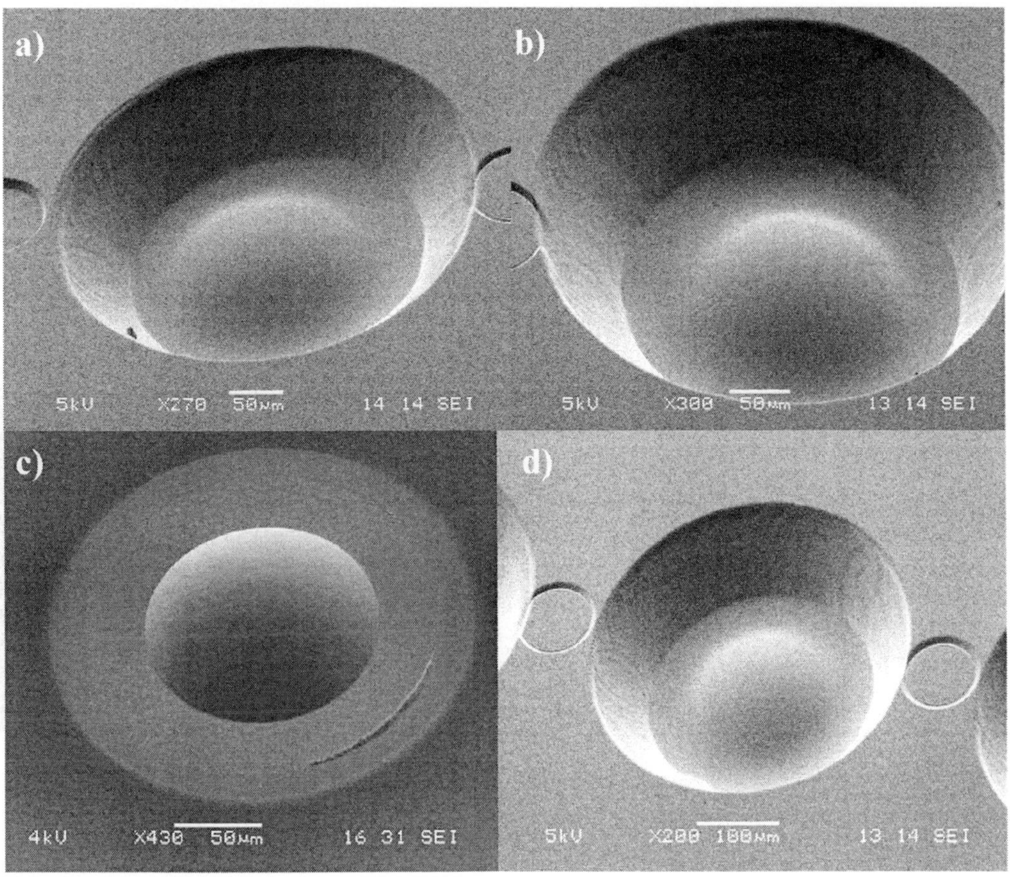


Figure 16. SEM images of microlenses ablated by ArF Laser at 345 mJ/cm$^2$ and 20 Hz using contour scanning in encapsulants: a) NOA 61, b) NOA 72, c) NOA 76, d) NOA 83H.

## ABLATION OF GLASS BASED MATERIALS

For decades, glass has been the most popular material for making optical components and devices. A wide range of glasses has been developed for the diverse needs of applications, from consumer products, such as lenses and components for cameras and projectors, binoculars, and telescopes, to high-tech industry applications, including microscopy, medical instrument, laser fusion, geodesy, reprographics, photonics, data discs, and image processing. Ablation of intricate glass-based microstructures is a value-added necessity in the development of the advanced industry applications. Excimer lasers provide an adequate and affordable approach for machining these microstructures to meet the expected demands. In this section, the basic ablation phenomena for glass materials are presented.

The laser workstation used for the ablation of glasses is similar to that shown in Fig. 1. The workstation is equipped with an ArF laser of Model LPX210, provided by Lambda Physik of Germany and a 7×7 fixed-array homogenizer, made by MicroLas of Germany, providing 12 x 23 mm uniform illumination at the mask. The workstation is enclosed with an inert gas to prevent $O_2$ absorption and mounted on a vibration isolation optical table to reduce vibration coupling between the mask and the substrate on the motorized stage. Since glasses ablated at $\lambda = 248$ nm by KrF lasers can easily induce microcracks in the ablated area, in the present study, only an ArF laser radiated at $\lambda = 193$ nm is used to avoid the possibility of cracks formation. Most of the glasses have adequate absorption at $\lambda = 248$ nm by KrF lasers. In operation, the operating parameters: the repetition rate, pulse number, and fluence, are adjustable and can be preset by the computerized controller, while the pulse duration is fixed at 20 ns FWHM.

The effects of changing major operating parameters, including the fluence, pulse number, repetition rate, and focal plane, on the ablated structures will be studied. Either a rectangular or a circular shaped stencil mask, made of 304 stainless steel, is used for the ablation study. The material selected to illustrate the ablation characteristics of glass is a borosilicate BK7 glass, which consists of mainly $SiO_2$ with boron and various other elements. In addition to its cost advantages, BK7 glass has low coefficient of expansion, approximately three times less than the other popular glass, such as soda-lime, which makes BK7 glass useful for heating and other thermal-environment packaging, without the risk of cracking due to thermal shock. Thermally stable up to 500 °C for continuous use and 600 °C for short periods. It is also more resistant to non-fluorinated chemicals and mechanically stronger than soda-lime glass. All of these properties make BK7 a popular material for optical applications

Figure 17 shows the experiment results of the relationship between the ablation depth (d) and pulse number for four fluences at a repetition rate of 10 Hz. As shown, for the range of fluence considered, d increases quite linearly with the pulse number, which is similar to those for PC shown Fig. 2. This implies that each pulse at a specific fluence (F) contributes equally to the ablation depth following the relationship shown in Eq. (5). Note that the coefficients of determination ($R^2$), for all curve fittings shown in Fig. 17, are higher than 0.99, which means the ablation depth correlates with the pulse number almost perfectly. Based on the data shown Fig. 17, the specific values of the ablation rate ($\Delta d$) can be found as 116, 129, 138, and 158 nm/pulse for the fluence at 1.4, 1.7, 2.1, and 2.4 J/cm$^2$, respectively. If $\Delta d$ is plotted against the incident fluence in a semi-logarithm scale as shown in Fig. 18, the corresponding fluence at zero ablation rates is $F_t$ while the effective absorption coefficient, $\alpha_e$, is the inverse of the slopes of the correlated curves. The respective values of $F_t$ and $\alpha_e$ for BK7 are found as 0.288 J/cm$^2$ and 13.54 μm$^{-1}$, in which $F_t$ is one-order of magnitude higher than and $\alpha_e$ is at the same order of magnitude of that reported in Table 2 for standard polymers. This implies that glasses need much higher energy density to be ablated but have similar optical absorption behavior as compared to that of polymers. Furthermore, the $\alpha_e$ obtained from Fig. 18 agrees very well with the value of 13.36 μm$^{-1}$ that is the data obtained from experimental extrapolation for BK7 glass at 193-nm wavelength (Tseng, Chen, and Ma, 2004). This may indicate that the relationship of Beer-Lambert law is reliable and accurate, and thus will be used for guiding further ablation of microstructures for different applications.

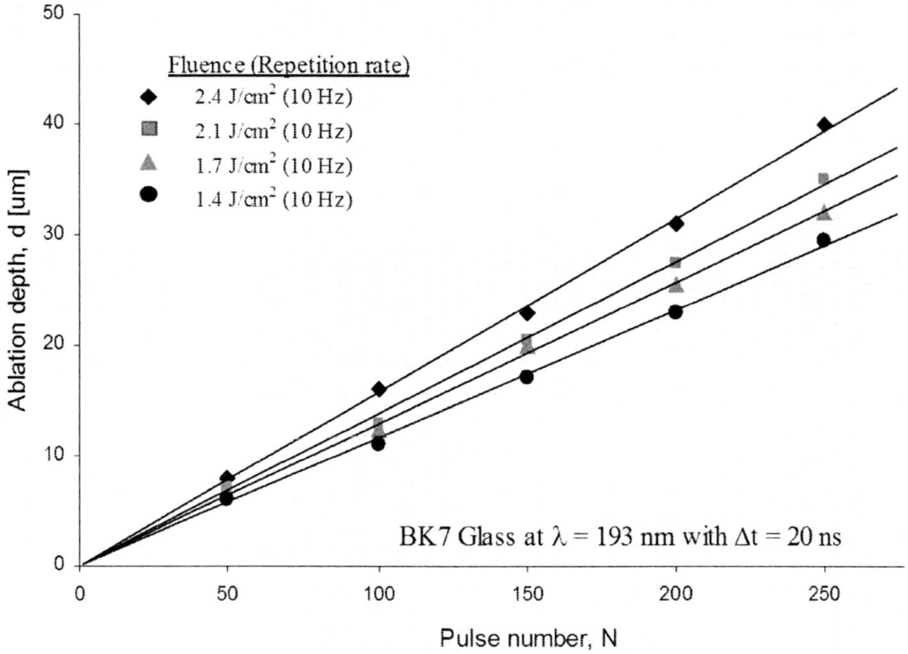


Figure 17. Correlations of ablation depth with pulse number at various fluences for BK7 glass by ArF laser.

For the sake of comparison, Fig. 18 also includes six additional sets of experimental data obtained from Ihlemann and Wolff-Rottke (1996) for a borosilicate glass and fused silica, Keiper et al. (2000) for a borosilicate Pyrex glass, Jackson, Metheringham, and Dyer (1995) for an Nd doped glass, and Awazu (2001) for a fused silica. Again, the corresponding threshold flence, $F_t$, can be found as the fluence at a zero ablation rate and the effective absorption coefficient, $\alpha_e$, is the inverse of the slopes of the correlated curves shown in Fig. 18. The respective values of $F_t$ and $\alpha_e$ are also reported in Table 4 with the coefficients $R^2$ for each curve fitting shown in Fig. 18. For all the data considered, except that by Keiper et al. (2000), the corresponding $R^2$ are all higher than 0.93. This indicates that the ablation depth correlates with the pulse number reasonably well and can be trusted for guiding future ablation operations for making microstructures. The correlation coefficient for the Pyrex glass data obtained by Keiper et al. (2000) is barely 0.58, which means that the data do not follow the Beer-Lambert law and the extrapolated parameter, $F_t$ and $\alpha_e$, should not be used further.

As indicated in Table 4, borosilicate glass is relatively easier to be ablated as compared with fused silica or doped glass. Based on the data of Nd doped glass, it also can be seen that, to ablate glass-based materials, ArF lasers should be much effective that the KrF lasers, since the threshold fluence with a KrF laser is more than three times higher than that of an ArF laser. Since different pulse durations (Δt) are used by different investigators, the effect of pulse duration should be considered in comparing these parameters. An empirical relationship between the threshold fluence and Δt: $F_t \sim (\Delta t)^{1/2}$ can be used for scaling $F_t$ for the duration time within the ns domain (Du et al. 1994). For example, in the two borosilicate glasses reported in Table 4, $F_t$ is 243 J/cm$^2$ for Δt =18 ns and 288 J/cm$^2$ for Δt = 20 ns, which implies that $F_t$ indeed increases with the duration time.

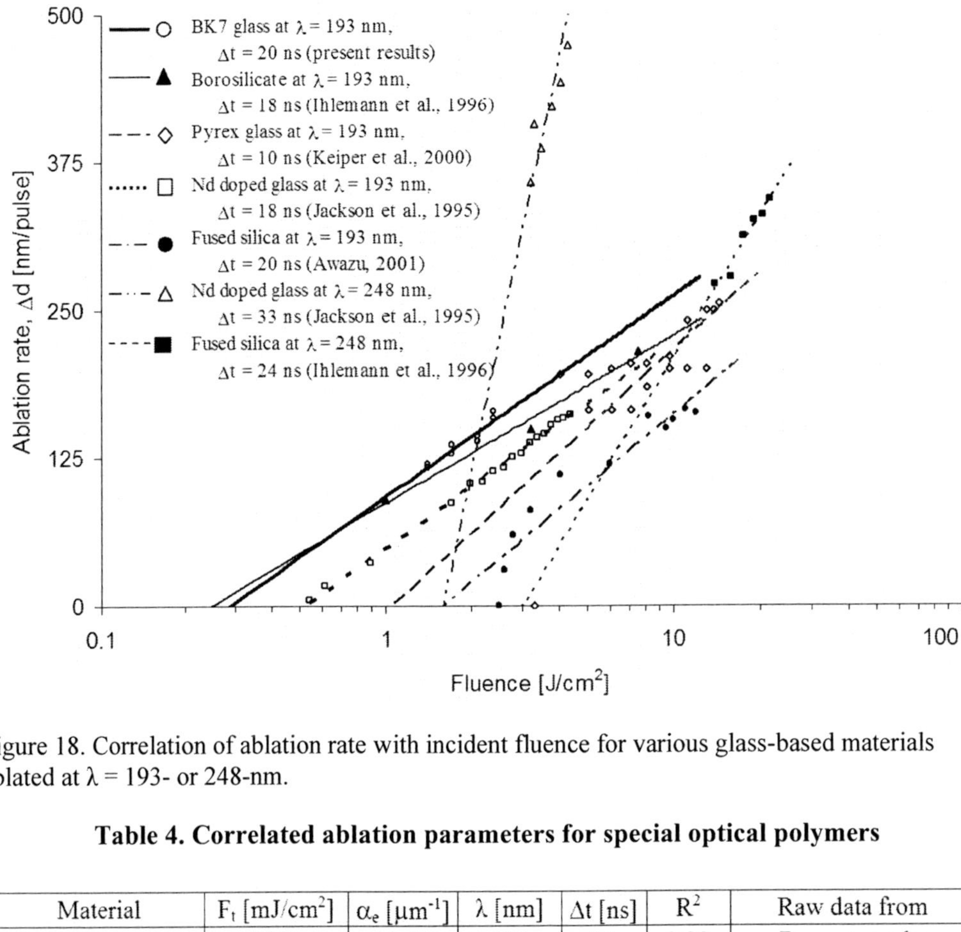


Figure 18. Correlation of ablation rate with incident fluence for various glass-based materials ablated at λ = 193- or 248-nm.

**Table 4. Correlated ablation parameters for special optical polymers**

| Material | $F_t$ [mJ/cm$^2$] | $\alpha_e$ [μm$^{-1}$] | λ [nm] | Δt [ns] | $R^2$ | Raw data from |
|---|---|---|---|---|---|---|
| BK7 glass | 288. | 13.54 | 193 | 20 | 0.93 | Present results |
| Borosilicate glass | 243. | 16.28 | 193 | 18 | 0.99 | Ihlemann et al. (1996) |
| Pyrex glass | 1638. | 8.443 | 193 | 10 | 0.58 | Keiper et al. (2000) |
| Nd doped glass | 529. | 13.22 | 193 | 18 | 1.00 | Jackson et al. (1995) |
| Fused silica ($SiO_2$) | 1287. | 12.87 | 193 | 20 | 0.93 | Awazu (2001) |
| Nd doped glass | 1599. | 1.999 | 248 | 33 | 0.97 | Jackson et al. (1995) |
| Fused silica ($SiO_2$) | 3034. | 5.725 | 248 | 24 | 0.94 | Ihlemann et al. (1996) |

# GLASS-BASED ABLATED MICROSTRUCTURES

Current microstructures made out of glass are mainly fabricated by photolithography and an etching process, which are only economically justified by mass production. Micromachining by lasers, or more specifically excimer lasers, is especially suitable for prototyping and for a small batch of glass-based product. In this section, the techniques to ablate arrayed and curved microstructures using an ArF laser are presented. The first two are arrayed structures used for optical waveguides and wave absorber, while the non-arrayed structures are fabricated for optical lenses and guides. The possible application for biomedical stamping is also discussed.

## Parallel Grooves for Optical Waveguide with Continuously Changing Gratings Profiles

The ablation technique for making glass-based parallel grooves is similar to that for ablating arrayed ferrule structures presented earlier (Fig. 9b). By controlling the movement of the substrate and the pulse number, a series of parallel trenches or grooves with a constant or variable pitch can be constructed. The parallel grooves or trenches structures shown in Figs 19a and 19b are respectively ablated by 100 and 250 ArF-laser pulses at a fluence of 2.4 J/cm$^2$ and a pulse duration of 20 ns. The corresponding groove mouth opening and array pitch are 20 and 40 μm, respectively, with an approximate depth of 16-μm for 100 shots and of 40-μm for 250 shots. The ablated mouth opening and array pitch are consistent with the dimensions designed in the mask. In additional to the depth of the groove, which is dictated by the pulse number as indicated in Eq. (5), the groove profile can also be changed by adjusting the fluence in ablation. As compared Fig. 19b with Fig. 19c, it can be seen that ablation at relatively high fluence (2.4 J/cm$^2$ in Fig. 19b) tends to form a U-shaped cross-section or profile, while a V-shaped profile can be obtained at lower fluence (1.7 J/cm$^2$ in Fig. 19c). The ablation results also indicate that the repetition rate has no noticeable effects on the ablated profiles from 2 to 10 Hz, the range studied.

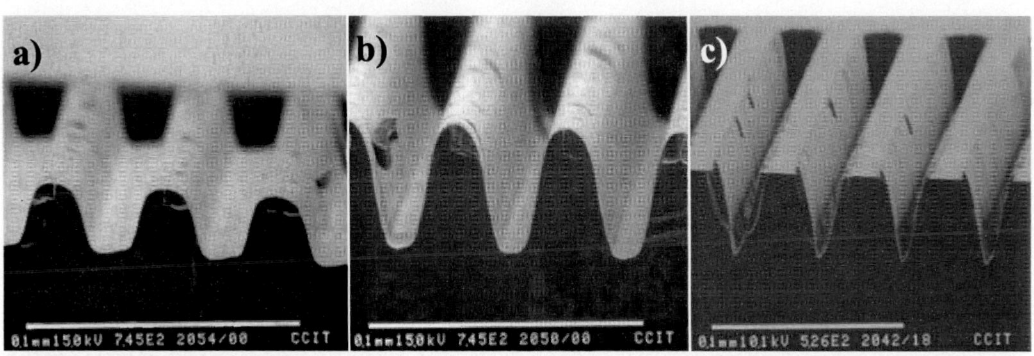


Figure 19. SEM images of parallel grooved microstructures ablated by ArF laser at: a) 2.4-J/cm$^2$ fluence with 100 shots, b) 2.4-J/cm$^2$ fluence with 250 shots, and c) 1.7-J/cm$^2$ fluence with 250 shots

The surface debris shown in Fig. 19 is the oxide layer frequently accumulated on the ablated surface. The debris can be etched out in a solution of 10% HF and 90% distilled water, for a period of ten to a few hundred seconds. This debris is also easier to break up at higher ablating energies and could cause micro-cracks near the ablated glass, especially at ablation wavelengths longer than 193-nm. In general, more debris is accumulated near the ablated region with higher fluences. Images obtained by an atomic force microscope (AFM), shown in Fig. 20, indicate that the ablated surface morphology is flatter and has less debris at lower fluences (0.6 J/cm$^2$) than that at higher energy density levels (2.4 J/cm$^2$) for the range of the fluence considered. This observation may be due to the fact that at higher fluences, a higher vapor or plasma pressure can build up inside the groove to sputter materials more rapidly along the periphery of the groove, re-melting the surroundings and ejecting more debris. In Fig. 20, both grooves are ablated with 5 pulses.

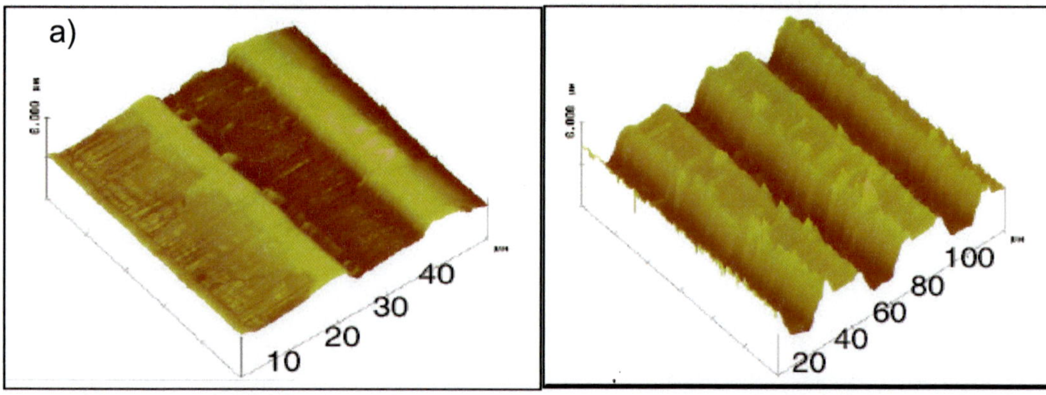


Figure 20. Effects of fluence on ablated surface morphology at: a) 0.6-J/cm$^2$ fluence with 5 shots, b) 2.4-J/cm$^2$ fluence with 5 shots.

The glass-based parallel grooved microstructure is an ideal microstructure for use as a waveguide (Levy-Yurista and Friesem, 2001). The waveguide consists of perturbing gratings that couple the guided wave to a free space diffracted order, as shown in Fig. 21. In the conventional waveguide, all grating elements are equal in depth and the coupling length is fixed by the decrease of perturbation. As a result, discontinuity scatter occurs at the beginning and end of grating. On the other hand, new couples can be designed to use gratings with tailored depth profiles to match the impedance at the beginning, so that the new design can maximize out coupling with a minimum length to increase its performance and size. The laser techniques presently developed can be conveniently used for making various waveguide structures and each grating element can be easily tailored to different depth profiles. Certainly, to fully demonstrate the versatility of the variable profiled waveguide proposed here, an experiment on light coupling and reduction of coupling loss or a numerical simulation of its performance is necessary.

It is to be noted that the ablated structure may not have the desired surface quality for waveguide applications. However, a thin glass film coating or deposition on a hot glass substrate subsequently with appropriate annealing can significantly improve its surface quality to make the ablated waveguide structure with satisfied optical quality. Also, the laser smoothing technique shown in Fig. 11 can be used, in which a thin surface layer covering the entire waveguide area is simultaneously removed by ablating the waveguide deeper into the substrate with a large projection beam covering the whole ablation surface. The effectiveness of any of these two techniques or surface improvement can be established by conducting more quantitative experiments to quantify the final surface quality and dimensional accuracy.

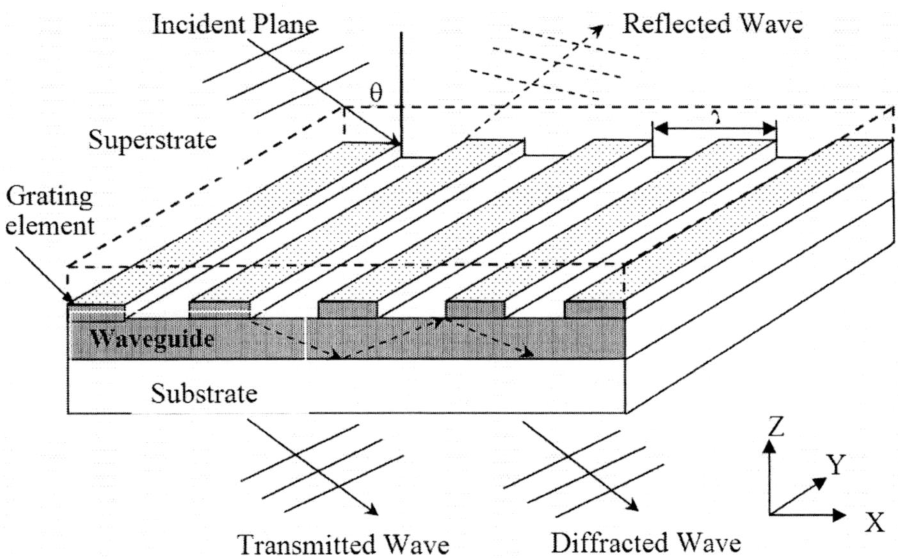


Figure 21. Conventional waveguide having equal grating elements.

## Arrayed Pillars for Microstamping and Wave Absorbing

Two steps are involved in making crossed arrays or checkered patterns. A set of parallel microgrooves similar to those shown in Fig. 19 can be ablated first. By rotating the grooved substrate or stage platform 90 degrees, a series of parallel grooves which are perpendicular to the grooves ablated in the first step can be formed; thus, a microstructure with checkered arrays can be obtained. The checkered-array pillars shown in Fig. 22a are ablated at the fluence, pulse number, and repetition rate of 1.4 J/cm$^2$, 300, and 10 Hz, respectively. The grooves are 20-μm wide, which is consistent with the projected beam size. As shown, the grooves are perpendicular to each other, and form a 2D array of pillars. The vertex or head of the pillars are flat, having a approximate 20 x 20 μm square cross-sectional area. The relatively low fluence (1.4 J/cm$^2$) used here is designed to avoid the possibility of microcracking, which can be induced occasionally at relatively high influences for ablating high-aspect-ratio structures as mentioned earlier.

The projection system has been set to have a 10x demagnification and the projection focusing lens used has a 14-mm focal length (Fig. 1). The sample is placed at the optimal focal plane ($\Delta f$ = 0), which is 1.4-mm, away from the focal point or 15.4-mm from the projection lens, as shown in Fig. 1. By shifting the focal plane (f) in the projection system, different shapes (vertex angles) of arrayed pillars can be produced. Figures 22b and 22c show two checkered-arrays ablated by the same procedure and same ablation condition as those used for ablating the flat-head array shown in Fig. 22a, except the focal plane being shifted from the original 1.40 mm to 1.60 mm ($\Delta f$ = +200 μm) for the case of Fig. 22b and from 1.40

mm to 1.70 mm (Δf = +300 μm) for that of Fig. 22c. As shown, the vertex shape or the pillar head geometry is clearly depicted and is changed from flat (Fig. 22a) to rough (Fig. 22b) and then smooth (Fig. 22c). The corresponding edge curvature of the pillar head increases from relatively small (sharp edge shown in Fig. 22a), to relatively large on the order of hundreds of μm (rough curved profile shown in Fig. 22b), and to even larger on the order of thousands of μm (smooth curved profile shown in Fig. 22c). This also implies that, if the focal plane deviates large enough, e.g. 300 μm, the square-cross section of the pillar head can shrink and converge into a sharp point and its corresponding sharpness can be controlled by the deviation or defocusing size (Δf). It is to be noted that the most uniform energy density is produced at the optimal focal plane (Δf = 0) to yield the best ablation effect, as shown in Figure 22a. Thus, more laser energy is needed for the ablation of the substrate located at non-optimal focal positions.

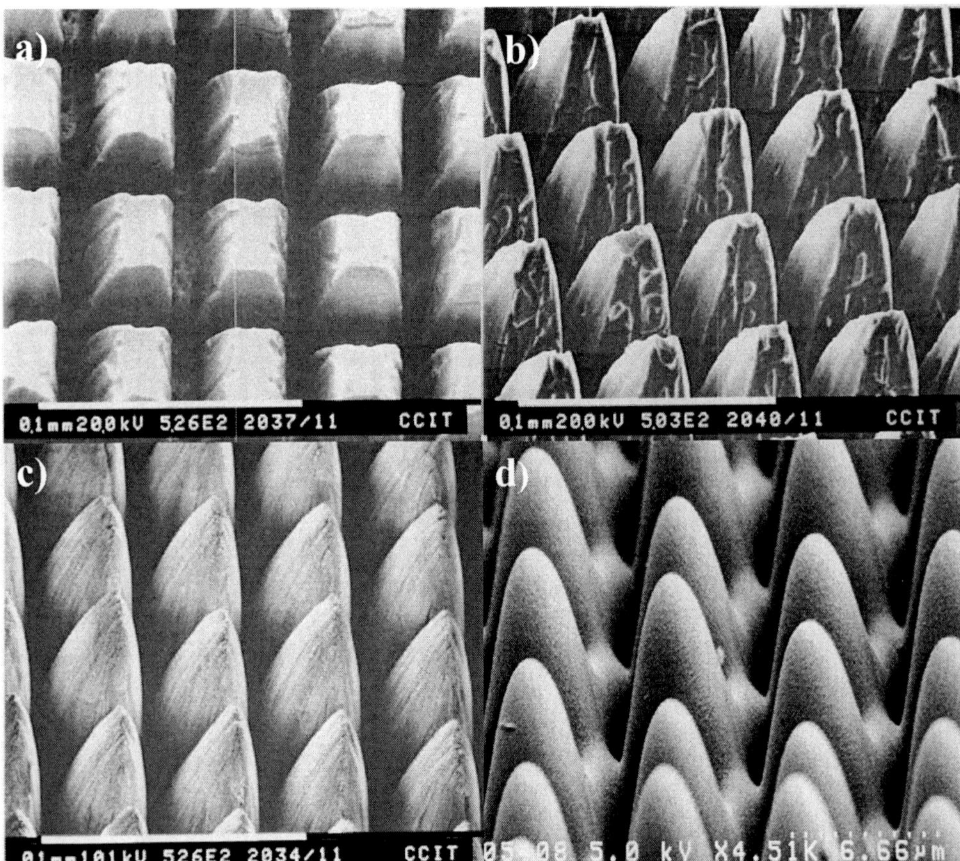


Figure 22. Arrays of micropillars with different vertex shapes ablated at 1.4-J/cm$^2$ fluence, 300 pulses, and 10 Hz repetition rate: a) flat head with Δf = 0, b) rough curved head with Δf = 200 μm, c) smooth curved head with Δf = 300 μm, and d) wave absorbing structure.

In additional to visible light, glasses are also transparent to allow a microwave to enter and to reduce its reflection back to the sources, while a flat metal can reflect most microwaves back to the source antenna. This makes the glass a good candidate for wave absorbing materials (Harakawa et al., 2002; Yokokawa et al. 2004). A wave absorb is used for suppress in unwanted reflection waves and has applications not only in military, such as radar absorbing structures used for the stealth of aircrafts, ships, and missiles, but also in commercial areas, including antenna pattern modification and electromagnetic interference reduction for telecommunications and information technology as well as for use in marinas, ships, and canals. Glasses with 1-D and 2-D periodic structures are often used for these applications (Baekelandt, Olyslager, and Zutter, 1995; Soh et al., 2003; Oh et al., 2004). With the technique developed for ablating 2D-arrayed pillars with curved heads, an arrayed glass structure with a waved surface can be fabricated for the use as a wave absorber shown in Fig. 22d. Since the waved glass surfaces can be ablated before or after the assembling process, so that it can be conveniently integrated into the structure to provide the required wave absorbing or reflecting properties to the products.

Furthermore, glass-based arrayed micropillars with flat heads can be applied for biomedical applications, including simultaneously printing biomolecules, such as proteins or other biologically active substances, onto a patterned bio-absorption surface for modern biomedical applications, including disease diagnosis, DNA sequencing, and drug screening and discovery (Renaultt et al., 2003). By parallel biological and chemical reactions, the arrayed flat-head pillars can be used as microstamps to process huge numbers of reagents in a short time (Tseng et al., 2002). The sizes of the stamps can be varied from 100 nm to 350 μm and they are normally made by the time-consuming semiconductor lithographic process (Morozov and Morozova, 1999; Tseng et al., 2002; Renaultt et al., 2003). The laser ablation technique developed provides a cost-effective alternative for making this type of arrayed microstamps. For further development of the ablation technique, the effects of surface properties of the glass-based probes on the effectiveness of transferring various biomolecules onto different patterned substrates in micro-bio printing processes should be studied.

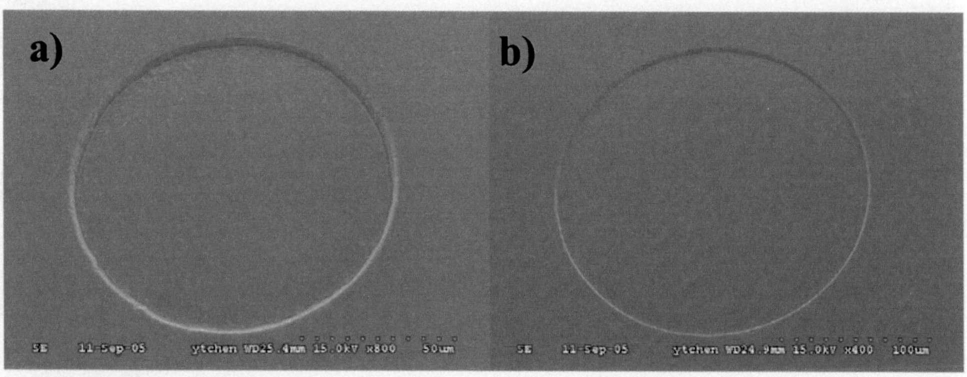


Figure 23. SEM images of circular cavities ablated by 10 ArF laser pulses at a fluence of 2.4 J/cm$^2$ and 5 Hz in BK7 glass from a mask opening size of: a) 1000-μm, b) 2000-μm.

## Curved Surface Ablation for Optical Lenses

The contour scanning ablation technique shown in Fig. 10 can also be used in ablation of glass for making curved surfaces. In ablating BK7 glass, a linear instead of rotational scanning velocity is adopted. A mask containing four circular openings, two with a diameter of 1000-μm and the other two being 2000-μm, is used to study the curved surface ablation technique. With 10x demagnification, the projected laser image on the substrates is 100 and 200 μm, respectively. The ablated circular cavities of 100 and 200 μm in diameter at the fluence, repetition rate, and pulse number of 2.4 J/cm$^2$, 5 Hz, and 10 shots are shown in Fig. 23.a and 23.b, respectively. If a linear scanning speed, which is a relative speed between the laser and substrate, is imposed, the ablated surface could become curved. Figure 24.a shows four channels with curved surface profiles being ablated at a linear scanning speed of 1 mm/s, in which the two channels on the left are resulted from the 1000-μm mask openings, while the other two channels on the right are attained by the 2000-μm openings. The curved-surface channels are approximately 100- or 200-μm wide, which are consistent to the projected respective bean sizes.

The curved profile has been measured by a 3D profile meter provided by SUN Precision of Korea (Model: SIS 1000) and is shown in Fig. 24.b for the 100-μm wide channel. The maximum depth is 6.34 μm and the corresponding standard deviation from a second-order polynomial curve is less than 0.2 μm. The size of the standard deviation, which is used to correlate the roughness of a curved surface, should be improved, so that the ablated curved configuration can be used as optical components, such as lenses and optical probes, and waveguides. Again, the laser smoothing technique shown in Fig. 11 can be used, in which a large projection beam covering the whole ablation surface is used to simultaneously remove a thin surface layer covering the entire area considered. Also the surface finish can be improved by an additive process, by coating or deposit a thin glass film on a hot glass substrate subsequently with appropriate annealing.

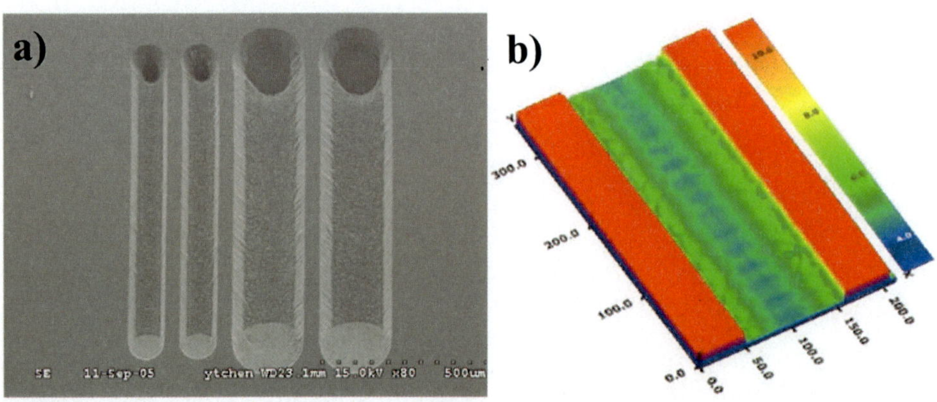


Figure 24. Curved channels ablated by ArF laser pulses at scanning speed of 1 mm/s with a fluence of 2.4 J/cm$^2$ and 5 Hz in BK7 glass using a circular opening mask: a) SEM image of four curved channel with a mask having two opening sizes, 1000- and 2000-μm, b) 3D profile image of channel from 1000-μm size opening..

It is also interesting to show the ablated channels using two different triangular-opening masks at the same ablation conditions as those shown in Fig. 24.a. The channels shown in Fig. 25.a are resulted from an equiangular-triangle mask, in which its base is perpendicular to the scanning direction, while the channels shown in Fig. 25.b use a right triangle mask, in which one of its non-hypotenuse sides is aligned with the scanning direction during ablation.

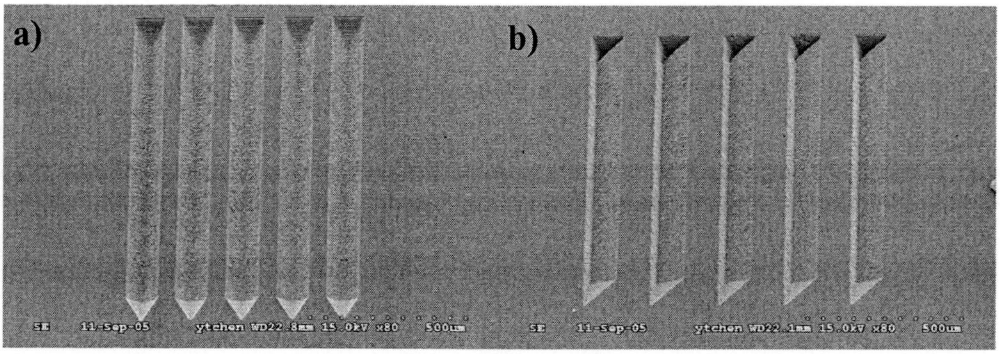


Figure 25. SEM images of curved channels ablated at scanning speed of 1 mm/s with a fluence of 2.4 J/cm$^2$ and 5 Hz in BK7 glass using: a) equiangular-triangle opening mask, b) right-triangle opening mask

## CONCLUSION

The feasibility of using excimer lasers for micromachining of transparent materials for optoelectronic applications has been illustrated and demonstrated. The ablation characteristics of two of the most important transparent materials, polymers and glasses, have been studied. Experiments on ablated simple cavities are first performed to provide the basic information and guidelines for the ablation of more intricate microstructures, which are often demanded by industry. The experimental study, using both ArF and KrF excimer laser, examines the ablation rates by varying several major operating parameters. The linear relationship between the ablation depth and pulse number has been observed for all the polymers and glasses considered. The relationship between the ablation depth and pulse fluence, which appropriately follows the Beer-Lambert law, has also been identified for most of the materials evaluated and can be used for guiding the ablation operation for high-precision microstructures. In fact, the precise ablation depth of a curved or flat surface can be achieved by adequately controlling the number of laser pulses and the intensity of the fluences. Several related issues, including surface roughness and residual debris, which directly impact on the precision and quality of the ablated microstructures have also been addressed. Although the pulse repetition rate has no significant effects on the ablation rate for most materials, it has a noticeable impact on the morphology of the ablated surface for both polymers and glasses. Results have indicated that increasing the repetition rate favors the morphology of ablated surfaces; however, sometimes the resulting debris or residue is hard to remove.

In polymers, the ablation behavior for both common and proprietary polymers is investigated. Experimental verification of cavity quality and influence of laser parameters on the ablation process are described systematically for the common polymers, including PC, PET, PI, PMMA, and PS. PMMA exhibits the lowest absorption, in particular at 248 nm. A thermal degradation pathway and low absorption are responsible for the corrugated cavity surface and increased ablation rate of PMMA with higher pulse repetition rate at 248 nm. Ablation in PC, PI, and PS gives the best cavity definition and a relative surface roughness as low as 0.5% is obtained without any beam-homogenizing optics. With the basic ablation information accumulated in the experiment, PC and PMMA are then selected as the base material for fabrication of micro-alignment and micro-optical components for photonic applications. Alignment grooves with a cross-section optimized for holding plastic optical fibers are ablated in PMMA substrates and employed for the assembly of a 2D fiber ferrule.

A scanning contour technique is also developed for the ablation of circular microlenses with desired surface profiles. For relaxed single-mode fiber-to-fiber coupling a suitable lens has designed and ablated in a small PC plate on top of the individual fibers. Equipped with the microlens, the relaxed lateral misalignment can be improved from 3.1 to 1.1 μm with little insertion loss penalty. Furthermore, excimer laser ablation is also applied for micromachining of a broad range of commercially proprietary polymers. Ablation of microholes and microlenses in cyclic olefin copolymers and several epoxies and acrylates is successfully demonstrated. The epoxy and acrylates based polymers are of particular interest to chip encapsulation, which opens the way to direct integration of micro-optics and micro-alignment features in the chip encapsulant.

In glass ablation, the technique to ablate different types of arrayed microstructures with flat and curved surfaces are described. The applications of several arrayed microstructures for optical waveguide, wave absorber, microstamper, and microlens are specifically illustrated. It has been found that glass is an excellent material for laser ablation of intricate optical microstructures, which are in demand by photonics and communications industries. However, micromachining of some special glass-based materials, such as fused silica or quartz is still not appropriate for excimer lasers, because these materials are basically transparent at the wavelength of excimer lasers. Recently, it has been discovered that, by applying organic solution, including pyrene or chlorine, to the surface, these materials can be ablated by the excimer laser at reasonable ablation rates (Wang, Niino, and Yabe, 2001; Vakanas, Tseng, and Winer, 2002). Further research and development efforts should be encouraged in this area to enhance the applicability of excimer lasers in micromachining.

Indeed, excimer lasers are effective tools for machining both polymers and glasses at micrometer scales. The excimer laser ablation offers quite a few advantages in machining polymers and glasses, such as very localized and well-controlled energy absorption with minimized heat affected zone, no mechanical contact with part, and no tool wear. Also, laser ablation is essentially a direct-write technology: it directly creates the pattern into a substrate without an intermediate resist layer. By scanning the laser beam over the part, complex patterns can be ablated without an expensive mask tool. Finally, since no clean-room environment is required during operation and its ability for on-line process monitoring, the laser ablation technique demonstated currently offers a viable alternative to the conventional semiconductor processes.

## ACKNOWLEDGEMENTS

The authors would like to thank the US National Science Foundation under Grant No. DMI-0423457and CMS-0115828, the Flemish IWT through the ITA-II, VLSI Photonics, and LIAM projects, and the ROC National Science Council under Grant No. NSC90-2811-E-002-007, for their financial supports. The authors would also like to express their gratitude to Dr. Ying-Tung Chen of Chung Cheng Institute of Technology of Taiwan for conducting useful experiments for the ablation of glass-based microstructures, to Yvan Van Gorp from Emerson & Cuming Benelux for providing encapsulant materials, and to Jan Watté from Tyco Electronics in Kessel-Lo, Belgium for supplying fiber-optic material and the fruitful discussions on fiber-optics. The assistance from Jong-Seung Park and KC Kruger of Arizona State University in preparing this manuscript should be specifically acknowledged.

## REFERENCES

Awazu, K. (2001), "Ablation of amorphous $SiO_2$ using ArF excimer laser," in *32nd Annual Boulder Damage Symposium- Laser-Induced Damaged in Optical Materials, Proc. SPIE*, Vol. 4347, pp. 169-176, *Int. Soc. Optical Eng.*.

Alexander, M.R., Short, R.D., Jones, F.R., Michaeli, W., and Blomfield C. J. (1999), "A study of HMDSO/O-2 plasma deposits using a high-sensitivity and -energy resolution XPS instrument: curve fitting of the Si 2p core level," *Applied Surface Science*, Vol. 137, No. 1-4, pp. 179-183.

Baekelandt, B., Olyslager, F., and De Zutter, D. (1995), "Comparison of two-dimensional periodic and finite absorber structures," in *Proc. 1995 IEEE International Symposium on Electromagnetic Compatibility*, pp. 573-577, IEEE.

Ball, Z., Feurer, T., Callahan, D., and Sauerbrey, R. (1996), "Thermal and mechanical coupling between successive pulses in KrF-excimer-laser ablation of polyimide," *Appl. Phys. A*, Vol. 62, p. 203-211.

Bityurin, N, Luk'yanchuk, B. S., Hong, M. H., and Chong, T. C. (2003), "Models for laser ablation of polymers," *Chem. Rev.*, Vol. 103, pp. 519-552.

Blanchet, G. B., Cotts, P., and Fincher, C. R., Jr. (2000), "Incubation: Subthreshold ablation of poly-(methyl methacrylate) and the nature of the decomposition pathways," *J. Appl. Phys.*, Vol. 88, pp. 2975-2978.

van den Boom, H.P.A., Li, W., van Bennekom, P.K., Monroy, I.T., and Khoe, G.-D. (2001), "High-capacity transmission over polymer optical fiber," *IEEE J. Select. Topics in Quantum Electron.*, Vol. 7 , pp. 461-470.

Burns, F. and Cain, S. (1996), "The effect of pulse repetition rate on laser ablation of polyimide and polymethylmethacrylate-based polymers," *J. Phys. D*, Vol. 29, p. 1349-1355.

Chen, Y. T., Ma, K. J., Zhou. J. G., and Tseng, A. A. (2005a), "Excimer laser ablation of glass-based arrayed microstructures for biomedical, mechanical, and optical applications," *J. Laser Applications*, Vol. 17, No. 1, pp. 38-46.

Chen, Y. T., Ma, K. J., Tseng, A. A., and Chen, P. H. (2005b), "Projection ablation of glass-based single and arrayed microstructures using excimer laser," *Optics & Laser Technology*, Vol. 37, No. 4, pp. 271-280.

Du, D., Liu, X., Korn, G., Squier, J., and Mourou, G. (1994), "Laser-induced breakdown by impact ionization in SiO2 with pulse widths from 7 ns to 150 fs," *Appl. Phys. Lett.*, Vol. 64, No. 23, pp. 3071-3073.

Dyer, P. E. (2003), "Excimer laser polymer ablation: twenty years on," *Appl. Phys. A*, Vol. 77, pp. 167-173.

Dyer, P. E. and Sidhu, J. (1985), "Excimer laser ablation and thermal coupling efficiency to polymer films," *J. Appl. Phys.*, Vol. 57, pp. 1420-1422.

Gorodetsky, G., Kazyaka, T. G., Melcher, R. L., and Srinivasan, R. (1985), "Calorimetric and acoustic study of ultraviolet laser ablation of polymers," *Appl. Phys. Lett.*, Vol.46, pp. 828-830.

Gower, M.C. (2001), "Excimer laser microfabrication and micromachining, in *Laser Precision Microfabrication (LPM2000), RIKEN Review*, No. 32, pp. 50-56.

Harakawa, K., Yamanaka, K., Kageyama, K., Masakage, M., Togashi, M., and Hashimoto, Y. (2002), "Development of transparent wave-absorbing glass." in *Proc. 2002 Asia-Pacific Microwave Conference (APMC 2002)*, Kyoto, Japan, November 19-22, *Institute of Electronics, Information and Communication Engineers (IEICE) of Japan*, Tokyo, Japan.

Hayden, C. J. (2003), "Three-dimensional excimer laser micromachining using greyscale masks," *J. Micromechanics and Microengineering*, Vol. 13, No. 5, pp. 599-603.

He, M., Yuan, X., Ngo, N. Q., Cheong, W. C., Bu, J. (2003), "Reflow technique for the fabrication of an elliptical microlens array in sol-gel material," *Applied Optics*, Vol. 42, pp. 7174-7178.

Holman, J.P. (2001), *Experimental Methods for Engineers*, 7th ed., McGraw-Hill, New York, NY.

Ihlemann, J. and Wolff-Rottke, B. (1996), "Excimer laser micro machining of inorganic dielectrics," *Appl. Surf. Sci.*, Vol. 106, pp. 282-286.

Jackson, S. R., Metheringham, W. J., and Dyer, P. E. (1995), "Excimer laser ablation of Nd:YAG and Nd:glass," *Appl. Surf. Sci.*, Vol. 86, pp. 223-227.

Janssens, K., Vincze, L., Vekemans, B., Adams, F., Haller, M., and Knöchel, A. (1998), "Use of lead-glass capillaries for micro-focusing of highly-energetic (0-60 keV) synchrotron radiation," *J. Analytical Atomic Spectrometry*, Vol.13, pp. 339-350.

Katsura, K., Usui, M., Sato, N., Ohki, A., Tanaka, N., Matsuura, N., Kagawa, T., Tateno, K., Hikita, M., Yoshimura, R., and Ando, Y. (1999), "Packaging for a 40-channel parallel optical interconnection module with an over-25-Gbit/s throughput," *IEEE Trans. Advanced Packaging*, Vol. 22, pp. 551-560.

Keiper, B., Exner, H., Loschner, U., and Kuntze, T. (2000), "Drilling of glass by excimer laser mask projection technique," *J. Laser Applications*, Vo. 12, No. 5, pp. 189-193.

Kuper, S. and Stuke, M. (1989), "UV-excimer-laser ablation of polymethylmethacrylate at 248 nm: characterization of incubation effects with Fourier transform IR- and UV-spectroscopy," *Appl. Phys. A*, Vol. 49, No. 2, pp. 211-215.

Lazare, S. and Granier, V. (1988), "Excimer laser light induced ablation and reactions at polymer surfaces as measured with a quartz-crystal microbalance," *J. App. Phys.*, Vol. 63, pp. 2110-2115.

Lee, Y.-C., Chen, C.-M., and Wu, C.-Y., (2005), "A new excimer laser micromachining method for 3D microstructures with continuous surface profiles," *Sensors and Actuators A*, Vol. 117, No. 2, pp. 349-305.

Levy-Yurista, G. and Friesem, A. A. (2001), "Dual Spectral Filters with Multi-Layered Grating Waveguide Structures," *Appl. Phys. B*, Vol. 72, pp. 921-925.

Liau, Z.L., Nam, D.W., and Waarts, R.G. (1994), "Tolerances in microlens fabrication by multilevel etching and mass-transport smoothing," *Applied Optics*, Vol. 33, pp 7371-7376.

Liu, Z., Feng, Y., and Yi, X. (2000), "Coupling effects of the number of pulses, pulse repetition rate and fluence during laser ablation of PMMA," *Appl. Surf. Sc.*, Vol. 165, No. 4, pp. 303-308.

Morozov, V. N. and Morozova, T. Y. (1999), "Electrospray Deposition as a Method for Mass Fabrication of Nano- and Multicomponent Microarrays of Biological and Biologically Active Substance," *Anal. Chem.*, Vol. 71, pp. 3110-3117.

Naessens, K., Ottevaere, H., Baets, R., Daele, P.V., and Thienpont, H. (2003a), "Direct writing of microlenses in polycarbonate with excimer laser ablation," *Appl. Opt.*, Vol. 42, pp. 6349-6359.

Naessens, K., Ottevaere, H., Van Daele, P., and Baets, R. (2003b), "Flexible fabrication of microlenses in polymer layers with excimer laser ablation," *Appl. Surf. Sci.*, Vols. 208-209, pp. 159-164.

Oh, J.-H., Oh, K.-S., Kim, C.-G., and Hong, C.-S. (2004), "Design of radar absorbing structures using glass/epoxy composite containing carbon black in X-band frequency ranges," *Composites Part B: Engineering*, Vol. 35, No. 1, pp. 49-56.

Pãtzel, R. (2002), "An introduction to excimer lasers," in *Photonics Handbook*, 44$^{th}$ ed., pp. 247-252, Laurin Publishing, Pittsfield, MA.

Renaultt, J. P., Bernard, A., Bietsch, A., Michel, B., Bosshard, H. R., Delamarche, E., Kreiter, M., Hecht, B., and Wild, U. P. (2003), "Fabricating Arrays of Single Protein Molecules on Glass Using Microcontact Printing," *J. Physical Chemistry B*, Vol. 107, pp. 703-711.

Schmidt, H., Ihlemann, J., Wolff-Rottke, B., Luther, K., and Troe, J. (1998), "Ultraviolet laser ablation of polymers: spot size, pulse duration and pulse attenuation effects explained," *J. Appl. Phys.*, Vol. 83, p. 5458.

Shuto, Y., Tohno, S., Yanagi, S., Ohno, M., and Nagase, R. (2004), "Optical characteristics and reliability of plastic ferrules for MU-type simplified receptacles," *IEICE Trans. Electronics*, Vol. E87-C, pp. 1302-1306.

Soh, T., Kondo, A., Toyota, M., and Hashimoto, O. (2003), "A basic study of millimeter-wave absorber for two frequency bands using transparent resistive films," in *IEEE Int. Symposium on Electromagnetic Compatibility*, Vol. 1, pp. 149-154, IEEE.

Srinivasan, R., Smrtic, M., and Babu, S. (1986), "Excimer laser etching of polymers," *J. Appl. Phys.*, Vol. 59, p. 3861.

Sugioka, K., Obata, K., Hong, M.H., Wu, D.J., Wong, L.L., Lu, Y.F., Chong, T.C., and Midorikawa, K. (2003), "Hybrid laser processing for microfabrication of glass," *Applied Physics A*, Vol. A77, No. 2, pp. 251-257.

Tseng, A. A., Chen, Y. T., and Ma, K. J. (2004), "Fabrication of high-aspect-ratio microstructures using excimer lasers," *Optics & Lasers Eng.*, Vol. 41, No. 6, pp. 827-847.

Tseng, F. G., Lin, S. C., Huang, H. M., Huang, C. Y., and Chieng, C. C. (2002), "Protein Micro Arrays Immobilized by μ-Stamps and -Protein Wells on PhastgelR Pad," *Sensors and Actuators B*, Vol. 83, pp. 22-29.

Vakanas, G. P, Tseng, A.A., and Winer, P. (2002), "Laser-assisted chemical etching for embedded microchannels and overhanging microstructures on Si/SiO2 substrates," *J. Laser Applications*, Vol. 14, No. 3, pp. 185-190.

Wang, J., Niino, H., and Yabe, A. (2001), "Etching of transparent materials by laser ablation of an organic solutions," in *Laser Precision Microfabrication (LPM2000), RIKEN Review*, No. 32, pp. 43-46.

Yokokawa, H., Kondo, A., Hashimoto, O., Soh, T., Ono, K., and Toyota, M. (2004), "A fundamental study of acoustic and EM wave absorber using resistive films and glass wool," *Electronics and Communications in Japan, Part I: Communications*, Vol. 87, No. 12, pp. 60-67.

Zhigilei, L. V., Leveugle, E., Garrison, B. J., Yingling, Y. G., and Zeifman, M. I. (2003), "Computer simulations of laser ablation of molecular substrates," *Chem. Rev.*, Vol. 103, pp. 321-347.

In: New Developments in Lasers and Electrc-Optics Research ISBN 1-59454-771-8
Editor: William T. Arkin, pp. 41-74 

*Chapter 2*

# HIGH DISCHARGE TUBE RESOURCE OF THE UV $Cu^+$ NE-CUBR LASER AND SOME APPLICATIONS*

***Nikolay K. Vuchkov***
Metal Vapor Lasers Laboratory
Institute of Solid State Physics, Bulgarian Academy of Sciences,
Tzarigradsko Chaussee 72, Sofia 1784, BULGARIA
E-mail: nvuchkov@issp.bas.bg

## ABSTRACT

The UV $Cu^+$ laser excited in a nanosecond pulsed longitudinal Ne-CuBr discharge with a resource above 700 hours is developed and investigated. A comparative study on the temporal behavior of the average output power of the laser with inserts made of a quartz and a ceramic tube is reported. Spectral investigations of the discharge are also made. The use of a ceramic tube results in an absence of the tube metalization, and allows to maintain a constant gas mixture composition. The influence of different additives – air, synthetic air, $N_2$, $O_2$, $CO_2$, He, Ar, Xe, and $H_2$, to the neon buffer gas on the average output power is investigated. It is obtained that addition of a small amount of hydrogen 0.02-0.04 Torr leads to an increase of more than twice of the average output power at a simultaneous laser oscillation of the five $Cu^+$ lines - 248.6, 252.9, 259.7, 260.0, and 270.3 nm. The use of the IC excitation scheme leads to a 30 % increase of the average output power in compare with the use of the Blumlein scheme. The laser is operated at a periodic refreshment of the buffer gas mixture - 16.7 Torr Ne and 0.03 Torr $H_2$. In this regime for a 700-hour laser action the average output power is maintained about 600 mW. That make this laser usable for a number of scientific applications in a laboratory conditions. The influence of UV 248.6 nm laser treatment on the characteristics of the conducting polymer PEDOT is investigated. The refracting index of liquid stilbene films is measure in this UV region. It is shown the possibility to make micro holes in different materials which are transparences for visible and IR light.

* This work was supported by the Bulgarian Science Fund, Grant F-1204 and by project NATO SfP 971989 "High beam quality UV lasers for microelectronics".

# I. INTRODUCTION

A number of ion metal vapor lasers, operating in the UV spectrum area, are well known. The highest average output power was obtained with the copper ion laser. For the first time UV laser oscillation was obtained in hollow cathode lasers [1-3 ], but they did not prove viable because of serious discharge tube problems. The obtaining of an UV $Cu^+$ laser oscillation in a Ne-CuBr nanosecond pulsed longitudinal discharge [4,5] opened new prospects for the UV $Cu^+$ laser utilization. An UV $Cu^+$ laser action with average output power of 1.3 W at multiline output (248.6 nm, 252.9 nm, 260.0 nm и 270.3 nm) and 0.85 W on the 248.6 nm laser line was reported in [6]. At a laser pulse duration (FWHM) of 10.5 μs and an excitation pulse repetition frequency of 25 kHz the peak pulse power on the 248.6 nm laser line was 3.25 W. At an active volume of 23 $cm^3$ a specific average output power of 56.6 $mW/cm^3$ was obtained at multiline output.

The results presented in [6] shows, that UV $Cu^+$ Ne-CuBr laser could be applied in a number of areas, such as microelectronics, microbiology and genetic engineering, as well as in injection-locking systems with a KrF excimer laser. This laser could be also used for solving of various scientific and practical problems, such as pumping of UV dye lasers, dimers or crystals for obtaining of a tunable generation in the 200 nm region by sum-frequency mixing, fluorescence-activated cell sorting, photolithography, etc. For these applications it is required, as for the other gas discharge lasers and especially the metal vapor lasers, the laser tube lifetime to exceed hundreds of hours. That is why it is necessary to find out the reasons reducing the discharge tube lifetime. Another necessary condition for the UV Ne-CuBr laser application is the obtaining of a narrow generation line – less than 0.5 pm, and a high beam quality. The high quality beam profile is needed for a high image quality, while the narrow laser line decreases considerably the chromatic aberrations.

The high output parameters in [6] were obtained by an inside tube diameter scanning - quartz inserts with different inside diameter were sleeved in the basic tube, also made of fused quartz. Maximal results were achieved at an inside insert diameter of 5.7 mm and discharge current pulse amplitude of 85 A, i. e. the discharge current density was 335 $A/cm^2$. The high current density leads to a strong sputtering of the insert wall because of the ion bombardment and hence impurities are introduced into the discharge mixture. In the investigated discharge tube a considerable metalization of the inside tube wall is observed. That reduces the tube lifetime – after 20-hour laser operation the UV output power decreases with an order of magnitude compared to the maximal value. By utilizing inserts to achieve the necessary active zone diameter the temperature of the inside tube wall increased considerably. That made difficult the Cu and Br atoms recombination in the interpulse period and increased the tube metalization. Simultaneously with the tube metalization an increasing heating was observed that necessitates a thermal insulation reduction. When the already metalized insert was replaced with a new one with the same inside diameter, at the beginning the output power was high and after 15-hour UV laser action started decreasing.

In spite of a number of advantages which make easier creation of the metal halide lasers they have some disadvantages, complicating their investigation and practical application. First of all, the laser active medium includes metal compounds and has considerably more complex composition in comparison with the elemental metal vapor lasers. For example, in the CuBr vapor laser except for the atoms of metal and buffer gas there is CuBr molecules and different products of the CuBr dissociation, such as bromine atoms and molecules. In relation to that some additional chemical reactions are possible in the plasma as well as in the material of the tube wall, electrodes, and other parts of the laser tube. Different chemical and physical processes in the CuBr vapor laser determine the laser tube lifetime. The most probable reason reducing the laser tube lifetime is the disturbance of the chemical equilibrium. At a partial recombination of the lasant except for its expenditure, some free bromine could be accumulate in the active volume as well as the residual bromine could react with the tube parts resulting in elemental copper deposition. That also disturbs the laser operation conditions.

Main physical factors, determining the laser tube lifetime, are very similar to those with the atomic CuBr laser operating on the two copper atom lines - 510 and 578 nm, and described in detail in [7,8]. The lifetime of the atomic CuBr vapor laser was investigated with a laser tube specially designed with discharge confining diaphragms and the lasant placed in sidearm reservoirs independently heated via external ovens. Copper porous electrodes, providing long laser operation without their deterioration, were used. It was found that the lasant expenditure was unessential because of the copper bromine recovery by recombination of the copper and bromine atoms in the area between the discharge confined by the diaphragms and the tube wall. It was obtained by measurements and estimations that the bromine was absorbed in the porous copper of the electrodes and the free bromine was not accumulated in the tube volume. A tube lifetime of the atomic CuBr vapor laser up to 1000 hours was obtained by such a laser tube construction.

Unfortunately, as it was shown in [5], when a laser tube with discharge confining diaphragms was used in the UV $Cu^{+}$ Ne-CuBr laser, a very low average output power was obtained (7 mW) in comparison with the use of a quartz tube without discharge confining diaphragms (270 mW). Moreover the utilization of a laser tube without diaphragms leaded to a considerable extension of the variation range of the operation parameters. However, at this tube design there is no distance between the discharge and the tube wall and therefore no area where the recombination of the free copper and bromine atoms would occur. It was observed the following laser tube performance: 1) larger lasant expenditure, that could be compensate with a larger amount of CuBr powder in the reservoirs; 2) copper deposition on the quartz tube wall. A significant metalization of the discharge tubes leaded to a decrease of the tube lifetime to about 20-25 hours. These results showed the necessity of a new solution of the problem with the UV $Cu^{+}$ laser lifetime. .

In the present paper a summary results concerning laser tube recourse and some applications are reported. A comparative investigation on the UV $Cu^{+}$ laser operation with the most inner insert made of quartz and ceramic respectively is made. Spectral studies on the cold zones and active zone of the gas discharge are carried out. A new laser tube design with ceramic insert and electrodes protected from deterioration caused by CuBr condensation on and around them is demonstrated. A new stabilized temperature regime of the active zone is realized. The influence of different additives to the neon buffer gas on the average output power is investigated. The use of the IC excitation scheme and the Blumlein scheme is compare.

The influence of UV 248.6 nm laser treatment on the characteristics of the conducting polymer PEDOT, the possibility to make micro holes in different materials which are transparences for visible and IR light and the measurement of the refracting index of liquid stilbene films in the UV region is carry out.

## II. Experimental Apparatus

The construction of the laser tube T1 is shown in Fig. 1. The basic tube with a 16-mm inside diameter and a 20-mm outside diameter is made of fused quartz. The active length is 100 cm. The CuBr powder is placed in seven quartz side-arm reservoirs. The reservoirs are heated with externally ovens.This tube is investigation in two modifications. At the first T1q three quartz tube inserts with inside diameters 12.5 mm, 8.4 mm, and 5.7 mm are sleeved in the basic tube. In each insert seven holes are made above each reservoir for CuBr vapor diffusion into the active zone. At the second studied tube modification T1c two inserts are used: a quartz insert with inside diameter 11.7 mm and a ceramic insert with inside diameter 7.3 mm. This way in T1q the quartz tube confines the active zone and in T1c - the ceramic tube.

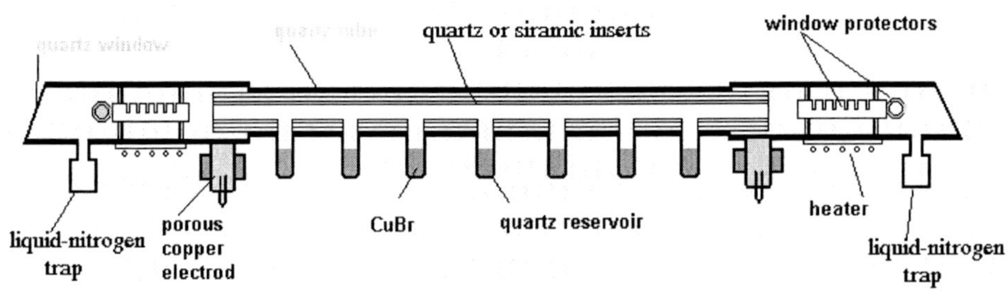


Figure 1. Construction of discharge tube

At the previous discharge tube construction [6] the active zone ended in front of the electrodes, and that results in condensation of the CuBr and the free $Br_2$, obtained at the CuBr molecule dissociation, on the electrodes and around them. That leads to: a deterioration of the porous copper electrodes, due to the bromine interaction with copper; a growth of copper dendrites from the electrodes to the active zone, which aperture the active zone cross-section; and to discharge instabilities because of the CuBr penetration into the electrodes. As a result the laser discharge conditions are disturbed and the discharge tube lifetime is reduced. In the new laser tube construction (Fig.1) the two ends of the quartz tube, forming the active zone, are placed in a such a way, that the CuBr and the free $Br_2$ obtained at the CuBr molecule dissociation condense outside the electrodes and their area [9]. That leads to an increase of the laser tube lifetime.

In the tube cold zones (between the electrodes and the Brewster windows) liquid nitrogen traps are placed. The purpose is to trap different impurities introduced into the discharge mixture as a result of the insert wall sputtering. In this way the free $Br_2$ molecules are also trapped as it is mentioned in [8].

The second tube T2 is with the construction similar of the T1 without liquid nitrogen traps in cold zones.The basic tube with an 11.8-mm inside diameter and a 14.8-mm outside diameter is made of fused quartz. The active length is 86.5 cm. The CuBr powder is placed in five quartz side-arm reservoirs. A ceramic tube insert with inside diameters 7.1 mm is sleeved in the basic tube. In the ceramic insert, five holes are made above each reservoir for CuBr vapor diffusion into the active zone. The discharge zone and CuBr reservoirs are in stainless steel enclosure (Fig.2a). On the bottom of the enclosure, there are long heaters just under the CuBr reservoirs. The temperature inside the enclosure is maintained by a temperature controller with a thermocouple.

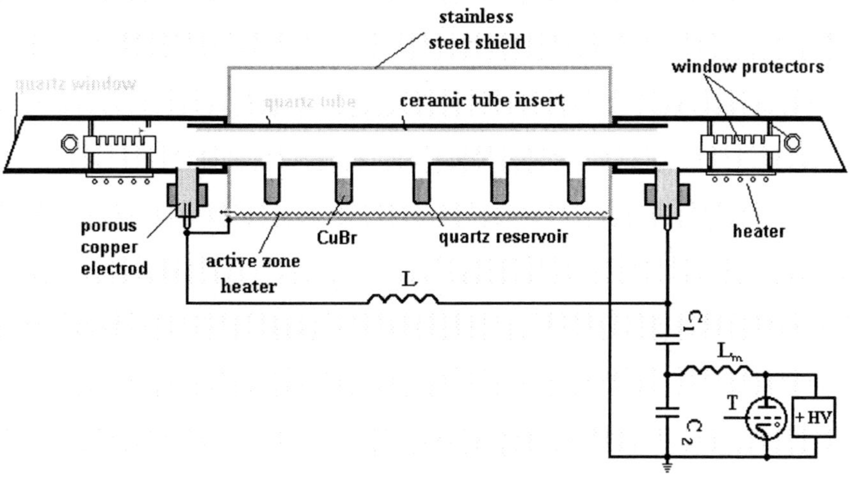


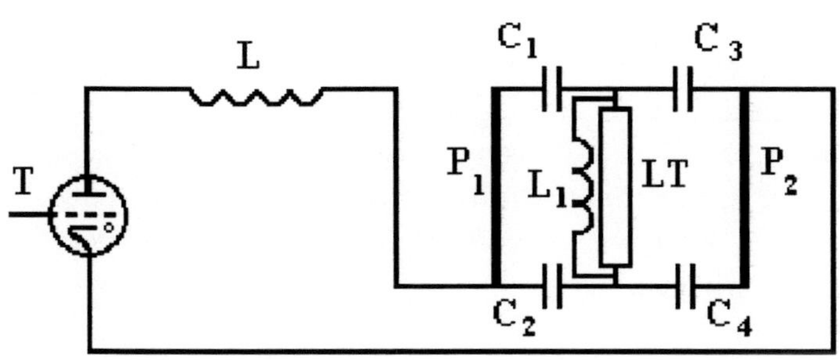


Figure 2 Schematic diagram of investigated discharge tube and electrical excitation schemes; a - Blumlein electrical scheme; b - Interacting Circuits (IC).

In all experiments the laser is excited by an electrical scheme with Interacting Circuits (IC) (Fig.2б). The IC excitation of the CuBr lasers, operating on the self-terminating copper atom transitions, was described in detail in [10.11]. In pargraf 2 a comparative investigation on the output characteristics and electrical parameters of the UV $Cu^+$ Ne-CuBr laser excited by the IC excitation and Blumlein (Fig.2a b) schemes are given. The Blumlein electrical scheme for the excitation of the CuBr lasers, operating on the self-terminating copper atom transitions, was described in detail in [12].

The ends of all laser tubes are set at the Brewster angle for 250 nm and quartz (SQ1) windows are sealed to them. The optic cavity of the laser studied is formed by two dielectric-coated mirrors (Rocky Mountain Instrument). The high reflector has a reflectivity of 99.95 % for the 235-270 nm spectral range and a radius of curvature of 2 m or 3 m. The transmission of the flat output coupler is 4 % or 6 % at 248.6 nm. The mirror separation is 1.7 m.

The spectral investigation is made with a Bentham M300 spectrometer, equipped with two 1800 groove/mm holographic gratings and a DH-2 photo-multiplier. The discharge spectra are recorded by a XY recorder-device. When it is necessary, an additional spectral selection of the laser lines is made via a quartz prism and a holographic grating. The average output power is measured by a calorimetric Scientech Vector S310 power energy meter. The waveforms of the discharge current and the tube voltage are detected by a Rogowski-coil probe and a Tektronix P6015 HV probe, respectively. The laser pulses are observed by an optic pulse detector for the UV spectral range. All pulses are displayed on a Tektronix 2455A oscilloscope with a 20 MHz cutoff filter.

# III. Experimental Results

## 1. Output Parameters and a Spectral Study of UV Cu+ Ne-CuBr Laser [13]

### *1.1. Temporal Behavior of the Average Output Power for the Laser Tube Construction T1*

At optimal Ne buffer-gas pressure of 19 Torr, from the laser tube T1q maximal average output power of 400 mW and 500 mW at multiline operation is obtained, at average input power of 1550 W and 1900 W, respectively. The excitation pulse repetition frequency is about 24-25 kHz. The output parameters are lower than these presented in [6] because of the difficulties in maintaining a homogenous CuBr vapor concentration along the discharge tube length. After about 15 hours of UV laser operation of the average output power decreases considerably to about 50 mW. Similarly to the results in [6], because of the high discharge current density (about 335 A/cm$^2$), a considerable contamination of the gas-discharge mixture is observed due to deterioration and metalization of the quartz insert inside surface.

In the absence of a distance between the discharge and the tube wall the recombination between the free Cu and Br atoms in the discharge afterglow is disturbed (there is no discharge-confining diaphragm, typical for the CuBr laser tube construction operating on the two Cu atom transition) and that leads to the following effects: 1) There is an increased consumption of CuBr, which can be compensated by more CuBr in the reservoirs; 2) A copper deposition on the internal tube wall, where the temperature is about 550 $^0$C is observed and the copper can not be evaporated. The use of a quartz tube with frosted inside wall as the most inner insert (the aim is to increase the inside tube wall surface) results in an insignificant tube lifetime increase (up to 20 hours). Elemental composition of metalized tube samples (after UV lasing of 20 hours) is studied by a scanning electron microscopy (SEM) and a X-ray diffraction analysis. The electron-microscopy photo of the samples shows that homogeneously distributed copper 0.5-1 μm crystals are observed on the dark surface. The analysis shows that copper particles have penetrated into the quartz.

With the discharge tube T1c (an outside quartz insert and the most inside ceramic insert with an inner diameter of 7.3 mm) at optimal Ne buffer-gas pressure of 16.5 Torr maximal average output power of 455 mW and 475 mW is obtained at multiline output, average input power of 1550 W and 1950 W respectively, and a pulse repetition frequency of about 24-25 kHz. It can be noted that for this inside diameter (7.3 mm) this is the highest obtained average output power. That could be due to the use of a smaller number of inserts. With the discharge tube T1c at a periodic refreshment of the buffer-gas a metalization of the ceramic insert inside wall is not observed for 100 hours UV laser action. For this 100-hour period the average output power is around 300 mW. In this tube growths of small pure copper dendrites are observed along the ceramic insert length – mainly around the holes above the CuBr reservoir. These growths lead to output power fluctuations because of the aperture obstruction. At a mechanic cleaning of the Cu dendrite growths the output power is restored. A destruction of these dendrite growths is also observed during the laser operation.

### *1.2. Spectral Investigation on the Gas Discharge*

A spectral investigation in discharge tube T1q is carried out for two essentially different regimes of the gas discharge: without any CuBr powder in the reservoirs, and with CuBr powder loaded in the reservoirs, at optimal reservoir temperature for the laser oscillation.

The purpose of the initial spectral measurement without CuBr is to find out if some impurities, including a large amount of free $Br_2$, are introduced into the discharge from the melted CuBr. The discharge spectrum from 200 nm to 750 nm is recorded in two cases: with nitrogen cooling (the traps mounted to the discharge tube cold zones are poured on with liquid nitrogen) and without cooling. A considerable change in different lines intensity is observed in the two cases.

In Fig. 3 the discharge spectrum from 300 nm to 330 nm is presented, where essential variations are observed without (curve 1) and with liquid nitrogen cooling (curve 2). The spectra are recorded under the following conditions: the discharge tube is filled up with Ne buffer-gas and the discharge switch on; after 2 hours of operation the spectrum is recorded; the traps are filled with liquid nitrogen and after a 30-minute cooling period the spectrum is recorded again (in the same way the next spectra measurements are made). From Fig. 3 it can be seen that the discharge spectrum is essentially different in the two cases. The strong spectral lines in the spectrum with cooling (curve 2) are identified as neon atom and ion lines. In the spectrum without cooling (curve 1) a molecular band is observed, which probably belong to CO, $CO^+$, $CO_2$, $CO_2^+$, $N_2$, NO, CN and $CN^+$ molecules.

In Fig.4 the waveforms of the tube voltage (Fig.4a) and discharge current pulses (Fig.4b) are shown for the two cases – without liquid nitrogen cooling (curve 1) and with cooling (curve 2). It can be seen from the figure that the utilization of the liquid nitrogen traps leads to a considerable change of the pulses.

After loading CuBr in the reservoirs it is found out that during the first 10 hours of the UV laser operation, when the average laser power is about 300-500 mW, the influence of the liquid nitrogen cooling on the output power is not so significant. After decreasing of the average output power to less than 30 mW for an UV laser action of about 25 hours, the liquid nitrogen traps are used again. The waveform and the amplitude of the tube voltage and discharge current pulses are recovered. The output power on the four UV $Cu^+$ laser lines increases about two times. The output power of the two Cu atom lines - 510.6 nm and 578.2 nm, increases with about 50 % (a similar behavior of the CuBr atom laser average output power was observed in [8,14], too). When He is used as a buffer-gas, the output power on the 780.8-nm $Cu^+$ lines increases with about 50 %.

In this case in the discharge spectrum some copper atom and ion lines with extremely high intensity are observed which leads to a damping of the rest of the spectral lines. In Fig.5a the spectrum from 300 nm to 315 nm is shown – without cooling (curve 1) and with cooling (curve 2). The observed lines are identified as copper atom lines. Because of the discharge current amplitude increase with the liquid nitrogen cooling, the intensity of most copper lines also increases and that makes the investigation on the spectrum changes with and without liquid nitrogen cooling very difficult. With the studied discharge tube geometry and at the operation Ne buffer-gas pressure a parasitic discharge is observed between the electrodes and the Brewster windows (cold zones). The spectrum change in exactly this zone is studied with and without the liquid nitrogen cooling. For this purpose the magnetically driven protectors of the Brewster windows are made opaque, in order to obstruct the active zone aperture.

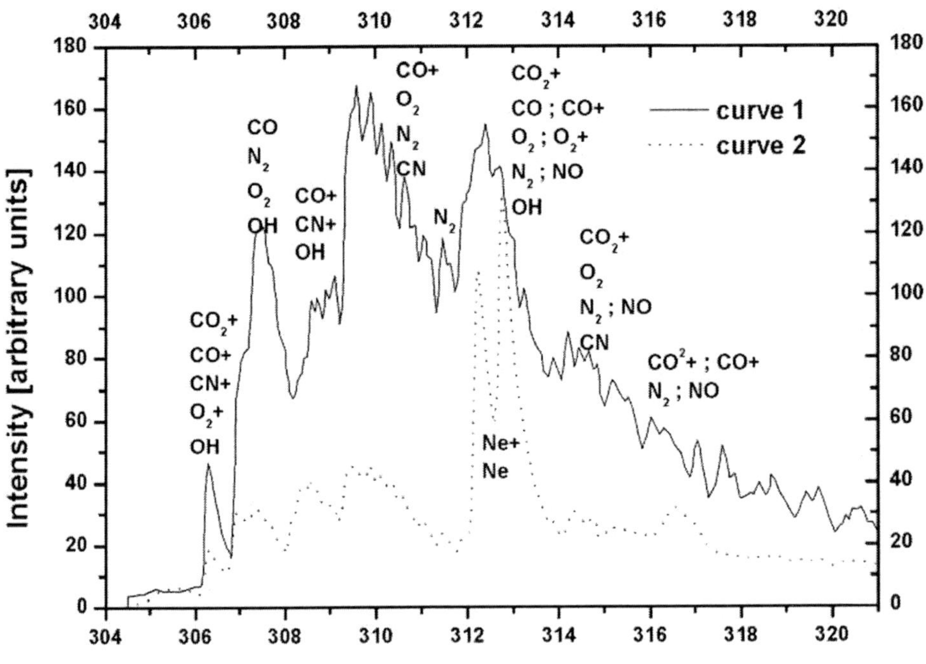


Figure 3. Active zone discharge spectrum from 300 nm to 330 nm without liquid nitrogen cooling (curve 1) and with cooling (curve 2) for the case without CuBr powder in the reservoirs of the laser tube construction T1q.

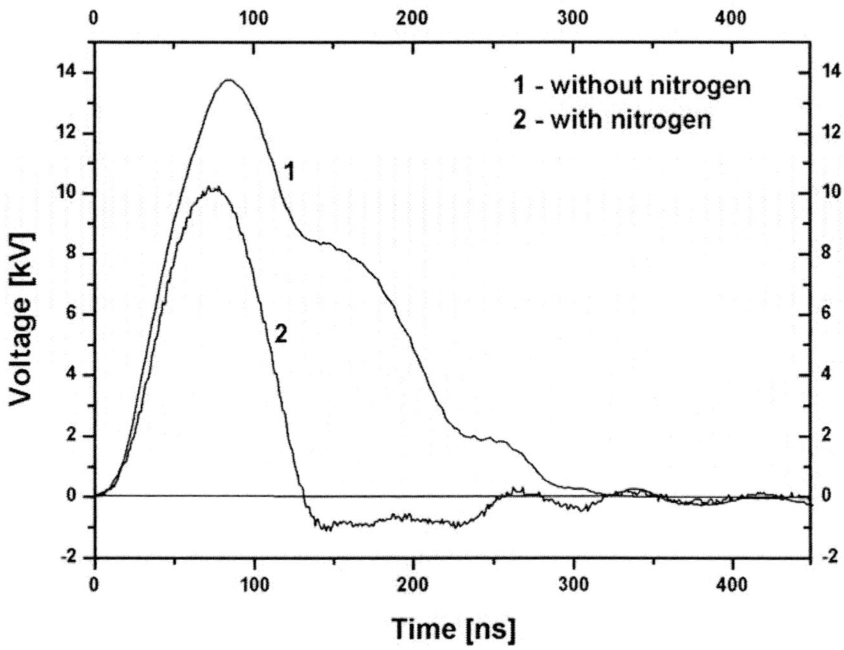


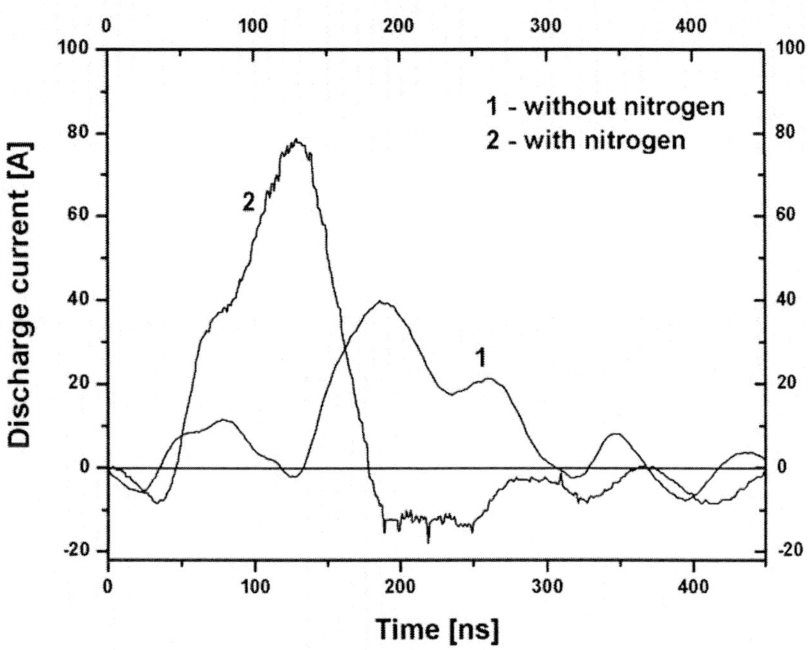


Figure 4. Waveforms of tube voltage (a) and discharge current pulses (b) without liquid nitrogen cooling (curves 1) and with liquid nitrogen cooling (curves 2).

In Fig. 5d b the spectrum of the cold zone discharge is presented in the spectral range 300-330 nm for the two cases - without (curve 1) and with the cooling (curve 2). The spectrum is recorded after the most inside quartz insert is already metalized (after 20-hour laser operation) and the average output power is 35 mW without cooling, and 70 mW with cooling. The discharge conditions and the reservoir temperature are optimal for a maximal output power. In this case, because of the considerably lower intensity of the spectral lines in comparison with the spectrum recorded from the active zone, the monochromator slits are widely opened and that results in worse resolution of the spectrum in Fig. 5b compared to that in Fig. 3.

In Fig.6 the spectrum of the cold zone discharge in the same spectral range for the tube T1c is presented for two cases – without liquid nitrogen cooling (curve 1) and with cooling (curve 2). The spectrum is recorded after 50-hour UV laser operation of the ceramic insert and the average output power is about 300 mW.

### *1.3. Discussion*

A comparative investigation on the output power temporal evolution with two different discharge tube – T1q and T1c is carried out. For each tube the optimal discharge conditions are determined, and average output powers of 500 mW and 475 mW are obtained, respectively. With the discharge tube T1q after about 20-hour UV lasing a strong metalization of the insert inner wall is observed, and the output power decreases to about 30 mW. The utilization of a ceramic tube as the most inner insert, on one hand does not reduce the maximal output parameters obtained – the average output power is 450-475 mW, and on the other hand, the laser tube lifetime increases considerably – for 100-hour laser operation the average laser power is around 300 mW. In this case no metalization of the ceramic insert inner wall is observed.

The undertaken analysis of the metalized quartz insert shows that copper species penetrate into the quartz substrate. One possible explanation of this fact is that the amorphous structure of the fused quartz under conditions of a strong ion bombardment and a high gas temperature allows the copper penetration under the quartz insert inner surface, and that leads to it metalization. It could be assumed that the metalized inner wall has a screening effect which results in a gas temperature increase and in worsening of discharge conditions.

When using a ceramic tube made of $Al_2O_3$ as the most inner insert no metalization of the inner wall is observed. It could be supposed that the crystalline structure of the ceramic prevents the copper penetration under the insert inner surface, and hence, the tube is not metalized.

From the spectral investigation carried out it can be concluded that at the optimal discharge current density – above 300 A/cm$^2$ and the high gas temperature some impurities are released from the quartz tubes, that leads to the discharge condition worsening, and hence, to a considerable decrease of the average output power and to a reduction of the laser tube lifetime – to about 20 hours. Our investigation shows that this problem could be solved by utilization of ceramic tubes as the most inner insert.

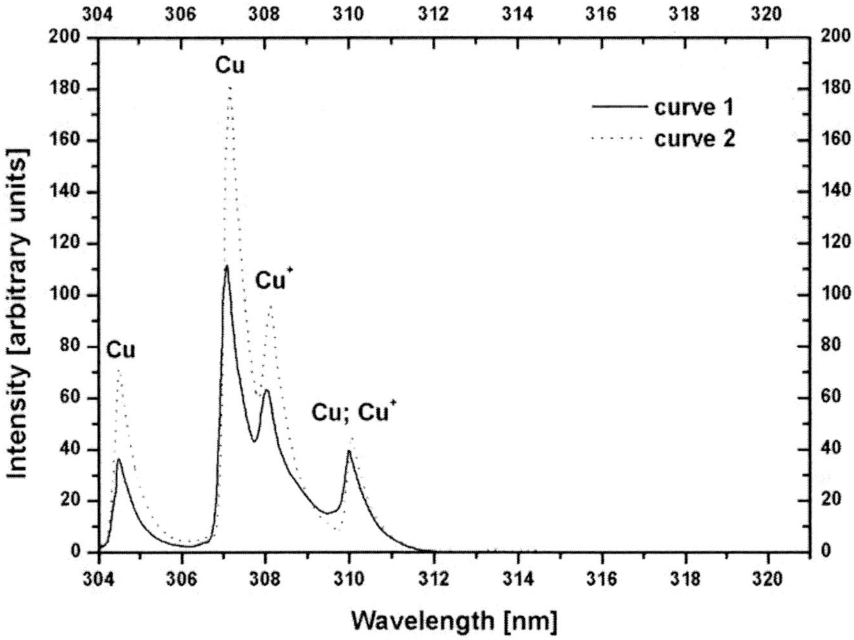


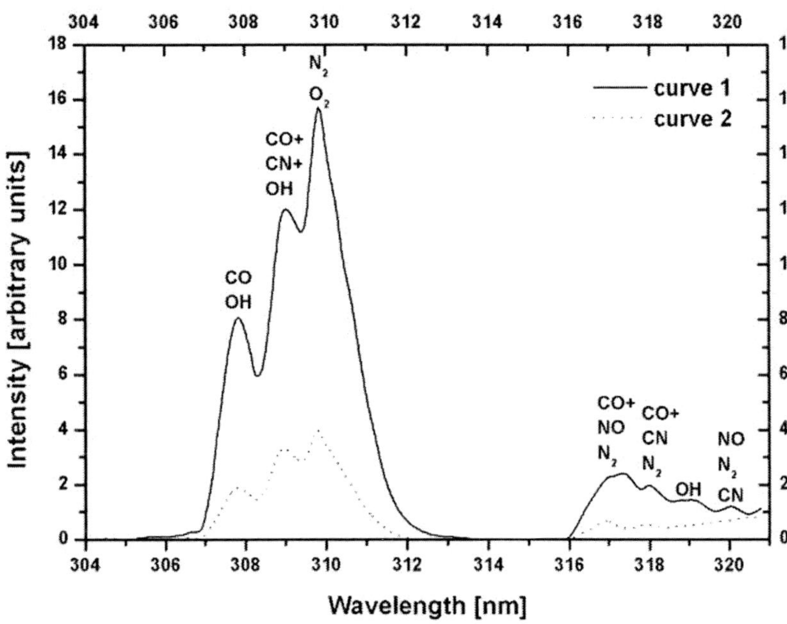


Figure 5. Active zone (a–slit width 0.1 mm) and cold zone (b- slit width 1 mm) discharge spectra without liquid nitrogen cooling (curve 1) and with cooling (curve 2) for the laser tube construction T1q.

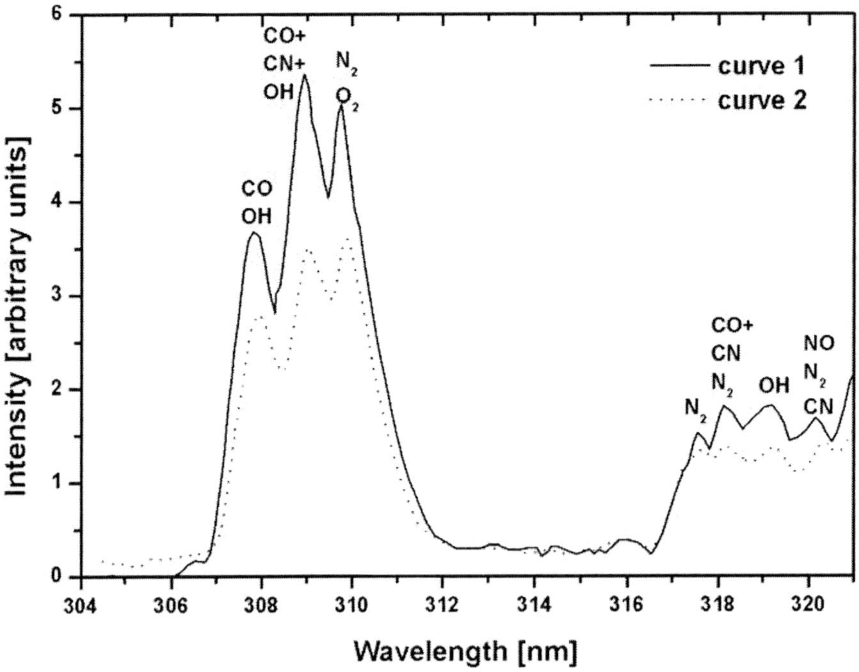


Figure 6. Cold zone discharge spectrum from 300 nm to 330 nm without liquid nitrogen cooling (curve 1) and with cooling (curve 2) for the laser tube construction T1c.

The results show that because of the high discharge current density and the high gas temperature the quartz tubes confining the active zone are metalized and some impurities are released into the gas mixture, that leads to average output power reduction from 500 mW to 30 mW for approximately 25 hours of UV lasing. The use of ceramic tubes results in an absence of tube metalization, unchanged gas mixture composition, and a relatively constant average output power – 300 mW, for 100-hour UV laser operation.

All next experiments are carried out with discharge tube T2 with ceramic tube confining the active zone.

## 2. Comparative Investigation on the Output Characteristics and Electrical Parameters of the UV $Cu^+$ Ne-CuBr Laser Excited by the Blumlein and IC Excitation Schemes

Electrical scheme of the pulsed excitation circuit is of great importance to the output parameters and the operation stability of the metal vapor lasers. The pulsed excitation scheme must deliver excitation pulses with an amplitude, duration, risetime and repetition frequency appropriate to create population inversion at each particular pulsed metal vapor laser. The possibility to change essentially or to govern intentionally the excitation pulse parameters by the excitation scheme is a principal method together with the gas mixture changing to increase the laser output characteristics. A new electrical scheme with Interacting Circuits (IC) was designed and developed for the excitation of the atomic CuBr vapor laser [10]. Its use for the excitation of a number of ion metal vapor lasers allowed us to obtain appropriate conditions for a laser oscillation on transitions with an upper laser level populated via different processes – charge exchange, Penning ionization, electron impacts and recombination [4,11,15,16]. As a result laser output parameters, superior or comparable to those reported in the references, were obtained.

For the IC scheme it is characteristic that neither of the tube electrodes is grounded and there is a metal plate under high potential to ground. That causes certain technical difficulties at the utilization of the IC scheme for the excitation of CuBr lasers for practical applications. Except for the above mentioned IC scheme for the excitation of the CuBr laser operating on the two Cu atom lines - 510.6 and 578.2 nm, a modified Blumlein scheme (LC inversion circuit) was used [12]. Both schemes leaded to an increase (to almost doubling) in the laser output characteristics of the atomic CuBr vapor laser in comparison with the standard pulsed excitation schemes with and without peaking capacitor.

The LC inverter uses the same voltage doubling principle as the Blumlein circuit, originally devised to achieve high voltages at the magnetron modulators (that is why this scheme is also called modified Blumlein scheme). The two capacitors $C_1$ and $C_2$ are resonantly charged to twice the power supply voltage(Fig. 2a). The triggering of the thyratron connected in parallel to the capacitor $C_2$ inverts the voltage across this capacitor and hence delivers a voltage of up to four times the power supply voltage across the discharge tube. Gas breakdown normally occurs before this voltage is reached and the capacitors discharge through the laser tube. The laser output characteristics could be improved using capacitors with different capacitance. It should be noted that the metal enclosure is connected to the grounded electrode and the capacitor $C_2$ and is used as a current return path as well as the thermal shield.

The IC scheme shown in Fig. 2b comprises two interacting circuits each consisting of a metal plate with a semicylindrical shape, two capacitors connected the plate ends to the electrodes and the laser tube. The metal plates are placed near the laser tube along its length. The metal enclosure used as a thermal shield for the laser temperature stabilization is electrically disconnected from the discharge tube and the excitation scheme. The four capacitors form a capacitor bank, which is resonantly charged in the usual manner from the HV rectifier. When the thyratron is switched the capacitors start discharging and a voltage appears on the tube electrodes – the electrode between the capacitors $C_1$ and $C_2$ is under high negative potential, while the one between $C_3$ and $C_4$ – under low positive potential. When the breakdown voltage is reached the discharge current starts. The change of the four capacitors nominal value (i.e. the capacitor bank optimization) gives us a possibility to control essentially the parameters of the tube voltage and discharge current pulses – the amplitude, duration, risetime and the delay between them. In this way operation conditions for a maximal output power on the corresponding laser transition are searched.

A steady-state laser operation on the five $Cu^+$ lines 248.6, 252.9, 259.7, 260.0 and 270.3 nm is obtained by both excitation schemes over the following range of the discharge conditions: the Ne buffer-gas pressure is varied from 15.5 to 17.5 Torr, the pulse repetition frequency - from 18 to 21 kHz, and the average input power – from 1.3 to 1.6 kW.

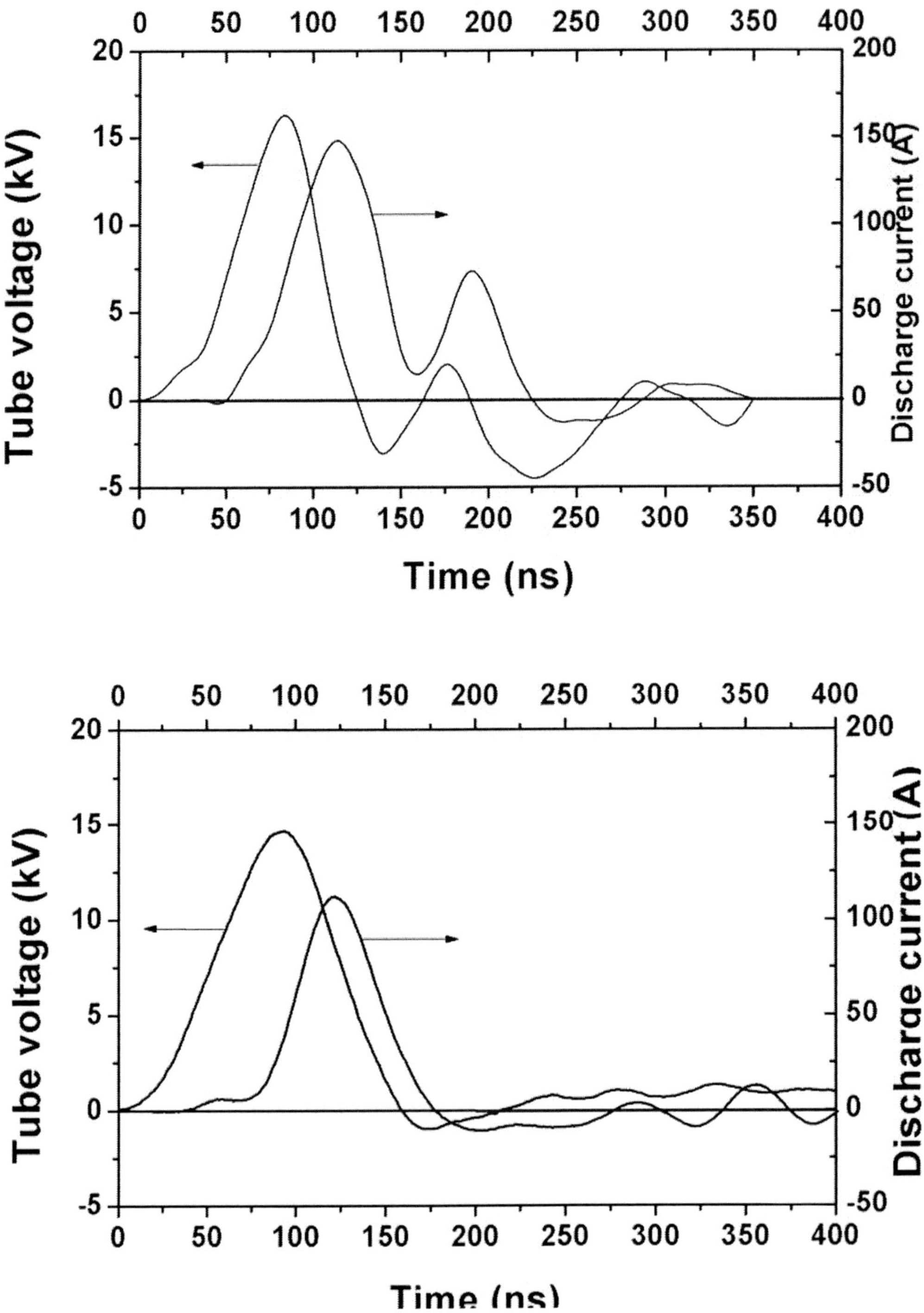


Figure 7. Waveforms of tube voltage and discharge current pulses; (a) for Blumlein and (b) for IC excitation schemes.

The optimal values (with regard to the maximal output power achievement) of the capacitors $C_1$ and $C_2$ in the Blumlein scheme are 500 pF and 235 pF, respectively. In the case of the IC excitation the optimal values of the four capacitors $C_1 \div C_4$ are 500pF, 235 pF, 235 pF and 2200 pF, respectively. The optimal temperature of the enclosure volume is about 495 $^0$C for the two excitation schemes. The optimal Ne pressure is 16.7 Torr for both excitation schemes.

At the optimal discharge conditions, the waveforms of the tube voltage and discharge current pulses are shown in Fig. 7 for the Blumlein (a) and IC (b) excitation schemes, respectively.

The results from the comparative investigation on the discharge and output parameters with the two excitation schemes are presented in the Table 1.

The comparison of the average output power obtained by the Blumlein and IC excitation schemes shows that the use of the IC excitation leads to a 30 % increase of the output parameters. It should be noted that the use of the IC scheme increases the thyratron lifetime, which is of great importance for the laser practical application.

**Table 1**

| Excitation Scheme | Optimal Capacitor Bank | Tube Voltage Amplitude | Discharge Current Amplitude | $P_{out}^{max}$ |
|---|---|---|---|---|
| Blumlein | $C_1 = 500$ pF, $C_2 = 235$ pF | 16 kV | 150 A | 0.9 W |
| IC | $C_1 = 500$ pF, $C_2 = C_3 = 235$ pF, $C_4 = 2200$ pF | 10 kV | 90 A | 1.2 W |

For both excitation schemes using well-known equations [11], the electrical pulses (Fig. 7 a and b) are mathematically processed in order to eliminate the inductive component of the voltage and to calculate the active voltage, the electric pulsed power deposited into the discharge, and the plasma resistance. In Fig. 8 the waveforms of the electric power pulse are presented for the Blumlein (solid curve) and IC (dashed curve) excitation schemes. On the basis of these results the electric pulse energy is calculated to be 53.3 mJ for the Blumlein and 48.5 mJ for the IC excitation schemes. This shows that the higher laser output parameters obtained with the IC excitation scheme are not due to a higher electrical efficiency of the pulsed generator.

## 3. Effect of Hydrogen on the Average Output Power of the UV $Cu^+$ Ne-CuBr Laser [17]

### *3.1. Experimental Results*

In paragraph 1 it is obtained, that some admixtures leads to change the laser operation regime. In the present paragraph an investigation of the influence of various gaseous additives to the neon buffer gas on the laser output power is reported. In the investigations the discharge tube T2 is used. The laser is excited by the IC scheme (Fig.2b).

Optimization of the discharge conditions – the enclosure temperature, neon buffer gas pressure, pulse repetition frequency, and average input power, is carried out in order to obtain a maximal average output power at multiline output on the five UV $Cu^{+}$ laser lines. At nominal values of the $C_1 \div C_4$ capacitors 500 pF, 235 pF, 235 pF, and 2200 pF, respectively, the determined optimal Ne pressure is 16.5 Torr, at enclosure temperature of 495 $^{0}$C. The maximal average input power and pulse repetition frequency, at which our studies are carried out, are 1.4 kW and 19.5 kHz, respectively.

During the investigation of the laser tube lifetime, following changes in the laser output power are observed depending on the tube decontamination. Clean-up of the discharge tube, which initially is at 1 atm air, is made using discharge and enclosure heaters, at a periodic neon buffer gas refreshing. Initially an increase in the output power up to a maximal value of about 700 mW is obtained, but with further decontamination the laser power decreases to about 350 mW. This laser performance is reproduced for the subsequent laser tube refilling with air at atmospheric pressure. These observations required further investigation of the influence of various gaseous additives to the neon buffer gas on the output power behavior. Spectral investigations on the discharge glow in the active and cold zones are also carried out.

The influence of small amounts (from $10^{-2}$ to over 1 Torr) of different gaseous air ingredients - $N_2$, $O_2$, $CO_2$, $H_2$, He, Ar and Xe, added to the neon buffer gas, is studied. With the exception of hydrogen additives, which will be considered in detail below, the addition of all remaining gases results in a reduction of the laser output power. It should be noted that the greatest power decrease is observed at the addition of argon – the average output power is reduced more than twice.

The addition of small amounts of atmospheric air (from $10^{-2}$ to 0.5 Torr) to the neon buffer gas leads to a 20 % increase in the output power, but the obtained results have poor reproducibility (different results are obtained for each particular measurement). The effect of the same amount of synthetic air (Messer-Himko Austria – 20 % $O_2$ in $N_2$, $CO_2 < 0.5$ vpm, NO $< 0.1$ vpm, water vapor $< 10$ vpm) added to the neon buffer gas is also studied. In contrast to the atmosphere air additives, the synthetic air addition always causes a reduction of the laser output power by nearly 20 %.

Hydrogen dosing in the discharge tube is realized via a hydrogen getter that ensures the necessary hydrogen purity. At the optimal neon buffer gas pressure of 16.5 Torr, variations in the laser output power resulting from the hydrogen addition (from 0.02 Torr to 1 Torr) are studied for two different values of the average input power. The dependence of the average output power (at multiline output on the five UV $Cu^{+}$ laser lines) on the hydrogen additive pressure is shown in Fig. 9. In the pressure range from 0.02 to 0.04 Torr $H_2$, the obtained average laser power is from two to five times higher depending on the average input power than the power obtained with pure neon buffer gas. The output power decreases at hydrogen partial pressure higher than 0.04 Torr.

The waveforms of the tube voltage, discharge current and $C_4$ capacitor current are detected without hydrogen (i.e. in pure neon) and at the optimal 0.03 Torr hydrogen pressure. Using well-known equations [11], the pulses are mathematically processed in order to eliminate the inductive component of the voltage and to calculate the active voltage, the electric pulsed power deposited into the discharge, and the plasma resistance. In Fig. 10a the waveforms of the active voltage and the discharge current pulses are presented with (dashed curves) and without (solid curves) hydrogen additives. The electric power pulse and the time-dependent plasma resistance are shown in Fig. 10b for both cases.

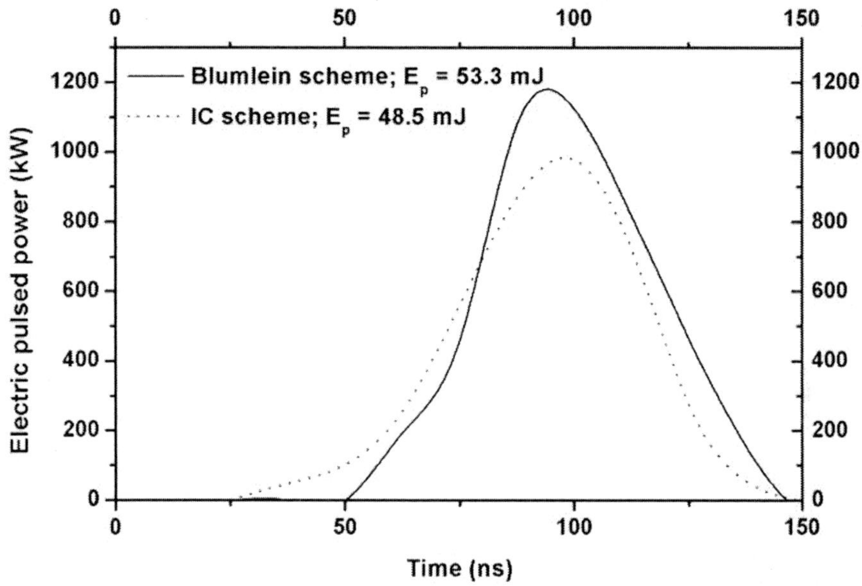


Figure 8. Waveforms of electric power pulse for Blumlein (solid curve) and IC (dashed curve) excitation schemes.

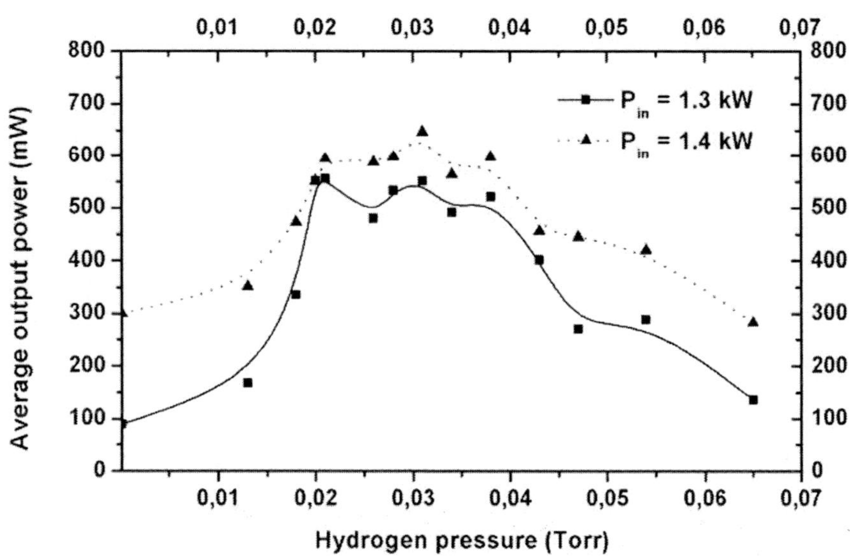


Figure 9. Dependence of UV multiline average output power on hydrogen additive pressure for different values of average input power.

In Fig. 11 the waveform of the laser power pulse for the 248.6-nm line is shown without hydrogen (solid curve) and at the optimal 0.03 Torr hydrogen pressure (dashed curve). The observed variation in the amplitude and duration of the laser pulse caused by addition of hydrogen corresponds to the increase in the average output power with hydrogen addition.

With hydrogen addition, some variations in the temperature regime of the laser operation are also observed. At constant heating conditions - enclosure heater power and average input power from the HV rectifier, the optimal temperature regime (i.e. maximal output power) in pure neon is achieved after 20-25 minutes of discharge operation. With hydrogen addition and at the same heating conditions this time interval is reduced to 10-15 minutes. It should be mentioned that after 30-40 minutes of discharge operation a temperature stabilization occurs in both cases.

The influence of small amounts of ethanol vapor added to the neon buffer gas on the average output power is also studied. Ethanol ($C_2H_5OH$) consists of 6 hydrogen atoms, two carbon atoms, and one oxygen atom. The purpose is to verify whether, despite the hydrogen additives the presence of carbon and oxygen would lead to an output power decrease as in the case of the $CO_2$ addition. The obtained results are displayed in Fig. 12. As could be seen from the figure a significant increase of the output parameters is observed when ethanol is added to the neon buffer gas. The optimal ethanol vapor pressure is about 0.01 Torr, and the maximal average laser power is 100 mW lower than in the case of the hydrogen addition.

### *3.2. Discussion*

Laser operation at two copper atom lines - 510.6 nm and 578.2 nm, is simultaneously obtained under discharge conditions which are optimized for the UV $Cu^+$ laser oscillation. The total average output power of these two lines reaches 1.5 W, using mirrors for the corresponding spectral area. In [18] the so-called strong effect of hydrogen was observed. The addition of 0.3 Torr hydrogen to the neon buffer gas caused a doubling of the average output power on the two atom laser lines in comparison with the case of pure neon. At low hydrogen partial pressure (below 0.05 Torr) no variation of the output power was observed in [18]. In the detailed investigation carried out in [19] it was shown that the increase in the average output power of the two copper atom laser lines is mainly connected with a decrease in the Cu ground state losses due to the electron impact ionization. The reduced losses were attributed to the dominant role of electron detachment from the negative hydrogen ions over the copper atom ionization at the beginning of the excitation pulse.

In the present experiments no increase in the average output power of the two copper atom lines is observed for hydrogen additive pressures up to 0.06 Torr. Furthermore, a reduction (up to 10 %) in the output power is observed, which is possibly due to changes in the temperature regime of laser operation. This allows us to conclude that the mechanism leading to an increase in the average output power at the UV $Cu^+$ lines with the hydrogen addition is not the same as in the case of the two copper atom lines. Moreover, the optimal hydrogen pressures differ considerably in the two cases.

In [5] using He buffer gas, laser oscillation on several IR $Cu^+$ lines was obtained in a nanosecond pulsed longitudinal He-CuBr discharge. The most intensive IR $Cu^+$ line is at 780.8 nm. The laser pulse of the 780.8-nm line starts at about 1 μs after the end of the discharge current pulse and has duration of about 20 μs. At optimal helium pressure the influence of a small hydrogen addition to the He buffer gas on the average output power of the IR $Cu^+$ laser lines is investigated. The average output power shows no change when hydrogen is added. This experimental result permits us to make the following assumption: 1) Hydrogen addition leading to a reduction in the pre-pulse electron concentration because of the $H^-$ formation does not lead to a higher ionization rate and hence to a higher population of the upper laser level; 2) In the discharge afterglow up to 20 μs any changes, which could result in an increase of the inversion population on the IR laser transition, do not occur.

From the electric pulsed power deposited into the discharge (Fig. 10b) the pulse energy is calculated. In the case of pure Ne discharge the pulse energy is 49 mJ, while with hydrogen additives - 47 mJ. This fact shows that within the experimental error at the electric pulses measurements the pulse energy is practically equal for both cases and the increase observed in the UV average output power is not due to a higher electric pulse energy introduced in the discharge.

The spontaneous emission variations of the 248.6-nm line are studied with and without hydrogen additives. No changes in the shape, amplitude, and duration of the signal are practically observed for the two cases. The influence of the high-reflectivity mirror (the reflectivity coefficient is 99.95 %) on the 248.6-nm spontaneous emission is measured. It is observed that this effect is enhanced by nearly 40 %, when hydrogen is added. As could be seen from Fig. 11 the laser pulse duration of the 248.6-nm line is larger with the hydrogen addition than in the case of pure neon. These two consequences of hydrogen addition – the enhanced influence of the high-reflectivity mirror on the spontaneous emission and in the laser pulse duration, show that the conditions for creation of inversion population on the 5s $^3D_1$-4p $^3F^O_2$ transition (248.6-nm laser line) are not only improved (higher inversion population) with hydrogen addition, but also have a larger duration. If the assumption made during the IR $Cu^+$ line investigation, namely that the hydrogen addition does not lead to an increased ionization rate of the buffer gas (neon in this case) and hence to a higher population of the upper laser level is accepted, it could be supposed that the higher inversion population is mainly due to a lower population of the 4p $^3F^O_2$ lower laser level. Furthermore, it was shown in [5,6], that there is a strong dependence of the lower laser level population on the admixtures (Br) and the laser tube diameter.

It was also shown in [5,6] that the population of the 4p $^3F^O_2$ lower laser level mainly depends on the population of the Ne and $Cu^+$ metastable levels. The mechanism, by which hydrogen atoms and/or molecules influence the population of all these levels, could not be determined from the present experiments, and will be object of future studies.

## 4. Discharge Tube Resource

In this paragraf the results concerning laser tube design, laser operation and some problems related to the lifetime of the UV $Cu^+$ laser excited in a nanosecond pulsed longitudinal Ne-CuBr discharge are presented. In the investigations the discharge tube T2 is used. The laser is excited by the IC scheme (Fig.2b).

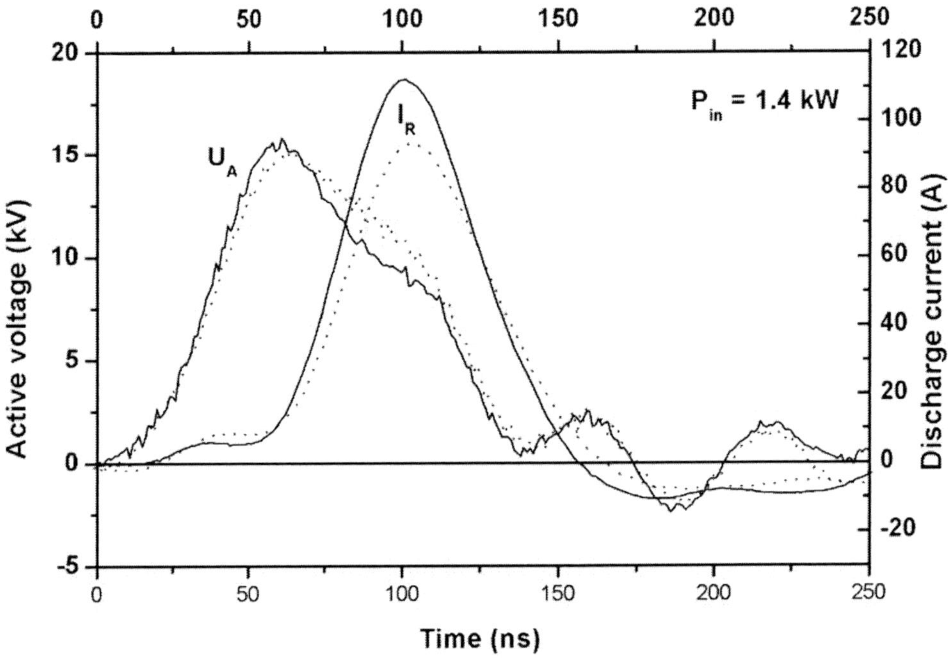


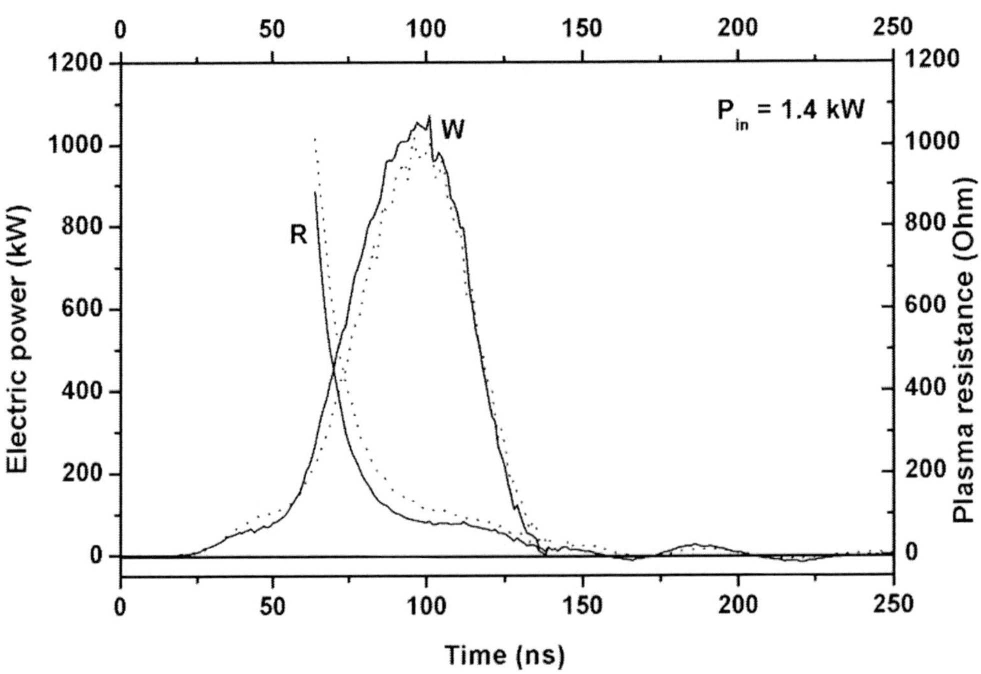


Figure 10. Waveforms of electrical pulsed characteristics: (a) active voltage $U_A$ and discharge current $I_R$; (b) electric power W deposited into the discharge and plasma resistance R. The solid curves are in 16.5 Torr Ne and the dashed ones are in 16.5 Torr Ne with 0.03 Torr $H_2$.

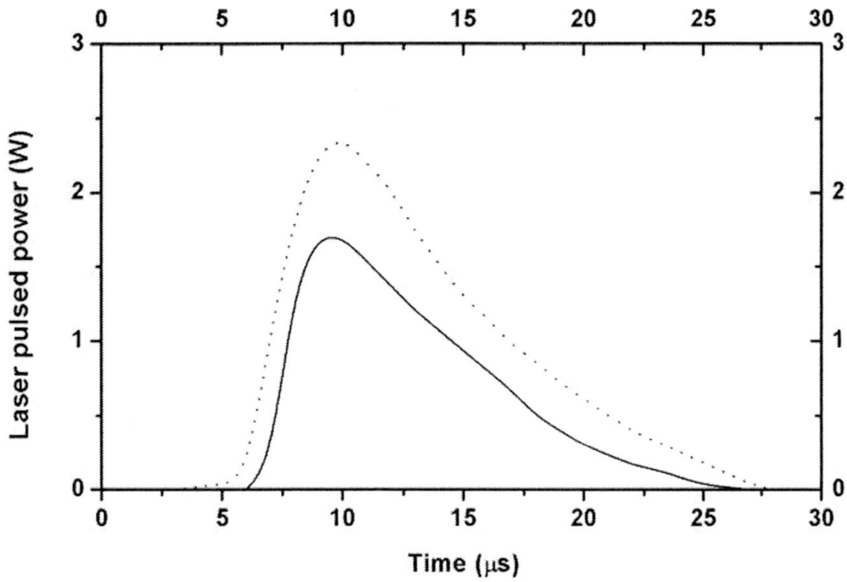


Figure 11. Waveform of laser power pulse of 248.6-nm line. The solid curve is in 16.5 Torr Ne and the dashed curve is in 16.5 Torr Ne with 0.03 Torr $H_2$.

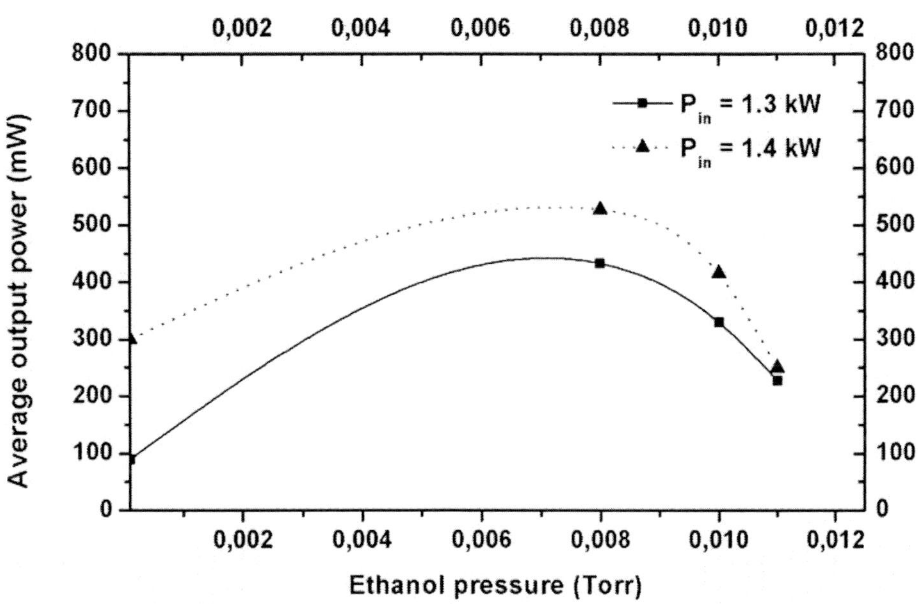


Figure 12. Dependence of UV multiline average output power on ethanol additive pressure for different values of average input power.

Initially a laser tube should be developed for the laser lifetime investigation. When a ceramic insert is used some problems arise from the considerably higher thermal expansion coefficient of the ceramic in comparison with the quartz. Temperature on the outside wall of the basic quartz tube is 580 $^{0}$C. Taking into account the temperature gradient in the quartz, the temperature on the inside wall of the basic quartz tube is estimated to be 600 $^{0}$C. It means that the average temperature of the ceramic insert is about 650 $^{0}$C at least. Using the linear thermal expansion coefficient of the ceramic and quartz, it is estimated that at the given tubes length and temperatures the quartz tube extension is about 0.3 mm, while the ceramic tube extension is about 6 mm. Because of this difference in the tube extension it is not possible to place immovably the ceramic insert into the basic quartz tube. A possibility for movement should be provided. At these conditions the ceramic insert shift inhibits CuBr vapor diffusion into the discharge zone (up to entirely ceasing). It should be noted that at the extension and shortening of the ceramic tube a rotation of this insert occurs. That leads to an entire obstruction of the reservoir inlets and to a laser operation termination. In order to overcome these problems the following design of the laser tube is devised: the five holes in the ceramic tube are lengthened up to 1 cm; according to the ceramic tube extension at its ends some restrictions are made; in order to prevent ceramic tube from rotating at one of its end a groove is made. A restrictor entering the groove is sealed to the quartz tube. So designed laser tube allows us to carry out the investigations on the conditions determining the laser tube lifetime.

The second problem which should be solved is the operational temperature regime of the laser. The optimal reservoir temperature for a maximal UV average output power is about 560 $^{0}$C [4]. For the atomic CuBr vapor laser it is shown in [8] that using laser tubes without discharge confining diaphragms and at temperatures higher than optimal one it is necessary to maintain a uniform temperature distribution along the reservoirs and the tube length. Otherwise, unlike the laser tubes with discharge confining diaphragms, it is observed a discharge constriction resulting in an output power decrease. In the studied UV $Cu^{+}$ laser because of the smaller tube diameter the problem with the temperature stabilization is considerably intensified. That demands the use of a thermostatic metal enclosure confining the active zone and CuBr reservoirs and equipped with long heaters. The corresponding distances and additional insulation are determined in such a way that in the operating regime enclosure heating together with the heat evolved into the discharge provide a higher tube wall temperature than the CuBr reservoir temperature as well as a uniform temperature distribution along the reservoirs and the tube length. That allows us to obtain a stable laser operation and to carry out the investigations on the conditions determining the laser lifetime.

For the used discharge tube with a 7.1-mm inside diameter an optimization of the discharge conditions is carried out in order to obtain a maximal average output power at multiline output. The optimal Ne pressure is 16.7 Torr. The enclosure temperature, at which a stable laser operation with a maximal output power is obtained, is 490 $^{0}$C. At an average output power exceeding 500 mW besides the known four $Cu^{+}$ laser lines 248.6 nm, 252.9 nm, 260.0 nm and 270.3 nm a laser oscillation on a fifth $Cu^{+}$ line with a wavelength of 259.7 nm is obtained for the first time in a nanosecond pulsed longitudinal discharge. At the optimal discharge conditions as well as at a pulse repetition frequency of 25 kHz and an average input power of 1.6 kW a multiline UV average output power of 1.2 W is measured.

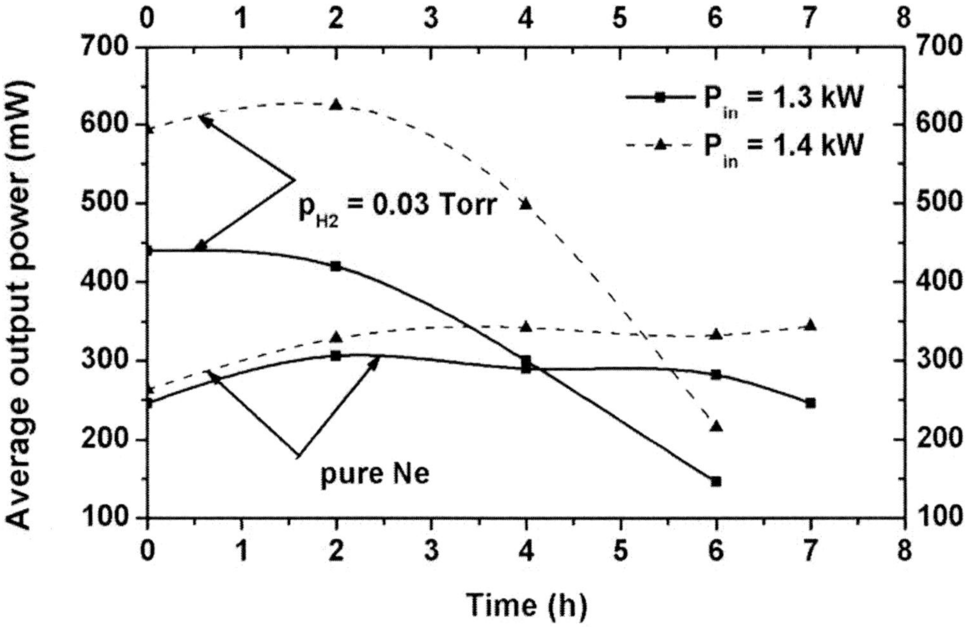


Figure 13. Time evolution of UV multiline average output power in sealed-off regime.

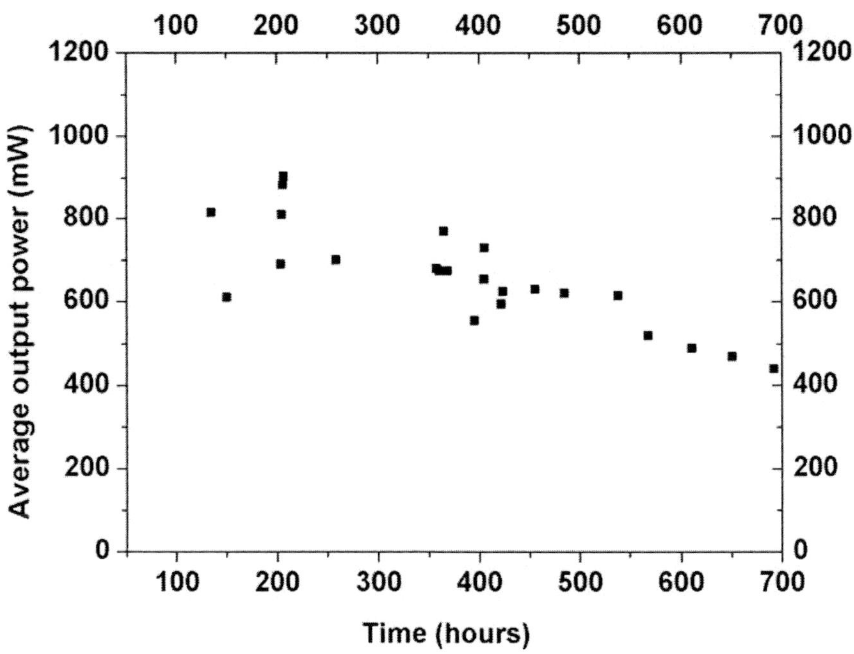


Figure 14. Time evolution of UV multiline average output power in expense the case of buffer gas refreshing.

The investigations on the discharge tube resource and laser lifetime are carried out at an average input power of 1.4 kW and a pulse repetition frequency of 19.5 kHz, at which the power of the additional enclosure heating is optimized for a stable laser action. At these conditions the multiline average output power is 650 mW.

In paragraph 3 it is shown that an addition of 0.03 Torr hydrogen to the neon buffer gas resulted in an increase of more than twice of the average output power. That imposes some changes in the laser lifetime investigation. In [8] the lifetime of the atomic CuBr vapor laser was investigated in a sealed-off regime, while in [20] – in a neon-gas flow regime at a low rate, that complicated the tube construction, but leaded to a stabilization of the atomic Cu laser action. The investigation on the UV $Cu^+$ laser is carried out at a periodic refreshment of the neon buffer gas with and without 0.03-Torr hydrogen additives. In Fig. 13 a typical variation of the average output power is shown at an UV laser operation in pure neon (curve 1) and with addition of 0.03 Torr hydrogen (curve 2) to the neon buffer gas. As can be seen from the figure for a 7-hour UV laser operation in pure neon the average output power is almost constant about 350 mW, whereas with the hydrogen addition – power decreases from 650 to 500 mW for a 6-hour operation. It should be noted that at a more prolonged operation – more than 10 hours, the laser output power decreases in both cases. Because of the considerable difficulties occurring at a laser operation in a neon flow regime possibly due to the relatively small tube diameter in comparison with the laser tube for the atomic copper vapor laser [20] as well as the problem with the maintenance of a mixture composition of 0.03 Torr $H_2$ in 16.7 Torr Ne in the gas flow regime, the investigations are carried out with a periodic refreshment of the gas mixture every fife hours of the UV laser operation. In this way it is determined not quite the laser lifetime, but rather the laser tube resource, i. e. how many hours the discharge tube can withstand the high gas temperature (above 2000 $^0$C), the high discharge current density (above 300 A/cm$^2$), and the certain CuBr lasant expense. Time evolution of the UV multiline average output power is shown in Fig.14 for 700-hour laser action at the following operation conditions. Gas mixture is changed every five hours. The highest value of the output power for this 5-hour period is given in the figure (power performance is the same as is shown in Fig. 13 – curve 2). Approximately at every 100 hours the laser tube is opened and cleaned up from the copper dendrites formed at the ceramic tube ends and in the zone between the basic quartz tube and the ceramic insert.

The obtained results demonstrate that the developed laser tube for the UV $Cu^+$ laser excited in a nanosecond pulsed longitudinal Ne-CuBr discharge can operate above 700 hours. For this period the maintenance of a relatively constant output power requires a periodic buffer gas mixture refreshing. That make this laser usable for a number of scientific applications in a laboratory conditions. Study on the laser lifetime problems is carried on.

## IV. APLICATIONS

### 1. UV-Laser Induced Modification of Poly-3,4- Ethylenedioxythiophene (PEDOT)

Poly-3,4-ethylenedioxythiophene (PEDOT) is one of the well known conducting polymers with unique properties enabling its use as active component in electronic, sensing and catalytic devices. Despite the numerous investigations on PEDOT, studies on the effect of laser irradiation on this polymer are still scarce. The recent investigations [21,22] show the influence of UV-laser treatment on the chemical, electrochemical and morphological characteristics of PEDOT as well as on the PEDOT activity with respect to copper electrodeposition.

PEDOT thin films are electrochemically synthesized on platinum electrodes by anodic polymerization from an aqueous microemulsion. By means of a mask, half of the polymer sample is irradiated by using the laser line of 248.6 nm at an exposition dose of 40 J/cm$^2$. In this way irradiated and non-irradiated parts of one and the same polymer sample are directly compared and eventual UV induced changes become identified.

The polymer coated electrodes are subsequently used for electrodeposition of copper species by galvanostatic and potentiostatic techniques [23-25]. This is one of the ways for obtaining metal/polymer materials with new or improved electronic, catalytic and sensing properties.

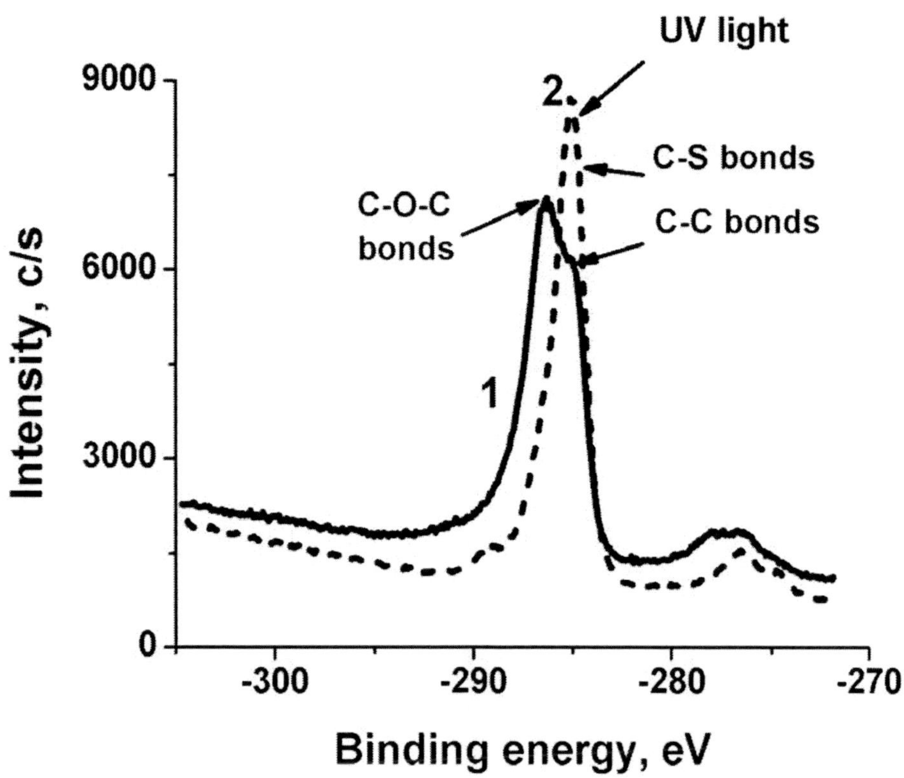


Figure 15. XPS C (1s) spectra of PEDOT layer before (curve 1) and after UV $Cu^+$ irradiation (curve 2).

The non-irradiated and irradiated parts of the polymer samples are characterized by means of different techniques: electrochemical measurements, scanning electron microscopy (SEM), X-ray photoelectron spectroscopy (XPS), energy dispersive X-ray (EDX) and electron spin resonance (ESR) spectroscopies. It is found that the laser treatment does not influence the electrochemical activity and the surface morphology of the PEDOT layers. Nevertheless, XPS analysis of irradiated PEDOT specimens shows a change in the oxidation states of the chemical elements (C, S, O) involved in the polymer (Fig.15). At the same time ESR results do not indicate any difference between non-irradiated and irradiated samples, giving evidence for preservation of the polymer paramagnetic system. Thus the laser induced oxidation states changes, detected by XPS, should not be related to damages in the bulk of the polymer material but rather to carbonization of the PEDOT layer surface.

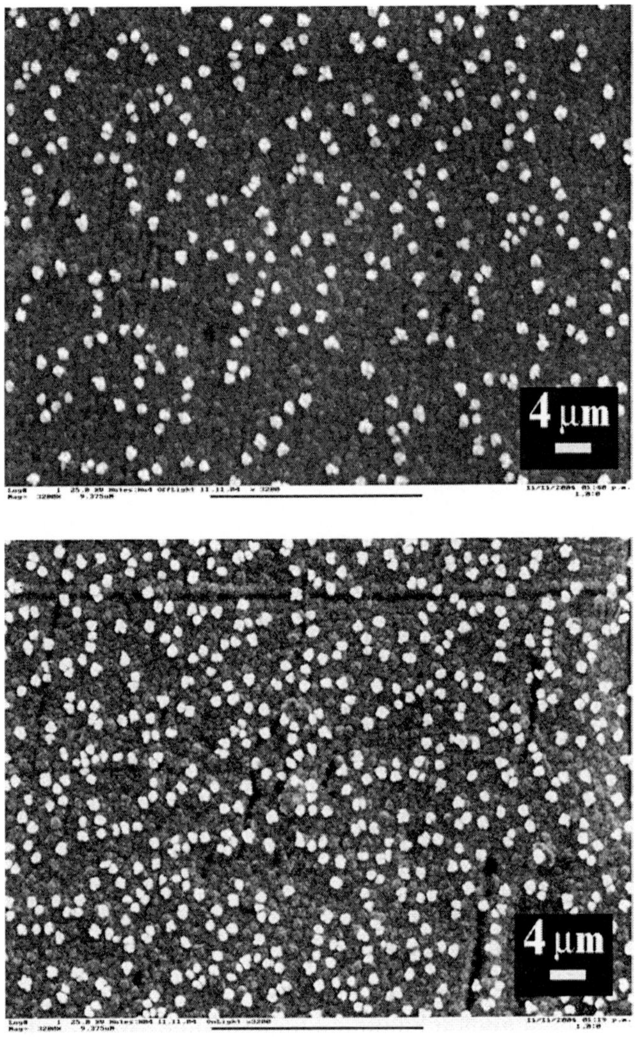


Figure 16. SEM image of Cu crystals deposited on (a) non-irradiated and (b) irradiated parts of a PEDOT layer.

It is established that the UV laser irradiation allows for significant activation of the copper electrocrystallization on PEDOT. The number of deposited metal crystals at the irradiated PEDOT samples drastically increases. Fig.16 a and b illustrate a twofold increase in the copper crystal number - from 2.8 $10^7$ $cm^{-2}$ (non-irradiated part) to 5.6 $10^7$ $cm^{-2}$ (irradiated part).

## 2. Refractive Index of Polymers Films

The refractive index of the polymer films containing trans-stilbene in polymethyl methacrylate (PMMA) deposited on quarts substrates is measured. 1,2-dichlorethan is used as a solvent. The film thickness is 3-10□m. The stilbene concentration to PMMA in the samples is varied from 0.125 to 3wt%. These thin azobenzene organic films, which have some unique properties, can be used as recording medium. In this sense, the optical investigations in the UV region are very promising.

The Cu-Br laser is used as a light source. In this case the laser operates at one $Cu^+$ line 248.6 nm (an output coupling mirror with reflectivity of 94 % for this line is used) and simultaneously at two copper atom lines - 510.6 nm and 578.2 nm.

Refractive index (RI) investigation of the material is performed by the method of the disappearing diffraction pattern (MDDP) [26,27]. This method belongs to the critical angle methods and is appropriate for measuring both liquid samples and solid films in near IR, VIS and UV spectral region [28,29].

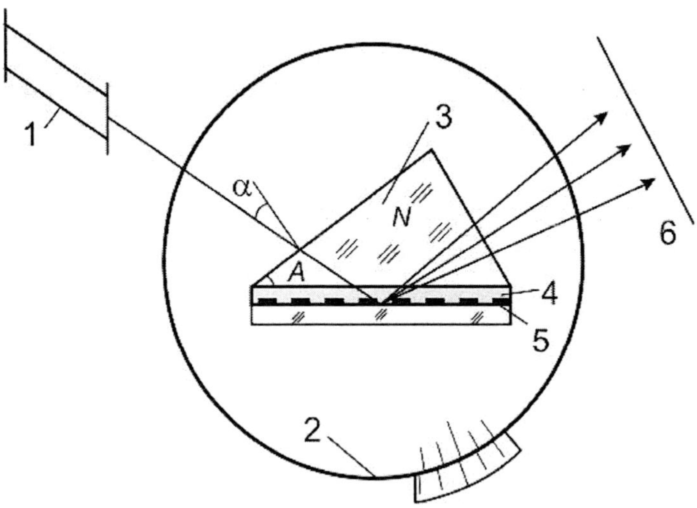


Figure 17. Experimental setup; 1- laser, 2- rotary table 3- prism, 4- investigated layer, 5- diffraction grating, 6- screen.

The experimental setup is shown in fig.17. The diffraction grating (5) is used to indicate the critical angle ($\alpha_{cr}$) for total internal reflection on the prism (3) and sample layer (4) surface. For small angles of incidence ($\alpha$) toward the prism-liquid interface, part of the light refracts into the liquid, reaches the grating and diffraction pattern can be observed in reflection. If the angle of incidence is equal or beyond the angle of total internal reflection, the light is totally reflected from the liquid and all diffraction orders disappear on the screen (6). Thus, measuring the angle of incidence (for which the diffraction pattern disappears from the screen) by the rotary table (2) the index $n$ is calculated according to:

$$n = N \sin\left( A \pm \arcsin \frac{\sin \alpha_{cr}}{N} \right)$$

where N and A are the RI of the prism and the refracting angle, respectively. The signs "+" and "-" correspond to the clockwise and anti-clockwise laser beam rotation with respect to the normal position to the front prism's face. The measurement error depends mainly on the uncertainty in the determined critical angle:

$$\Delta n = N\cos\left(A + \arcsin\frac{\sin\alpha_{cr}}{N}\right)\left(N^2 - \sin^2\alpha_{cr}\right)^{-1/2}\cos\alpha_{cr}\Delta\alpha_{cr}$$

For small angles $\alpha_{cr} \leq 10°$, $\Delta n \sim \cos A\Delta\alpha_{cr}$. For $\Delta\alpha \sim 5\,'$ and $A = 60°$, $\Delta n \leq \pm 1.10^{-3}$.

The obtained data are presented in Table2.

**Table 2**

| Stilbene, % | 578.2 nm | 510.6 nm | 248.6 nm |
|---|---|---|---|
| 2 | 1.453 | 1.459 | 1.462 |
| 3 | 1.456 | 1.459 | 1.463 |
| Refractive index of the prism, *N* | 1.742 TF4 | 1.755 TF4 | 1.508 (fused silica |

## 3. UV $Cu^+$ Ne-CuBr Laser for Material Treatment

High precision micromachining requires an UV laser in order to achieve the necessary resolution. The photon energy is sufficient to produce photochemical ablation rather than melting and subsequent vaporization. The UV laser ablation produces clean precision cutting and drilling with a minimum of thermal or mechanical damage on the target.

Excimer lasers have long been used for surface photoablation owing to the non-thermal character of the laser-matter interaction [30]. On the contrary, the infrared lasers ($CO_2$ and YAG), which are widely used for industrial applications like cutting, drilling or welding, are limited by a more thermal type of interaction since the heat is the main source of matter decomposition, and because of diffusion it does not permit ablative etching with a resolution approaching the micron range. UV beams, in addition to heatless etching, offer the advantage of focusing with less limitation due to diffraction. The beam divergence, and thence the focal spot radius, is roughly proportional to the wavelength.

At present excimer lasers are the main industrial UV laser source used for micromachining. However, the low PRF of these lasers means that they are not well suited to machining applications requiring high production rate serial processing. Moreover, the wide laser linewidth leads to chromatic aberration and requires achromatic optic using.

The high PRF, high beam quality, and narrow linewidth makes the UV $Cu^+$ laser a possible source for many UV micromachining applications such as drilling micro-orifices in polymers and other organic materials at high production rates. Furthermore UV $Cu^+$ laser, which has demonstrated high beam quality at considerable output fluxes, could meet the parameters for high precision removal by photoablation of material with a high absorption coefficient in the UV (organic materials, polymers, semiconductors) via photoablation.

In this paragraph, results of photoablation microdrilling in a variety of transparent materials – glass, polymers, and ZnSe semiconductor by a UV $Cu^+$ Ne-CuBr laser are presented.

Schematic diagram of the experimental setup is shown in Fig. 18. The laser radiation has the following parameters: wavelength - 248.6 nm; average output power - 500 mW; pulse repetition frequency - 20 kHz; laser pulse duration - 10 μs (FWHM); laser pulse energy - 25 μJ; peak pulse power - 2.5 W. The photographs are captured with a CCD camera Hitachi VK-C150ED as it is described in [31].

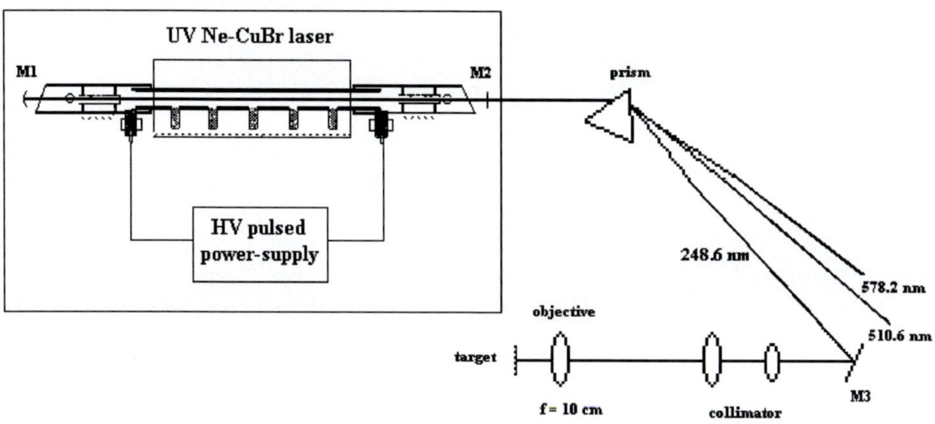


Figure 18. Schematic diagram of experimental setup.

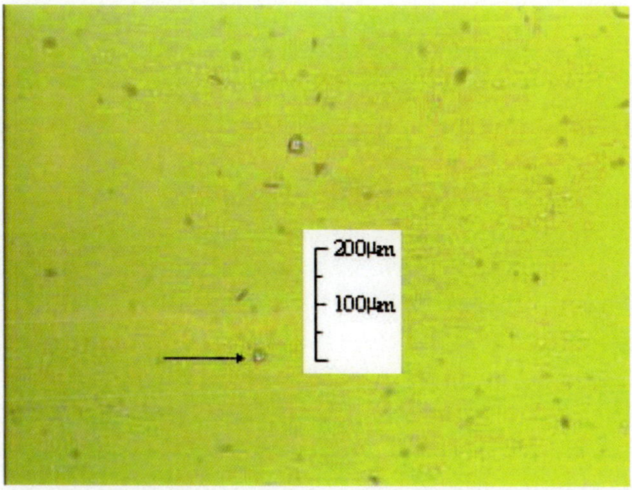


Figure 19. Glass microdrilled at 248.6 nm with $Cu^+$ Ne-CuBr vapor laser.

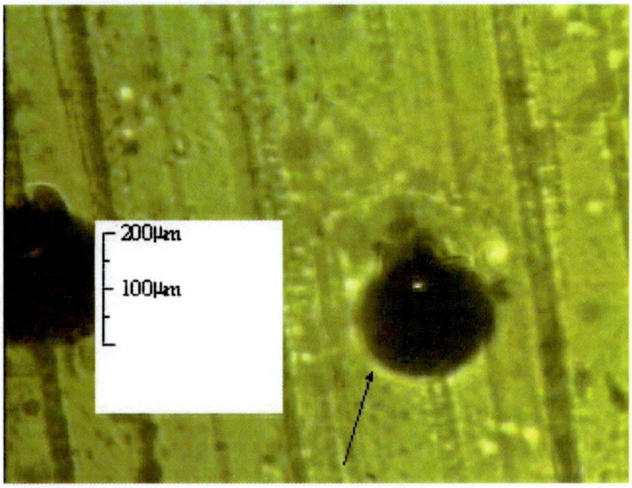


Figure 20. Polymethyl methacrylate (PMMA) microdrilled with $Cu^+$ Ne-CuBr vapor laser, which gives more thermal decomposition and significant HAZ.

The beam divergence for a cavity formed by a concave high reflector and a flat output coupler could be calculated via:

$$\theta_{1/2} = \frac{2\lambda}{\pi d} + \frac{\lambda}{\pi \rho_0},$$

where the first term describes the diffraction-limited divergence (d is the narrowest aperture diameter d = 7.1 mm), and the second term – the geometric divergence ($\rho_0$ is the waist radius which coincides with the Gaussian beam radius on the output coupler). For a curvature radius of the high reflector $R_1$ = 2 m the total divergence is 355 μrad (a sum of the diffraction-limited divergence 22 μrad and the geometric divergence 333 μrad). These divergence angles are estimated for a single transverse mode lasing. At the multimode laser operation, that is observed in our experiment, the divergence is higher and the measured beam divergence is 2.5 mrad.

For a divergence reduction a collimator, consisting of two convex lenses, with magnification about 3.5 is used. The laser beam is focused by an objective with f = 10 cm. To find the best position of the focus, the target is manually aligned in the laser beam direction via micro-translation screw. The utilization of the collimator leads to a decrease in the laser beam divergence to about 90 μrad, but also to a decrease in the output parameters (about two times) because of the absorption losses in the two lenses.

In Fig. 19 micro-hole drilled in glass is shown. As can be seen the hole diameter is 15 μm, while the Heat Affected Zone (HAZ) is 25 μm. Micro-hole drilled in polymethyl methacrylate (PMMA) is presented in Fig. 20. The hole diameter is 10 μm, but has moderate reproducibility (could reach 125 μm). The Heat Affected Zone (HAZ) is about 200 μm. The presence of significant HAZ in the PMMA case shows a different ablation mechanism, which is characterized mainly with a more thermal matter decomposition, possibly due to the moderate absorption coefficient. Micro-holes drilled in the wide-band semiconductor ZnSe are shown in Fig. 21 (a) and (b). The hole diameter varies from 50 to 125 μm, while the HAZ size – from a HAZ absence to 250 μm. The large HAZ is probably due to the laser pulse of long duration - 10 μs (FWHM), which leads to undesirable heating of the glass substratum and thence to a considerable damage of the semiconductor layer.

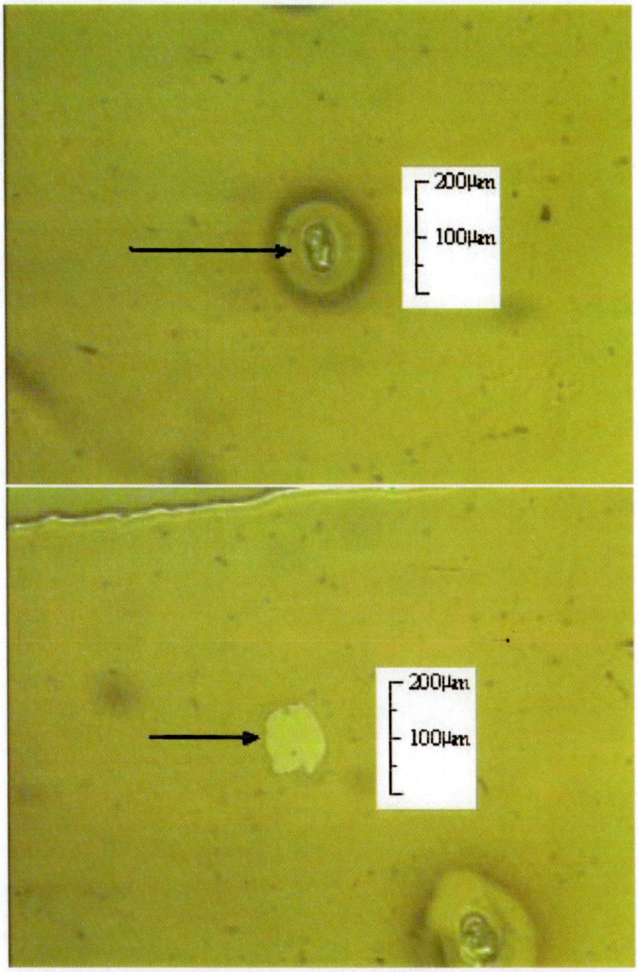


Figure 21. Wide-gap semiconductor ZnSe drilled with $Cu^+$ Ne-CuBr vapor laser.

The results obtained show that the UV $Cu^+$ Ne-CuBr vapor laser can be successfully used for micro-machining of transparent materials and wide-gap semiconductors via photoablation.

## IV. Conclusions

1. Because of the high discharge current density and the high gas temperature the quartz tubes confining the active zone are metalized and some impurities are released into the gas mixture, that leads to average output power reduction from 500 mW to 30 mW for approximately 25 hours of UV lasing.

2. The use of ceramic tubes results in an absence of tube metalization, unchanged gas mixture composition, and a relatively constant average output power – 300 mW, for 100-hour UV laser operation.
3. A comparative investigation on the output characteristics and the electrical parameters of the UV $Cu^+$ Ne-CuBr laser excited by the Blumlein and IC excitation schemes is reported. An optimization of the UV multiline average output power with respect to the discharge conditions is carried out for both excitation schemes. It is obtained that the use of the IC excitation scheme leads to a 30 % increase of the average output power at a simultaneous lasing on the five $Cu^+$ lines - 248.6, 252.9, 259.7, 260.0 and 270.3 nm, in comparison with the Blumlein scheme.
4. The effect of different additives – air, synthetic air, $N_2$, $O_2$, $CO_2$, He, Ar, Xe, and $H_2$, to the neon buffer gas on the average output power of an UV $Cu^+$ Ne-CuBr is investigated. It is found that the addition of small amounts of hydrogen, 0.02-0.04 Torr, leads to an increase by more than twice of the average output power for simultaneous laser oscillation at the five $Cu^+$ lines.
5. A laser tube construction for the UV $Cu^+$ Ne-CuBr laser with a resource exceeding 700 hours is developed and investigated. The laser is operated at a periodic refreshment of the buffer gas mixture - 16.7 Torr Ne and 0.03 Torr $H_2$. In this regime for a 700-hour laser action the average output power is maintained about 600 mW.
6. The influence of UV 248.6 nm laser treatment on the characteristics of the conducting polymer PEDOT is investigated. The refracting index of liquid stilbene films is measure in this UV region. It is shown the possibility to make micro holes in different materials which are transparences for visible and IR light.

## REFERENCES


[1] J. R. McNeil, G. J. Collins, K. B. Persson, and D. L. Franzen, "Ultraviolet laser action from Cu II in the 2500-A-region," *Appl. Phys. Lett.*, vol. 28, pp. 207-209, 1976.

[2] R. Solanki, W. M. Fairbank, JR., and G. J. Collins, "Multiwatt Operation of Cu II and Ag II Hollow Cathode Lasers," *IEEE J. Quantum Electron.*, vol. QE-16, No. 12, pp. 1292-1294, 1980.

[3] J. R. McNeil, W. L. Johnson, and G. J. Collins, "Ultraviolet laser action in He-Ag and Ne-Ag mixture," Appl. Phys. Lett., vol. 29, No. 3, pp. 172-173, 1976.

[4] N. K. Vuchkov, K. A. Temelkov, N. V. Sabotinov, " UV lasing on $Cu^+$ in Ne-CuBr pulse longitudinal discharge," *IEEE J. Quantum Electron.*, vol. 35, pp. 1799-1804, 1999.

[5] Nikolay K. Vuchkov, Krassimir A. Temelkov, Peter V. Zahariev, and N. V. Sabotinov, Optimization of an UV $Cu^+$ laser excited by pulse longitudinal Ne-CuBr discharge, *IEEE J. of Quantum Electron*, vol.37, No 4, 2001, pp. 511-517.

[6] Nikolay K. Vuchkov, Krassimir A. Temelkov, Peter V. Zahariev, and N. V. Sabotinov, Influence of the Active Zone Diameter on the UV Ion Ne-CuBr Laser Performance, *IEEE J. of Quantum Electron*, vol.37, No 12, 2001, pp. 1538-1546.

[7] D.Astadjov, N.Vuchkov, G.Petrash and N.Sabotinov - *Kvant.Elektron.* 11 (1984) 808, Investigation on the reasons limiting the CuBr laser lifetime.

[8] D.Astadjov, N.Vuchkov, G.Petrash and N.Sabotinov - *Metal Vapor and Metal Halide Vapor Lasers*, Nova Science Publishers, New York, Series, Ed. N.Basov (1989), pp. 183-249, A study of the causes limiting the operating life of copper bromide lasers.

[9] Nikolay K. Vuchkov, Krassimir A. Temelkov, Peter V. Zahariev, and Nikola. V. Sabotinov, "Laser tube for ultraviolet copper laser", *Published Applications for Inventions* BG 11/2003.

[10] N.Vuchkov, D.Astadjov and N.Sabotinov - *Optic.Quantum Electr.* 23 (1991) S549-S553, A new circuit for CuBr laser excitation.

[11] N.Vuchkov, D.Astadjov and N.Sabotinov - *IEEE .Quant.Electron.* 30 (1994) 3, 750-758, Influence of the excitation circuits on the CuBr laser performance.

[12] K. D. Dimitrov, T. S. Petrov, N. V. Sabotinov, "Long-live copper bromide vapor laser with cold point", *in Proc. of LTL'2001*, pp. 139-141, 2001.

[13] N. K. Vuchkov, K. A. Temelkov, P. V. Zahariev, N. V. Sabotinov, Output parameters and a spectral study of UV $Cu^+$ Ne-CuBr laser, *Optics and Laser Tech.*, vol.36, pp.19-25, 2004.

[14] NATO SfP 972685, *6 month progress report*, November 2000.

[15] N.K.Vuchkov, K.A.Temelkov, P.K.Telbizov, M.G.Grozeva and N.V.Sabotinov, - He-$Cd^+$ 325 nm line laser with radio-frequency transverse and pulsed longitudinal excitation. *Proc.SPIG 1998*, Contributed papers, p.529-532.

[16] N.K.Vuchkov, D.N.Astadjov - *Optics and Laser Technology* 27 (1995) 407-408, IC-excitated strontium recombination laser.

[17] N. K. Vuchkov, K. A. Temelkov, N. V. Sabotinov, Effect of hydrogen on the average output power of the UV Cu+ Ne-CuBr laser *IEEE J. of Quantum Electron*, vol.41, No1, pp.62-65, 2005.

[18] D.Astadjov,. N Sabotinov and N.Vuchkov - *Opt. Commun.* 56 (1985) 279, Effect of hydrogen on CuBr laser power and efficiency.

[19] D.Astadjov, N.Vuchkov and N.Sabotinov - *IEEE J.Quant.Electron.* QE-24 (1988) 1927, Parametric study of the CuBr laser with Hydrogen additivies.

[20] Y. Wang, W. Sun, Z.Yao, B. Shi, "A practical CuBr laser with flowing buffer gas", *Optics Commun.*, 55, No 5, pp. 345-346, 1985.

[21] M. Ilieva, V. Tsakova, N. Vuchkov, K. Temelkov, N. Sabotinov, 5th Stranski-Kaischew Surface Science Workshop SK-SSW'05, Pamporovo, Bulgaria 19-25 February, 2005, *Book of abstracts*, p. 58.

[22] M. Ilieva, V. Tsakova, N. Vuchkov, K. Temelkov, N. Sabotinov, *Synthetic Metals*, submitted.

[23] M. Ilieva, V. Tsakova, *Synthetic Metals*, 141 (2004) 281.

[24] M. Ilieva, V. Tsakova, *Synthetic Metals*, 141 (2004) 287.

[25] M. Ilieva, V. Tsakova, *Electrochimica Acta*, 50 (2005) 1669.

[26] Sainov S., Dushkina N., "Simple Laser Microrefractometer", *Appl. Opt.*, v.29, p.1406 (1990).

[27] Sainov S., "Laser Refractometer", *Rev. Sci. Instr.*, v. 62, p.3106 (1991).

[28] Sainov S., Sarov Y., and Kurtev S., "Wide -spectral-range laser refractometer", *Appl. Opt.*, v.42, p.2327 (2003).

[29] Sarov Y., Sainov S., Kostik I., Sarova V., and Mitkov S., "Automatik VIS-near IR laser microrefractometer", *Rev. Sci. Instr.*, v. 75, p. 3342 (2004).

[30] Lazare S, Lopez J, Weisbuch F. High-aspect-ratio microdrilling in polymeric materials with intense KrF laser radiation. *Appl. Phys. A* 1999;69:1-6.

[31] Katranchev BP , Naradikian HM, Keskinova EN, Petrov MP and Marcerou JP. The elekcroconvective dendrites in nematics with short range smectic order liquid crystal – 4,n-alkyloxybezoic acids. *Liq. Cryst* 2004;31(12):1663–76.

In: New Developments in Lasers and Electrc-Optics Research ISBN 1-59454-771-8
Editor: William T. Arkin, pp. 75-116 

*Chapter 3*

# IMPROVEMENT OF IMAGES RECONSTRUCTED FROM 3D COMPUTER-GENERATED HOLOGRAMS

*Kenji Nagashima**
Kyushu University, 9-1 Siobaru, Minami-ku 4, Fukuoka, 815-8540, Japan

## ABSTRACT

We review our works concerning improvement of reconstructed images from computer-generated holograms for 2D, 3D Fourier-transform and Fresnel types of holograms.

We show experiments of reconstruction of 2 level phase quantized 2D Fourier transformed holograms produced by using iterative method, error diffusion method and a combination method of the iterative and the error diffusion methods. In particular, use of a combination method was effective for image improvement of 2D objects, but in 3D objects, single use of the iterative method was more effective in improving the reconstructed images.

Since 3D computer-generated hologram that is created from the iterative operations requires much computing time, the 3D hologram using 1D Fourier transform operations was proposed and performed. As a result, similar reconstructed images as 3D holograms were obtained from the holograms made with 1D Fourier transform operations. In the case of the off-axis Fresnel hologram, to improve the reconstructed images, holograms with high construct diffraction patterns were formed by using histogram method. From reconstruction experiments of the holograms, clear images were obtained and 3D images were seen directly with eyes.

## 1. INTRODUCTION

Holography is especially excellent in display, since it can reconstruct the real 3D images. In many field, computer-generated holography[1,2,29] as well as optical holography is investigated extensively.

---

* The author is now with Daiichi Institute of Technology, 10-2 Tyuou 1, Kokubu-shi, Kagoshima-ken, 899-4395, Japan.

Computer-generated holograms have more advantages than the holograms produced by optical devices. Objects used for making holograms need not actually exist if the objects can be described in mathematical terms. The interference patterns made by numerical calculations can be represented by binary form and the patterns can be printed, while an optical hologram is produced by recording the interference pattern of an object wave and a reference wave. It is impossible to make the same computer-generated holograms as the optical holograms, since the latter have a great amount of information. Therefore, computer-generated holograms with the least sampled points are needed for their uses and consequently, the simplified holograms have been studied and reconstruction experiments were successful[3-6].

In order to improve the images reconstructed from computer-generated holograms, the employment of iterative methods is useful for the hologram improvement. Iterative methods are generally classified into two groups: the iterative Fourier method[3-6,14] and the iterative quantization method[9-13,15-21]. The phase information of complex amplitudes of objects is more important than the amplitude and it is possible to make hologram with use of phase terms alone[7,8]. From our experiments, good images can be obtained from 2 level phase quantized holograms produced by means of the iterative Fourier method[14]. The iterative method requires much computing time since this method consists of iterations of Fourier and inverse Fourier transforms, the same operations must be repeated many times until the desired results are obtained. The iterative method has the properties of rapidly improving the image quality within first several iterations, but only a little improvement can be seen in the image after a large number of iterations are performed. On the other hand, there is an error diffusion method that can improve the holograms through only one process. The application of the error diffusion method to amplitude holograms has succeeded [22-25,27,28]. We have also performed the error diffusion experiments of holograms having 2 phase quantization[26].

Most iterative methods excluding our method[9-21] are used to reduce the error between the desired quantization values and the calculated values of objects and further image enhancement cannot be expected even if iterative operations are performed infinitely. From this point of view, it is supposed that more improved reconstructed images may be obtained by means of using first a few iterative operations and secondly using the error diffusion operations. On the basis of this concept, a combination method of an iterative and error diffusion methods was used to improve the holograms in a short time[26]. Image enhancements using these methods have been performed for 2D objects. Since in the case of 3D holograms, whether image enhancement using the iterative or the error diffusion methods is possible or not remained unknown, we have performed the experiment of the hologram of the 3D objects consisting of several cross sections.

When the improvement operation used to decrease the quantization error is performed for Fourier transformed amplitude distributions of the 3D objects, all images obtained from the holograms synthesized by the iterations cannot be improved at the same time. This is because the variation in complex amplitudes due to improvement of one image can vary the magnitude of the other complex amplitudes. In other word, the amplitude intensities and phases of the total complex amplitudes may be disturbed even if the error that results from the quantization of a Fourier transformed amplitudes of one cross section can be decreased considerably by using the other methods. In our method presented here[31], holograms of 3D objects consisting of several cross sections were generated by maintaining the constant spectral intensities of the complex amplitudes obtained from the Fourier-transformed objects in each cross section.

In research on 2D and 3D computer-generated holograms, it is important to shorten the computing time for making holograms. The total operating time for 3D hologram making process of 3D object composed of several cross sections is increased by increasing the number of the cross sections since the 2D Fourier transform must be iterated with the amount of the whole cross sections. To shorten the hologram synthesizing time, the decrease of the Fourier transform process that requires much computing time is desirable. However, the decrease of the computing time cannot be expected due to difficulty of further improvement for the fast Fourier transform(FFT) technique which is used for Fourier transform of the input object. Therefore, decrease of iterative numbers of Fourier transform operations may be most suitable.

The amount of calculations for complex amplitudes in a hologram plane can be decreased, if 3D holograms are generated by using 1D Fourier transform operation only. The 2D Fourier transforms are also needed to iterate the calculation of the several cross sections of the 3D object, but if the transforms are exchanged with the iteration of 1D Fourier transform, then the computing time required for making holograms may reduce approximately to half. We, therefore, have performed the experiment of computer-generated holograms using 1D Fourier transform alone and compared these properties with normal holograms made by 2D Fourier transform[32].

From the experiments using 1D Fourier transform method, we have obtained similar reconstructed images as those from 2D Fourier transform. However, the holograms created by the 1D Fourier transform method frequently reconstruct the images whose intensities are not uniformly distributed. We, therefore, have investigated the method that the image with uniform light intensity can be reconstructed from the hologram made by 1D Fourier transform[35]. The method that we propose has 2 step processes: one is making hologram with addition of compensation factors to the input object arrays and the other is improvement operations of the images reconstructed by means of the iterative method.

The computer-generated holograms formed by 1D-FFT have the three-dimensional horizontal parallax but the vertical parallax is not felt in the reconstructed images. This phenomena is not disadvantageous in 3D visual display since movements of human heads occur mainly in the horizontal direction and the eyes are at the same height. In optical holography, images from Fresnel types of holograms can be seen directly with eyes and consequently, are suitable for the display applications. Especially, the rainbow hologram that can be reconstructed with white light may be suited to this purpose.

For this purpose, computer-generated rainbow holograms, namely, 1D Fresnel off-axis holograms were investigated[36]. In the experiments, holograms of 3D objects were produced with the calculation of the optical transform process without using 1D- FFT algorithm and the improvement studies of the holograms were performed by using a histogram method. From the experiments, clear images were obtained from the holograms whose interference patterns were increased in contrast by the histogram method[36].

In the first two chapters, improvement of the reconstructed images from the 2 level quantized 2D Fourier transformed holograms produced by using iterative method, error diffusion method and their combination method are presented. The next chapter describes reconstruction of 3D holograms generated by using an iterative method. The next two chapter show a 3D hologram using 1D Fourier transform operations and improvement of reconstructed images from this hologram. The final chapter describes the improvement of 1D Fresnel off-axis holograms.

## 2. Improvement of Images Reconstructed From Computer-Generated Holograms Using An Iterative Method

In computer-generated holograms the amplitude information in the Fourier plane, is less important than the phase information. Furthermore, the amplitude information can be discarded without causing any remarkable change in the reconstructed images if the objects have random phases. This fact is realized as kinoform[7] and phase holograms[8]. However, in each case, it is desirable that the amplitude of the Fourier spectrum is suppressed to make it uniform especially, in a hologram having only a few steps of quantized phase information[3-6]. This amplitude at the hologram plane needs to be distributed as uniformly as possible.

In order to obtain the uniformly spaced amplitude, we can multiply the original image $f(x,y)$ by a random phase factor $\exp[i\varphi(x,y)]$ before we calculate the Fourier transform. The amplitudes of the Fourier transformed data seem to be relatively uniform but they generally take various values. To minimize the difference in each amplitude value, the phase angle of the original image must be shifted to the optimal angle. However, the magnitude of the complex amplitude resulting from the Fourier transform cannot be arbitrarily controlled since the phase angle is chosen through a random number generated automatically by computer.

Thus, we have conducted an experiment based on a new technique by which the value of the amplitude in the Fourier spectrum can be uniformly distributed. The procedure is as follows: first, replace the magnitudes of the Fourier transformed values with a constant value like kinoform[7]. Secondly, after the inverse Fourier transform of the hard clipped complex amplitudes has been performed, the complex amplitudes obtained are exchanged for the amplitudes of the original image without changing the phase information in the transformed values. Namely, the modified complex amplitudes used as new image data have amplitudes equal to the original image, while the phases are considerably different. As the next step, the above procedure must be repeated many times.

Let $f(x,y)$ and $\varphi(x,y)$ denote the transmittance of the original image and the phase factor at the spatial coordinates in the object plane $(x,y)$, respectively. The 2D Fourier transform of $f(x,y)\exp[\varphi(x,y)]$ is given by

$$F(u,v)\exp[i\Phi(u,v)] = \iint f(x,y)\exp[i\phi(x,y)]\exp\left[-\frac{i2\pi}{\lambda f}(xu+yv)\right]dxdy \tag{1}$$

where $\lambda$ is the wavelength of the light and $f$ is the focal length of lens used to produce the optical Fourier transform. To equalize all amplitudes of the Fourier spectra, the amplitude $F(u,v)$ must be exchanged for the constant value $A(u,v)$. That is, $A(u,v)\exp[i\Phi(u,v)]$ is obtained. The inverse Fourier transform of the equation is

$$A'(x,y)\exp[i\phi'(x,y)] = \iint A(u,v)\exp[i\Phi(u,v)]\exp\left[-\frac{i2\pi}{\lambda f}(ux+vy)\right]dudv \tag{2}$$

$A'(x,y)$ in (2) may be quite different from $f(x,y)$ because the amplitude $A(u,v)$ takes a constant value. In our method, the amplitude $A'(x,y)$ must be replaced by the original amplitude $f(x,y)$

so as to equalize both intensities. As a result, $f(x,y)\exp[i\varphi'(x,y)]$ is obtained as new image information.

The second operation begins with the Fourier transform of $f(x,y)\exp[i\varphi'(x,y)]$, and the same procedure above is repeated until the amplitude in the Fourier spectrum becomes uniform over the plane. After many iterations, the amplitudes in the hologram plane may be more uniform than the first Fourier transformed amplitudes. The final values obtained through the iterative operation are used to construct the hologram pattern.

The 2D object used as an original image comprises the letters OPTICS, which are formed by bright areas on an opaque background. The image is represented by 128×128 sampled points corresponding to the descrete value $A_{mn}$. Then, random numbers having the value from 0 to 2π, are assigned to the phase variables $\theta_{mn}$. $A_{mn}$ multiplied by a random phase factor $\Psi_{mn}=\exp(i\theta_{mn})$ is calculated by using the FFT algorithm. This relation is written as

$$B_{pq}\Psi_{pq} = \sum_{m,n=0}^{N-1}\sum A_{mn}\psi_{mn}\exp\left[-\frac{i2\pi}{N}(mp+nq)\right] \tag{3}$$

As previously mentioned, we need to change the amplitude to a constant value such as unity. The complex value having the constant amplitude is, then, computed by the inverse discrete Fourier transform. It can be written in the form

$$A'_{mn}\Psi'_{mn} = \sum_{p,q=0}^{N-1}\sum B_{pq}\Psi_{pq}\exp\left[\frac{i2\pi}{N}(pm+qn)\right] \tag{4}$$

The amplitude of the reproduced image is equalized to that of the original image $Amn$ and the new image array $A_{mn}\Psi'_{mn}$ is again computed by the Fourier transform. Thus, a series of processes is repeated. When the process meets the desired iteration, the phase component $\Psi'_{mn}$ is quantized and display by various devices.

Figure 1 shows the hologram patterns made by a line printer. Fig.1(a) is a hologram constructed from the FFT without conducting iterative operations and Fig.1(b) has been produced by 20 iterations. The pattern in Fig.1(b) appears to be similar to that of Fig.1(a). However, it is clear from careful observation that the position of each dot is considerably different from the position shown in Fig.1(a).

Images optically reconstructed from the holograms in Fig.1 are shown in Fig.2. From the reconstructed images, Fig.2(b) is much improved in image quality compared to Fig.2(a).

Similar improvement is also shown in the image reconstructed by computer (see Fig.3). These results indicate that use of the iterative method can enhance the image quality. We cannot prove that by using this method complex amplitudes having uniform spectra can always be obtained, but improved images can certainly be reconstructed from holograms produced by this method.

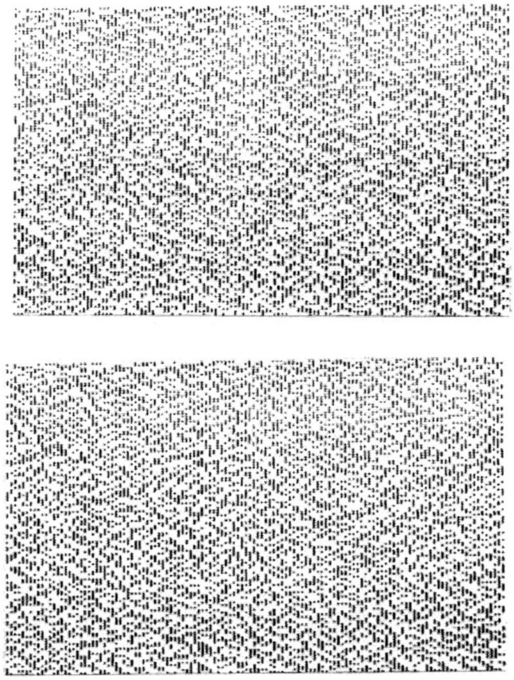

a

b

Figure 1. 2D Fourier transform holograms constructed by a line printer. Holograms made (a) without repeating the operation of image improvement and (b) with 20 iterative operations.

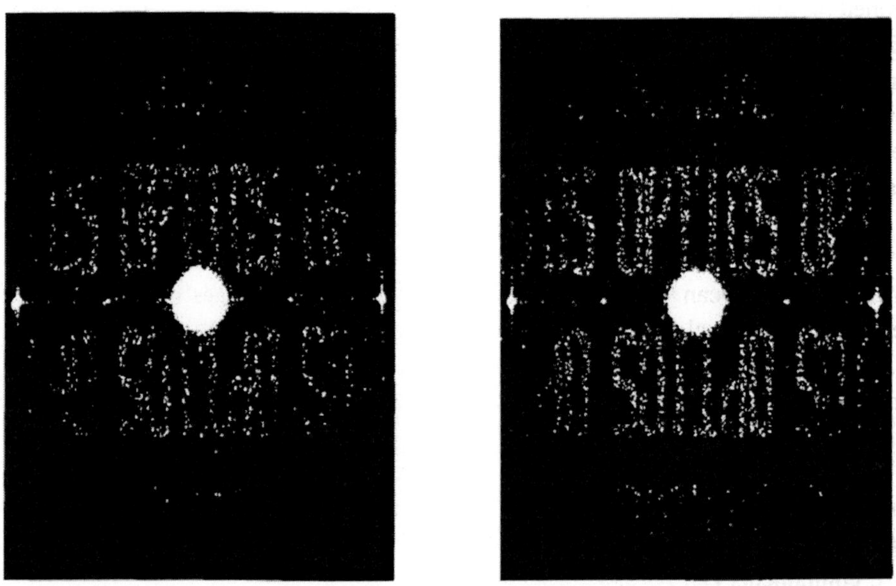

a

b

Figure 2. Optically reconstructed images. Images obtained (a) from the hologram in Fig.1(a) and (b) from the hologram in Fig.1(b).

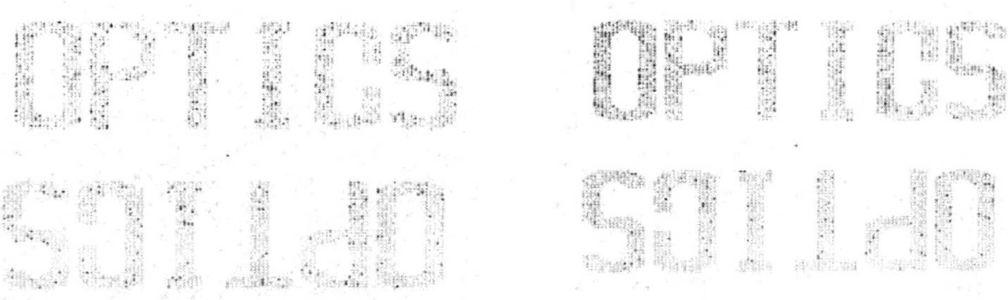

a b

Figure 3. Images reconstructed numerically by computer. Images obtained (a) from the hologram in Fig.1(a) and (b) from the hologram in Fig.1(b).

To estimate the deviation of the reconstructed value from the original value, we define the following relation

$$D = \frac{1}{N^2} \sum_{x,y=0}^{N-1} \sum | r(x,y) - f(x,y) |^2$$
$$= \frac{1}{N^2} \sum_{x,y=0}^{N-1} \sum | d(x,y) |^2 \quad (5)$$

where $f(x,y)$ is the magnitude of the original image and $r(x,y)$ is the reconstructed image obtained through the iterative operation. A small value of $D$ is desirable since the deviation $D$ indicates the total difference between the original image and the reconstructed image. Fig.4 represents the deviation $D$ as a function of the number of iterations. The curve rapidly decreases in height after two or three iterations and gradually approaches a constant value after many iterations. The value of $D$ is not equal to zero even after a large number of the iterations, which suggests that an image identical with the original cannot be produced even if the method is repeated infinitely.

A comparison of the reconstructed image and the original image of a rectangular object is shown in Fig.5. It can be seen from Fig.5 that the discrete values obtained by 30 iterative operations can more closely approach the original value than one operation. As can be seen in Fig.4, it is impossible to remove completely the difference between both values. Consequently, the optimal number of iterations may be decided from the computing time and the image quality required.

A line printer or an X-Y plotter can be draw hologram patterns, but it is difficult for an X-Y plotter to produce the density variation with lines alone. Therefore, it is not the magnitudes but the phases of the complex amplitudes that are quantized to four levels. The phase components can be displayed by an X-Y plotter without difficulty because quantized phase are represented as a shift of position corresponding to their magnitudes. Details of the hologram making process are described in a previous paper(6).

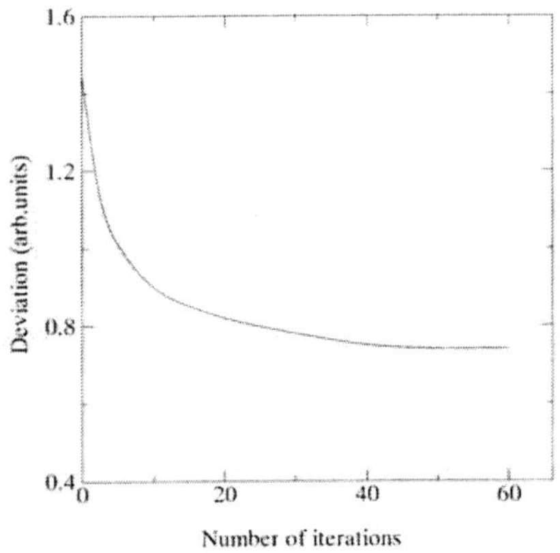


Figure 4. Deviation of a reconstructed image from the original image.

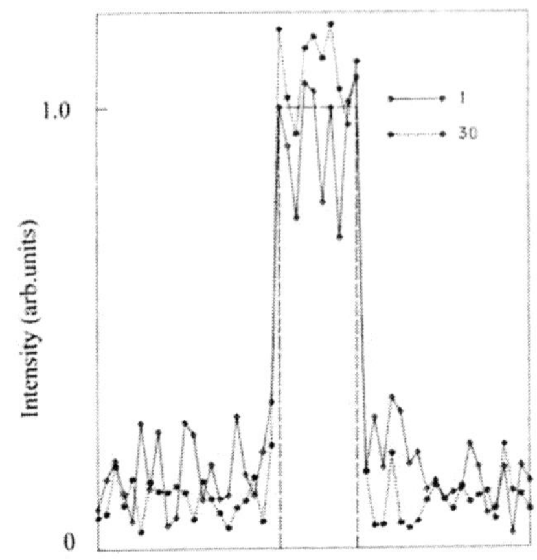


Figure 5. Comparison of the reconstructed and original images. The image is a part of the letter T.

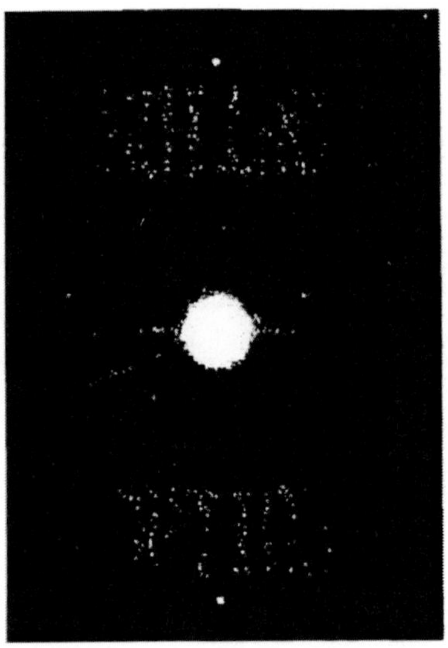

Figure 6. Optically reconstructed images from the holograms using an X-Y plotter. Images obtained (a) without repeating the operation and (b) with 20 iterative operations.

Images optically reconstructed from holograms made by an X-Y plotter are presented in Fig.6. Fig.6(b) represents the image obtained from the line hologram constructed with 20 iterative operations and Fig.6(a) is the image obtained without repeating the operation. The quality of the former image is better than the latter. The improvement on Fig.6 is greater than that in Fig.2. The reason for this lies in the difference of the number of phase quantization levels in each hologram. That is, the holograms used for the reconstruction of Fig.6 have four quantization levels, while the holograms of Fig.1 have only two quantization levels.

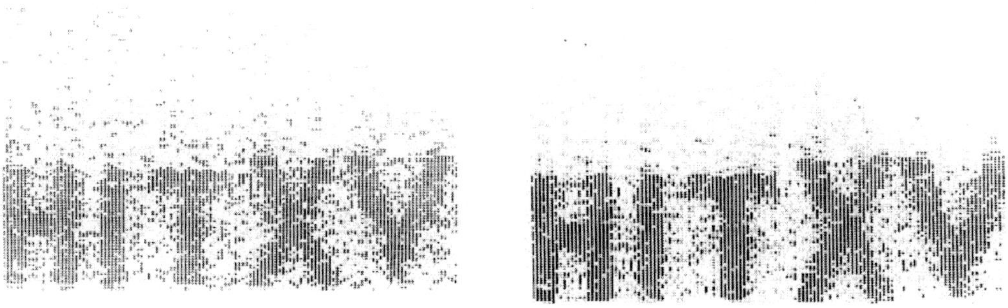

Figure 7. Reconstruction of the hologram obtained from a random phase object. Images obtained (a) without repeating the operation and (b) with 20 iterative operations.

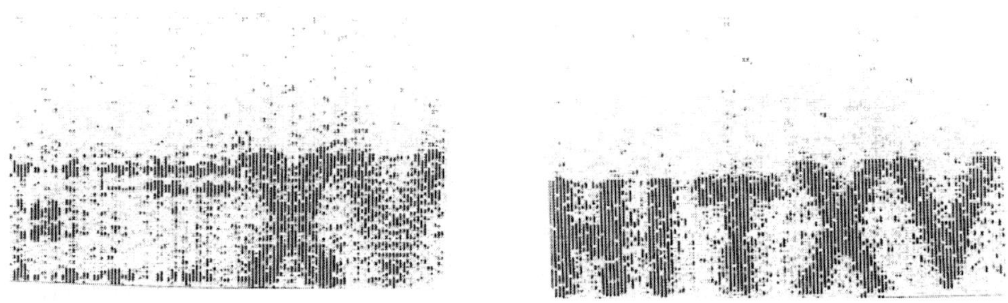

Figure 8. Reconstruction from an object without random phase. Images obtained (a) without repeating the operation and (b) with 20 iterative operations.

If the phases of input objects are randomly distributed, the images can be reconstructed from a part of the Fourier transformed data of the objects. This is observed in optical holograms made using diffused light. Figs 7(a) and 8(a) are images obtained without using the iterative method. Figs. 7(b) and 8(b) show the images having been improved by performing 20 iterative operations. The only difference between Figs. 7 and 8 is that the latter does not use the random phase data as an input object. Of course, reconstruction from a single iteration is disadvantageous but the image obtained after 20 iterative operations is similar to Fig.7 in which the original data with random phases was used. In this case, the magnitudes of the Fourier transformed values are forced to take a value of unity, with the aim of the investigation concerning the decrease of spectral information.

Since the method consists of iterative processes based on the Fourier transform and inverse Fourier transform operations, the computing time is necessarily increased. However, this may not become week point due to development of rapid CPU.

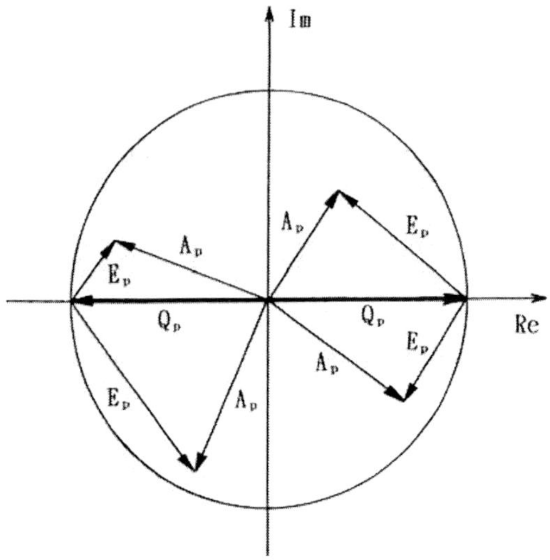


Figure 9. Two-level phase quantization with error diffusion. Ap, sampled complex amplitude; Ep, error resulting from quantization; Qp, quantized value.

## 3. Reconstruction of Holograms Generated By Using A Combination of an Iterative and Error Diffusion Methods

By using iterative method[9-21] good images can be obtained from two-level phase quantized holograms produced by means of the iterative Fourier method[14]. Such methods require much computing time since the same operations are repeated many times until the desired results are obtained. There is an error diffusion method[22-28] that can make improved holograms through only one process. It is supposed from properties of these methods that more improved images may be obtained by using a combination of the iterative method and the error diffusion method. Therefore, to obtain the improved images in a short time in phase hologram, the combination of the iterative Fourier method and the error diffusion method was performed[26].

Error diffusion method is useful for amplitude holograms[13] since complex amplitudes in the hologram plane can easily be transformed to the real values which are desirable for accomplishing the error diffusion operation and for displaying hologram patterns. The error diffusion can also be applied to Fourier-transformed complex values of the objects[18]. However, display as a hologram pattern is difficult, while the two-level quantized phase hologram[3-6] has the property desirable for the application of the error diffusion method. Here, the error diffusion method for the two level phase holograms is considered.

Figure 9 shows two level quantization with the error diffusion algorithm. In the figure, an arbitrary sampled value of the complex amplitude *Ap* is normalized to the maximum amplitude value of 1 and quantized to *Qp* taking constant amplitudes of 1 and the phase of 0 or $\pi(-\pi)$. When *Ap* is less than *Qp*, error *Ep* always arises with *Qp*. This error is added to the complex values at the points around the sampled point. Similar operations are conducted throughout the hologram plane. Some examples of the error distribution are illustrated in

Fig.10. Errors due to quantization of a sampled value *k(i,j)* of the complex amplitude *Ap* obtained from the two dimensional Fourier transform of the object are diffused to neighboring points, as shown in Fig.9.

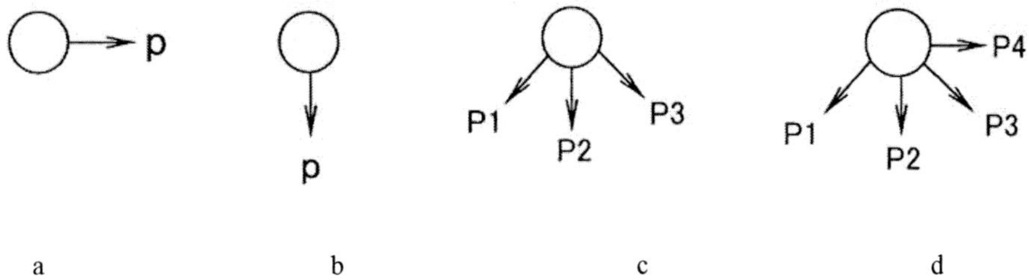


Figure 10. Distribution of error arising from quantization.

Diffusion directions are

(A) $(i, j+1)$

(B) $(i + 1, j)$

(C) $(i+1, j-1), (i+1, j), (i+1, j+1)$

(D) $(i+1, j-1), (i+1, j), (i+1, j+1), (i, j+1)$

The value of the diffusion rate $p$ is equal to 1 in types A and B in the figure. In types C and D, the following conditions are required

$$p1 + p2 + p3 = 1$$

$$p1 + p2 + p3 + p4 = 1$$

First, an experiment that chooses the optimal error diffusion from the four types shown in Fig.10 that can yield the most improved images from the hologram obtained has been performed. The input object used in the experiment is the 2D image with a three stepped intensity variation of 1,2 and 4. An input plane consisting of 128×128 pixels and the original image is placed in the area of 32×32 pixels.

Fourier transformed complex values of the object are quantized to binary states by the method shown in Fig.9 and errors due to quantization are distributed to the points around the sampled points. The value of the next point accumulated with the error arising from the former operation is again quantized and the new error due to quantization is also diffused. These operations are repeated throughout the hologram plane and after finishing the operations, the values quantized to binary states are displayed as a hologram pattern through substituting black and white for the binary states. Examples of holograms generated by using the four types of the error diffusion are shown in Fig.11.

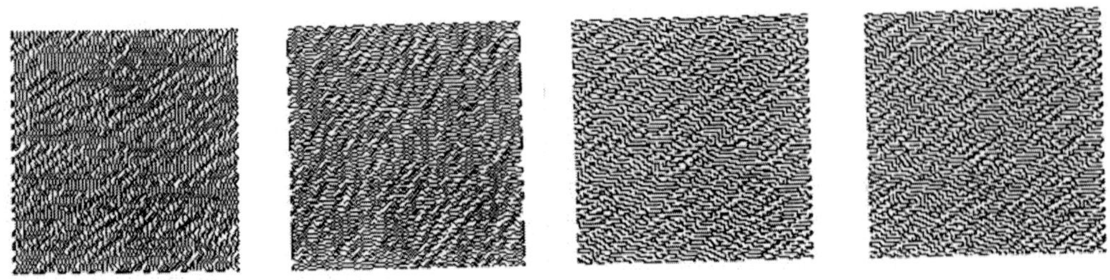


Figure 11. Holograms produced by using the error diffusion method. (a),(b),(c) and (d) are produced by using A,B,C and D shown in Fig.10.

The values of the diffusion rate $p$ used for the error diffusions presented in Fig.9 are as follows

(A) $p=1$

(B) $p=1$

(C) $p_1=0.517,\ p_2=0.368,\ p_3=0.115$

(D) $p_1=0.5,\ p_2=0.3,\ p_3=0.1,\ p_4=0.1$

The values of $p$ in the type C error diffusion were decided by referring to the optimal error diffusion for computer generated holograms proposed by Barnard[23] and the values for type D were also chosen from rough approximations of type C.

**Table 1. Brightness of various error diffusions**

| Type of error diffusion | Relative image brightness |
|---|---|
| A | 1.0 |
| B | 1.0818 |
| C | 1.1939 |
| D | 1.3895 |

It is shown in the figure that if the scan of the error diffusion is in the vertical or horizontal directions, then the patterns extending to the horizontal or vertical directions appear in the holograms and that, if the diagonal scan such as $p_2$ and/or $p_3$ is used for the error diffusion operation, then diagonals are added to the vertical and horizontal patterns.

Figure 12 shows the numerically reconstructed images from the holograms shown in Fig.11. Clearly, the noise has a relation to the hologram patterns. For example, noise in the vertical and horizontal directions is seen in Figs. 12(a) and (b) and noise in the diagonal direction due to $p_1$ and $p_3$ is seen in Figs. 12(c) and (d). Of course, variations of the values of $p_1$ and $p_3$ can affect the total form of the hologram patterns and then, in reconstruction, the area of the noise generation is varied together with the fluctuation of the diffusion rate.

Figure 12(d) seems to be the best image among all the reconstructed images, but it is difficult to estimate the image quality by observation alone. Brightness $B$ of the images was introduced to estimate the image quality numerically. Brightness $B$ is calculated by integrating the complex image amplitude $f(x,y)$ over the reconstruction area. This is defined as

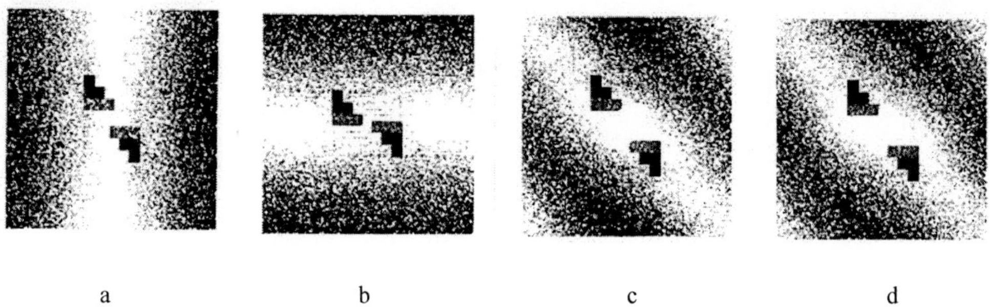

Figure 12. Numerical reconstruction of the holograms. (a),(b),(c) and (d) are reconstructions of (a),(b),(c) and (d) shown in Fig.11.

$$B = \iint | f(x,y) |^2 dxdy \tag{6}$$

The above brightness must be calculated for the images reconstructed from holograms of the objects with the same size and intensity. If these conditions cannot be satisfied, it becomes impossible to compare the qualities of the various images by using the above defined brightness $B$. The brightness of the images is listed in Table 1. As can be seen in Table 1, the use of type D error diffusion shown in Fig.10 leads to the highest image brightness. In addition, it is clear that the image intensity is related to noise intensity. Since the image energy in the input plane and the transform domains is identical (it can be derived from a generalization of Parseval's relationship) the noise intensity is decreased if the image intensity is increased. From the results, type D has been chosen as the most suitable error diffusion.

In the next experiment, holograms using the combination of the iterative method and the error diffusion method were produced. The iterative method was employed in the first process and in the following process the error diffusion method was employed. Since phase type holograms that we proposed can be rapidly improved through small amounts of iterations, the iteration number was set to 20 from the previous result[14]. When the iterative method is employed, we need not pay attention to the position of the original image, whereas the position of the image is important in the error diffusion method because characteristic noise due to the error diffusion is created in the reconstruction plane. If the original object is not properly placed in the input plane, reconstructed images may be overlapped with the noise yielded because of the error diffusion.

The position and size of the original image, therefore, must be chosen by taking the output noise into consideration and, moreover, the diffusion rate $p$ must be fixed on the optimal value that improved images can be obtained. The same object and arrangement used in the type D error diffusion previously mentioned have been employed in this experiment. From the experiments, when using the type D diffusion, $p_1$, $p_2$, $p_3$ and $p_4$ take the same value of 0.25, the best images were obtained. This value was used in the error diffusion of the combination method. In this case, noise appears around the central portion where the images exist. In Fig.13, holograms produced without iterative operations with 20 iterative operations and with the combination method are shown.

a b c

Figure 13. Holograms produced, (a) without iterative operations, (b) with 20 iterations in the iterative method, and (c) with combination of the iterative method and the error diffusion method.

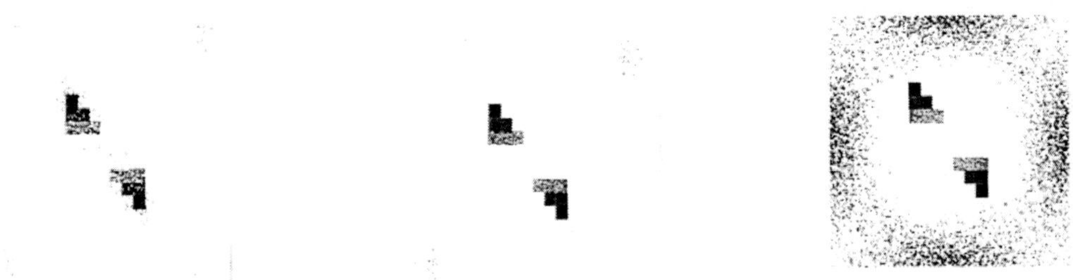

a b c

Figure 14. Numerical reconstruction of holograms. (a),(b) and (c) are the images obtained from the holograms of (a),(b) and (c) shown in Fig.13.

Figures 14(a),(b), and (c) show, respectively, the image numerically reconstructed from the hologram created with two level phase quantization without the iterative operation for image improvement, the image from the hologram created with the correcting operation of 20 iterations and the image from the hologram created with error diffusion after 20 iterative operations. Although noise around the reconstructed images is observed in Fig.14(c), it is obvious from Figs.14(b) and (c) that the improved image is obtained from the hologram produced by using the combination method, but noises are seen around the image. The image obtained from the final operation has a higher quality than that obtained with more than 360 iterations in the iterative method. The iteration number that the quality of the reconstructed image obtained by employing the iterative method is approximately equal to that obtained by employing the combination method remains unknown since we have not performed the iterative experiment over 360 iterations.

In the first process, an iterative operation is used and in the next process, an error diffusion method modified for phase type holograms was employed in the experiment. In the former, the phase errors of holograms can be decreased by forming the complex values into those with uniform amplitude intensities. In the latter, the phase errors that arise from the two level quantization of the complex amplitudes obtained through the iteration operations can be decreased by diffusing around the sampled point. From experiments, it became clear that the combination method can lead to obtaining more improved images than by using the single method and that the computing time required for image enhancement can be reduced since this combination method requires small amounts of iterations. In this case, characteristic noises are created in the images reconstructed from the holograms produced by the error diffusion process. Since an overlap of the image and the noise is disadvantageous for hologram reconstruction, the positions of the original images in the input plane are important for image enhancement.

## 4. Enhancement of Images Obtained From Reconstruction of 3D Computer-Generated Holograms

By using a combination method of an iterative and error diffusion methods more improved images were obtained in 2D holograms[26]. In the case of 3D holograms, according to our experiments, the iterative method was dominant in these methods. In the experiments, holograms of 3D objects were generated by maintaining the constant spectral intensities of all the complex amplitudes obtained from the addition of Fourier transformed images in each cross section.

A lens cannot take the exact Fourier transform of a 3D object, but the concept of Fourier transform holograms can be generalized to 3D objects[29]. Let a 3D object be located near the front focal plane of the lens, and let $g(u, v, z)$ denote the original image at the spatial coordinates. Then the Fourier transformed light amplitude distribution at the back focal plane of the lens is

$$G(x,y) = \iiint g(u,v,z)\exp\left(-\frac{i\pi}{\lambda f^2}\right)z(x^2+y^2) \times \exp\left(-\frac{i2\pi}{\lambda f}\right)(ux+vy)dudvdz \tag{7}$$

where $\lambda$ is the wavelength of the light and $f$ the focal length of the lens. If we can approximate the 3D object by a sum if equally spaced cross sections perpendicular to the z axis, the above integral is written as

$$G(x,y) \approx \sum_{k=0,1,2...}\left\{\exp\left(-\frac{i\pi}{\lambda f^2}\right)k\delta z(x^2+y^2)\iint g(u,v,\delta z)\exp\left(-\frac{i2\pi}{\lambda f}\right)(ux+vy)dudv\right\} \tag{8}$$

where $\delta z$ is a constant.

In our method, to create the 3D holograms, the 2D Fourier transformed image complex amplitude in each cross section is computed and the complex amplitudes of all cross sections are summed up. Since our aim is to obtain the uniformly distributed complex amplitudes in the hologram plane, the total complex amplitudes are normalized to a unity intensity. If the complex amplitudes with constant intensities are quantized in phase, then the reconstruction errors will be decreased more than in amplitude quantization.

In fact, it is difficult to obtain the complex amplitudes with completely uniform intensities by iterative operations. However, we assumed that the best reconstructed images may not be obtained from the holograms generated with such complex amplitudes, but that better images are obtained. With these assumptions, experiments for image enhancement in 3D computer-generated holograms have been performed. Details concerning the operations for image enhancement will be mentioned later.

In computation, the Fourier transform of an image *g(u, v)* in each cross section is calculated. Let the Fourier transform of the image be

$$Gn(x,y)\exp[i\phi(x,y)] \quad \text{(n=0,1,2....)} \tag{9}$$

According to Eq.(2), total complex amplitudes $Gs(x,y)\exp[i\phi s(x,y)]$ are given by

$$Gs(x,y)\exp[i\phi s(x,y)] = G_0(x,y)\exp[i\phi_0(x,y)] + G_1(x,y)\exp[i\phi_1(x,y)] + ... \tag{10}$$

To equalize the amplitude intensity of $Gs(x,y)\exp[i\phi s(x,y)]$, the amplitude $Gs(x,y)$ is exchanged for the constant value A, and we obtain

$$\mathrm{A}\exp[i\phi_s(x,y)] \tag{11}$$

The original images of the cross sections are reconstructed numerically by using above complex amplitude (5).

To reconstruct the individual image in the cross section, the inverse Fourier transform of the normalized complex amplitude (5) is performed. Original images obtained with the inverse Fourier transform are

$$g'(u,v)\exp[i\phi_n'(u,v)] \quad \text{(n=0,1,2...)} \tag{12}$$

In the next step, the transformed images are compared with the original images, and the amplitude intensities are exchanged for those of the original images. Through the exchange, the same images as the original images except for phase variations are obtained. That is, $g'(u,v)$ of the complex amplitude (6) is exchanged for the original amplitude $g(u,v)$. Then, we can write the expression of the complex value as

$$g(u,v)\exp[i\phi_n'(u,v)] \quad \text{(n=0,1,2...)} \tag{13}$$

The modified complex amplitudes used as new image data have amplitudes equal to those of the original images, but the phases are considerably different.

The Fourier transform of expression (7) is calculated again. Similarly, the same operation as mentioned above is iterated until the image improvement is accomplished or the iteration reaches a desired number. If uniformly distributed complex amplitudes are obtained by the iterative operations, then the images from the holograms made with these complex amplitudes will be enhanced. Since, however, it is impossible to obtain the fully uniformly distributed complex amplitudes in the hologram plane with iterative operations, iterations must stop when they reach a desired number. Holograms were generated with two level quantization in phase components of the complex amplitudes.

a b c

Figure 15. 3D original object consists of images in equally spaced cross sections. (a) image at a focal plane, (b) image shifted 15 mm, and (c) image shifted 30 mm from the focal plane.

The 3D object used for the experiments is shown in Fig.15. It consists of three cross sections with plane objects and two light intensities are placed at the different positions in the cross sections. Figure 15(a) is the original image at the focal plane, Fig.15(b) the image of a plane shifted 15 mm, and Fig.15(c) the image shifted 30 mm from the focal plane, respectively. The positions of the original images in the cross sections were chosen carefully.

In order to investigate the performance of the iterative operations in the images reconstructed from the holograms, the iterative number was firstly decided by estimating the value of mean-square-error $e^2$ which is defined as[13]

$$e^2 = \frac{\sum_{n=0}^{n-1}(|f_n|^2 - |f_0|^2)^2}{\sum_{n=0}^{n-1}|f_n|^2} \tag{14}$$

where $f_n$ is the complex amplitude of the image and $f_0$ is the intensity of the original data. We used the error $E$ instead of the value of $e^2$. The error $E$ is given by

$$E = e_n^2 / e_0^2 \quad (n=0,1,2...) \tag{15}$$

where $e_n^2$ is the value of mean-square-error in $n$ iterations and $e_0^2$ is the value of 0 iteration.

Figure 16 shows the error versus the iteration number of images in three cross sections located at different distances in the direction of the z axis. From the figure, it is clear that the image qualities vary with iteration number. The image qualities at the same iterations also varied in each image, and the images distant from the focal plane are more improved than that at the focal plane. Note that many iterations are not always suitable for best image reconstruction, and that furthermore, the enhancement of one image causes decrease in quality of the others.

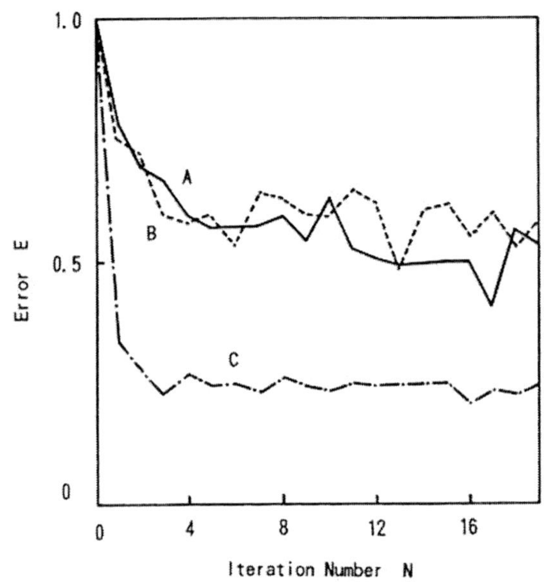


Figure 16. Error E versus the iteration number N. A,B, and C in the figure represent the images at the focal plane, shifted 15 mm and 30 mm from the focal plane, respectively.

Holograms were made with the objects shown in Fig.15, added to random phases. The holograms with and without iterative operations are shown in Figs.17 and 18, respectively. The iteration number of 3 used for producing the hologram patterns was chosen by referring to the curves in Fig.16. The difference in the hologram patterns with the iterative technique is observed in Figs. 17 and 18. A numerical reconstruction from the hologram of Fig.17 is shown in Fig.19. A similar reconstruction from Fig.18 is shown in Fig.20. The images shown in Figs.19 and 20 are located at the focal plane, and at a plane shifted 15 and 30 mm from the focal plane, respectively. In reconstruction of computer-generated holograms created with the 3D objects having several cross sections, twin images are usually obtained with the conjugate symmetry properties in the direction of the z axis with respect to the focal plane. At the cross sections distant from the focal plane, twin images are generated at the front and back planes and at equal distances from the focal plane. If the images from holograms are reconstructed at the same place, it is difficult to distinguish them because of image overlap. Consequently, the original images in the cross sections must be placed at different positions where the reconstructed images cannot overlap each other.

As can be seen in Figs. 19 and 20, in the case of reconstruction without iteration, relatively clear images appear in the focal plane and images in other planes are degraded considerably, whereas with three iterations, the image in the focal plane improves considerably, and, in addition, the other two images are more improved ( in particular, see the letter H in Fig.20). Since it is unlikely that all images will have the least values of error $E$ simultaneously with more than three iterative operations, we should choose the optimum iteration by observing the reconstructed images. Furthermore, from the experiments, this optimum iteration number varies with the objects used.

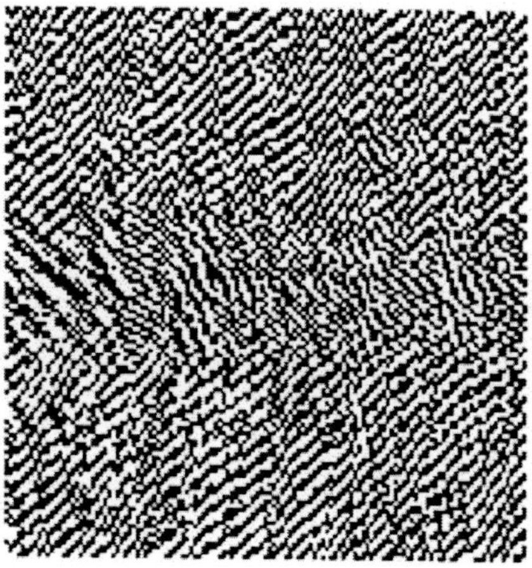

Figure 17. Hologram of the 3D object shown in Fig. 15, made without iterative operations.

Figure 18. Hologram of the 3D object shown in Fig.15, made with three iterative operations.

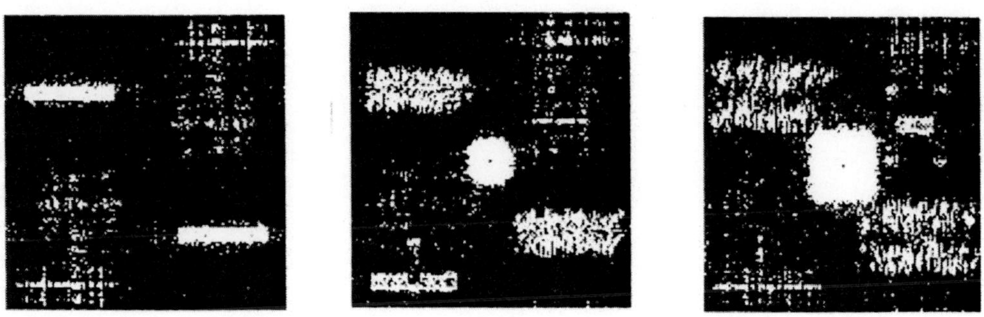

a b c

Figure 19. 3D images obtained by numerical reconstruction of the hologram shown in Fig.17. (a) reconstructed image at the focal plane, (b) image shifted 15 mm, and (c) image shifted 30 mm from the focal plane. The bright spot at the optical axis is removed from the figure.

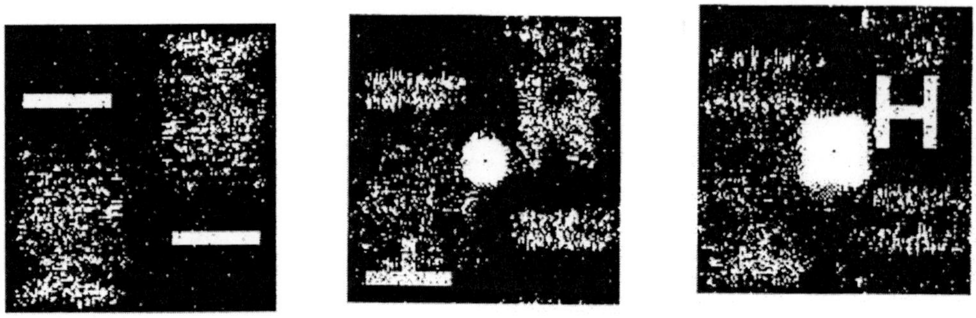

a b c

Figure 20. 3D images obtained by numerical reconstruction of the hologram shown in Fig.18. (a) reconstructed image at the focal plane, (b) image shifted 15 mm, and (c) image shifted 30 mm from the focal plane. The bright spot at the optical axis is removed from the figure.

From the reconstruction of holograms made from the uniformly distributed complex amplitudes used with the iterative method, we have obtained the improved images in all planes of the 3D objects, with fewer iterative operations, for example, with only three iterations.

## 5. 3D Computer-Generated Holograms Using 1D Fourier Transform Operations

In computer-generated holograms, it is important to shorten the computing time required for making holograms. The total operating time for 3D hologram making process is increased by increasing the number of the cross sections since the 2D Fourier transform must be iterated by the number of the whole cross sections[31]. For shortening the hologram synthesizing time, the decrease of the Fourier transform process is desirable, so that the least use of Fourier transform operations may be most suitable. The amount of calculations for complex amplitudes in a hologram plane can be decreased and then, the computing time required for making holograms may reduce approximately to half, if 3D holograms are generated by using 1D Fourier transform operation[32].

Here, we consider computer-generated Fourier transform holograms. Let a 3D object be located near the front focal plane of the lens, and let $f(x,y,z)$ denote the original object. Then, the Fourier transformed light amplitude distribution at the back focal plane of the lens is

$$F(u,v) = \iiint f(x,y,z)\exp(-i)(\pi z/\lambda f^2)z(u^2+v^2)$$
$$\exp(-i)(2\pi/\lambda f)(xu+yv)dxdydz \tag{16}$$

where $\lambda$ is the wavelength of the light and $f$ the focal length of the lens. If we can approximate the 3D object by a sum of its equally spaced cross sections perpendicular to the z axis, the above integral is written as

$$F(u,v) = \sum_{k=0,1,2..} \exp(-i)(\pi/\lambda f^2)k\Delta z(u^2+v^2) \iint f(x,y,k\Delta z)\exp(-i)(2\pi/\lambda f)(xu+yv)dxdy \tag{17}$$

where $\Delta z$ is a constant, and

$$F(u,v) = \sum_{k=0,1,2..} \exp(-i)(\pi/\lambda f^2)k\Delta z(u^2+v^2) \sum_x \sum_y f(x,y,k\Delta z)\exp(-i)(2\pi/\lambda f)(xu+yv) \tag{18}$$

From Eq.(18), it is obvious that the 3D holograms can be generated by the addition of the 2D Fourier transformed image complex amplitudes in each cross section. Next, we attempt to create a equation of a 3D hologram with 1D Fourier transform terms.

If the 2D discrete Fourier transform term in Eq.(18) can be exchanged for 1D Fourier transform, the Eq.(18) will be represented by the form with the single sum. Let $g(x,y,z)$ be the original complex amplitude. Then, the discrete Fourier transform of the object is expressed as

$$G(u,y) = \sum_{k=0,1,2..} \exp(-i)(\pi/\lambda f^2)k\Delta z(u^2) \sum_x g(x,y,k\Delta z)(-i)(2\pi/\lambda f)xu \tag{19}$$

where 1D Fourier transform is calculated in direction of x axis. From the Eq.(19), 3D computer-generated holograms can be created with 1D Fourier transform operation though the complex amplitude $G(u,y)$ in Eq.(4) is not equal to $F(u,v)$ in Eq.(18). Since calculation time of the 1D discrete Fourier transform in Eq.(19) is approximately a half of the 2D discrete Fourier transform in Eq.(18), total computing time is also decreased. It is noted that the complex amplitudes calculated from both Eqs. (18) and (19) are not identical. Some effects of the different complex amplitudes on reconstruction are considered below.

Equation(19) shows the possibility of a 3D hologram creation by 1D Fourier transform iterations. Matters expected from the two kinds of holograms made by using the Eqs.(18) and (19) are

1. If Eq.(18) is used, it may be possible to reconstruct the images from the parts of the holograms, since the informations of the sampled points in a 2D object plane are distributed all over the hologram plane.
2. Since in the 3D holograms created by the method using only 1D Fourier transform operations, all informations concerning the objects are distributed to the one dimensional direction of x or y axes corresponding to horizontal or vertical directions, the transformed complex amplitudes of the data arrays in directions of x or y axes do not affect the complex values arranged in other x or y axes directions.

In other words, the above express the characteristics of the 1D and 2D Fourier transform operations. We can suppose from above mentioned matters that the properties of the images reconstructed from two kind of holograms are also different each other. These differences are expected to be similar in property to those between the general Fresnel[33] holograms and rainbow holograms[34].

3D objects used in this experiment consist of 3 cross sections with plane ring images on the surfaces and 3D object consists of equally spaced 3 cross sections at a distance 0, 10 mm and 20 mm from the focal plane. There are 2 rings with different diameters on a plane with 256×256 points and the rings on the plane do not overlap each other. Figure 21 shows the images used as input object. In the experiment, holograms have been created by using Eqs.(18) and (19). Hologram patterns were drawn by the 2 quantized phase method and hologram reconstruction was performed numerically by computer.

Figure 22 is a hologram pattern obtained from 2D Fourier transform method using Eq.(18), and Fig.23 is a hologram obtained from Eq.(19). Spaced two hologram patterns are shown in Fig.23. This space is corresponding to the part without objects, as is shown in Fig.21, so that the hologram pattern is not created in this place by 1D Fourier transform in the direction of x axis. Images reconstructed from the hologram in Fig.22 are shown in Fig.24, and Fig.25 shows the images from the hologram in Fig.23. Figures 24(a),(b)and (c) which are reconstructed from the hologram in Fig.22, show the images in the focal plane and in the 2 planes distant from the focal plane, respectively.

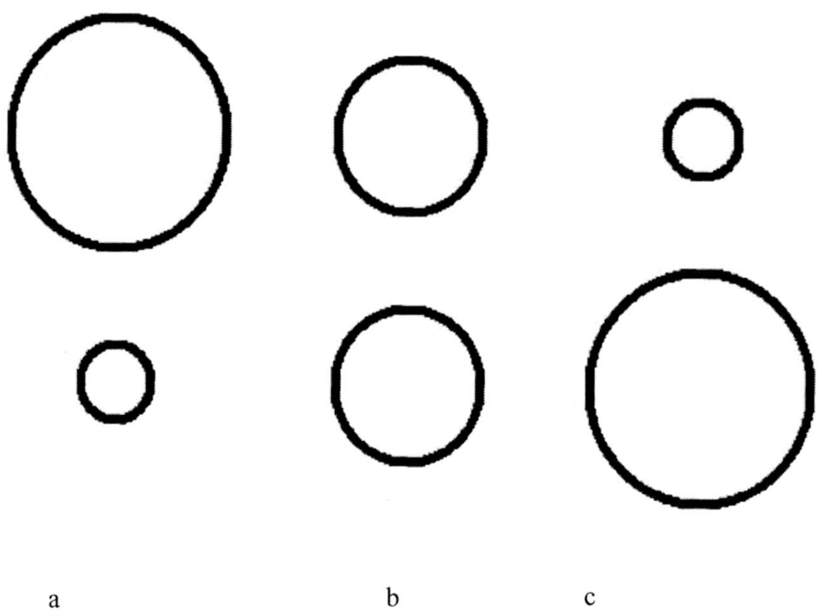


Figure 21. Three images used as input data of a 3D object. The 3D object consists of equally spaced three cross sections at a distance (a) 0 (focal plane), (b) 10 mm and (c) 20 mm from the focal plane, respectively.

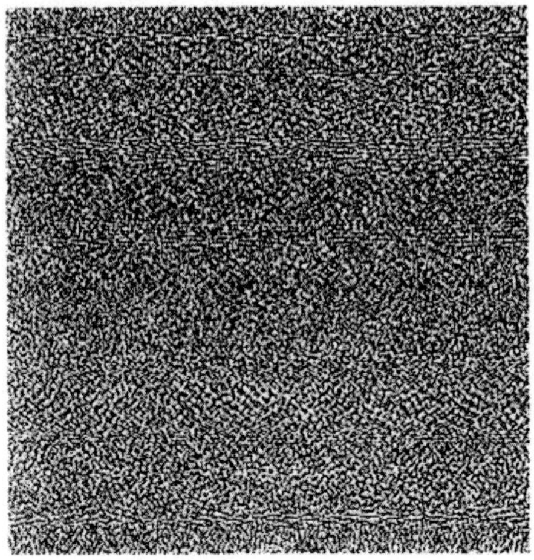

Figure 22. Hologram created by using iterative 2D Fourier transform method.

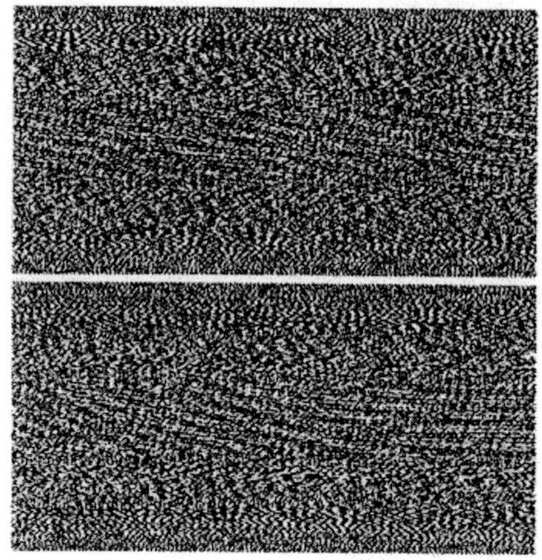

Figure 23. Hologram created by using iterative 1D Fourier transform method.

Similar results were obtained from Fig.23. Figures 25(a), (b) and (c) show the images reconstructed from Fig.23 and show the images in the focal plane, in the planes distant from the focal plane, respectively. It is clear that the holograms created with two methods can reconstruct the similar images, since Fig.24 is nearly equal to Fig.24 in image quality. However, the directions of the twin images in Fig.24 are considerably different from the directions of the images in Fig.25. The twin images which are seen in the focal plane in Fig.24(a), appear in the symmetric planes with respect to the central point (light axis), while in Fig.25(a) appear in the symmetric planes with respect to vertical line. Positions of twin images are symmetric with respect to the focal point, and for this reason, the images reconstructed in the plane apart from the focal plane are blurred completely, though twin images in Figs.24(a) and 25(a) are seen clearly on the focal plane.

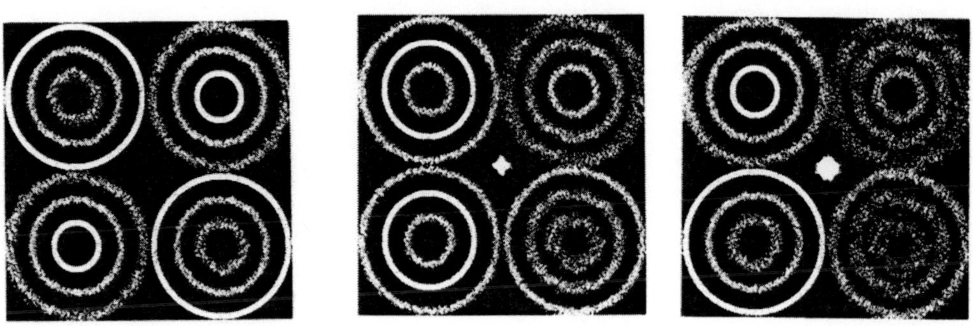

a b c

Figure 24. Images reconstructed from the hologram shown in Fig.22. Reconstructed images correspond to the input images shown in (a) Fig.21(a), (b) Fig.21(b) and (c) Fig.21(c), respectively.

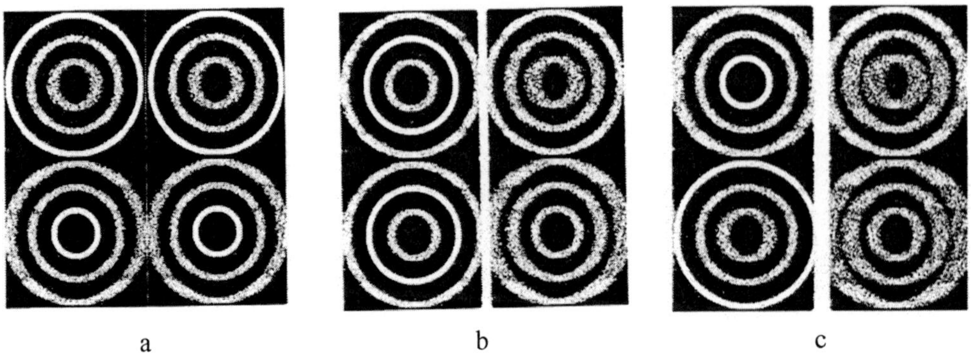


Figure 25. Images reconstructed from the hologram shown in Fig.23. Reconstructed images correspond to the input images shown in (a) Fig.21(a), (b) Fig.21(b) and (c) Fig.21(c), respectively.

The relative positions of the images reconstructed from holograms made by 2D and 1D Fourier transforms are illustrated in Figs.26 and 27. For example, in 2D Fourier transform method, the images in P1 and P1' planes are symmetric with respect to the focal point on the P0 plane, and in addition, the direction of the image in P1 plane is opposite to the direction of the image in P1' plane, but in 1D Fourier transform method, the images have same directions. Next, from image quality if P1 plane is in focus, then P1' is out of focus and therefore, the images in the P1' is blurred. Conversely, when the P1' plane is focused on, the image in the P1 plane is more blurred than that in P2', since distance between P1 and P1' is longer than that between P2' and P1', and the image in P2 is the most blurred. Further, in 2D Fourier transform method one of the pair images takes opposite direction to the other, as shown in Fig.6, whereas in 1D Fourier transform method the directions of a images is the same to the other and twin images appear in the planes with respect to the focal point and vertical or horizontal lines. Of course, all the twin images except for those in the focal plane, appear in distant planes from the focal plane each other.

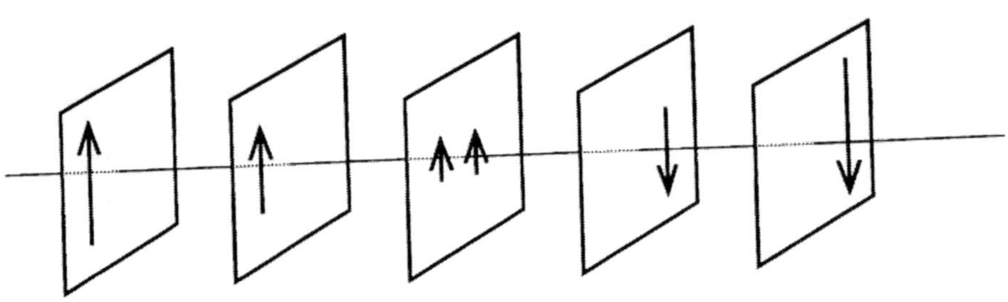

Figure 26. Schematic illustrations of the spaces of the three twin images reconstructed from the hologram shown in Fig.22.

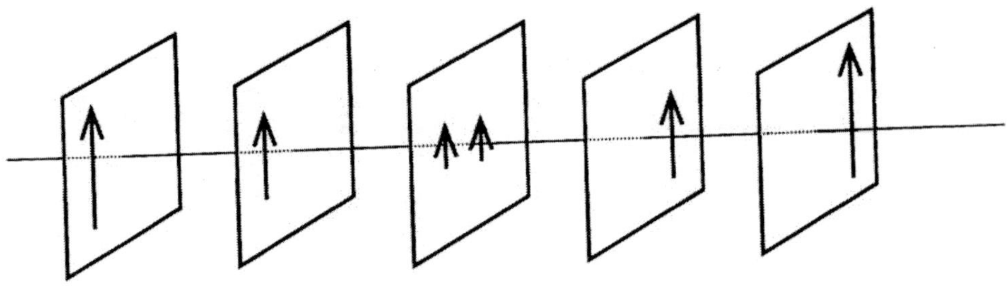

Figure 27. Schematic illustrations of the spaces of the three twin images reconstructed from the hologram shown in Fig.23.

As mentioned before, there is a serious difference between the images reconstructed from the holograms made by the two methods. Especially, in reconstruction from holograms of the objects with 2D light intensity variations, made by 1D Fourier transform method, the same restoration as that from 2D Fourier transform method may not be obtained, since complex amplitudes of Fourier transformed object arrays of vertical or horizontal directions in the hologram plane are independent of the Fourier transformed arrays of the other vertical or horizontal directions.

In the computer-generated holograms of the 3D objects consisting of some cross sections, a method that synthesize the holograms with iterations of 1D Fourier transform operations was considered for the purpose of shortening computing time and the hologram made by using this method was compared with the hologram created with the iterations of 2D Fourier transform operations. As a result, similar images were reconstructed from the holograms synthesized by using two methods, but properties of reconstructed images were slightly varied. That is, the directions of the twin images reconstructed from holograms made by the iterations of 2D Fourier transform method are reversed each other, while the twin images from the 1D Fourier transform method have the same directions. These differences were clearly shown in the figures obtained from the experiment and moreover, there was a quality difference in the images reconstructed from the two types of holograms. Especially, in the 1D Fourier transformed phase type hologram, it is difficult to equalize the density variation of reconstructed images to that of original input images. This difficulty will be removed in the next chapter. However, the most different point in the two methods is the time required in the computations. This point of view, the method using the iterations of 2D Fourier transform requires twice the time of the method using the 1D Fourier transform operations, but it should be noted that the image properties vary with the individual method. If input objects have complex forms and/or various light intensities, then the degraded images will be obtained from the holograms of 1D Fourier transform method. Restorations of these image degeneration are important when 1D method is used instead of 2D method in making 3D computer-generated holograms.

# 6. Improvement of Reconstruction in 3D Computer-Generated Holograms Using 1D Fourier Transform Operations

Since computing time for a 3D computer-generated hologram is increased by increasing the numbers of the cross sections of the object, it is effective for using 1D Fourier transform operations. As mentioned before, we have proposed the computer-generated holograms using 1D Fourier transform and compared the properties with normal holograms made by 2D Fourier transforms[32]. From the experiments using 1D Fourier transform method, we have obtained similar reconstruction as those from 2D Fourier transform. However, the light intensities of images from holograms created by the 1D Fourier transform method are not uniformly distributed. We, therefore, have investigated the method that the image with uniform light intensity can be reconstructed from the hologram made by 1D Fourier transform. The method that we propose has 2 step processes. One is making hologram with addition of compensation factors to the input object and the other is improvement of the images reconstructed by means of the iterative method[9-21]. In this chapter, experimental results using the proposed method are described.

Let a 3D object be located near the front focal plane of a lens. If we can approximate the 3D object by a sum of its equally spaced cross sections perpendicular to the z axis, then the 2D discrete Fourier transform term can be exchanged for 1D Fourier transform and the complex amplitudes will be represented by the form with the single sum in the hologram plane. Let $g(x, y, z)$ be the complex amplitude of the original object. Then, the discrete Fourier transform of the object is expressed as [32]

$$G(u, y) = \sum_{k=0,1,2..} \exp(-i)(\pi/\lambda f^2)k\Delta z(u^2) \sum_{x} g(x, y, k\Delta z)(-i)(2\pi/\lambda f)xu \tag{20}$$

where $\lambda$ is the wavelength of the light and $f$ the focal length of the lens and $\Delta z$ is a constant. Here, 1D Fourier transform is calculated in direction of x axis. From the Eq.(1), 3D computer-generated holograms can be created by 1D Fourier transform operation but the complex amplitude $G(u, y)$ in Eq.(1) is not equal to the complex amplitude obtained from 2D Fourier transform. Of course, properties of holograms that are made with 1D Fourier transform are different from those made with 2D Fourier transform.

The sum of the amplitude intensity of the arrayed image reconstructed from the hologram of a line object in direction of x axis must be identical to that of 1D complex amplitude in the hologram plane, since the image energies are identical in the object plane and the transform domain as mentioned before. The energy equivalence causes the generation of the irregular density patterns in the image field, even if the original object has uniform intensity amplitudes.

Since in the 3D holograms created by the method using 1D Fourier transform operations, all information concerning the objects are distributed to the 1D direction of x or y axes corresponding to horizontal or vertical directions, the transformed complex amplitudes of the object arrays in direction of x or y axes do not affect the other complex amplitudes arranged in the same direction. Therefore, even if the objects have the same light intensities over all parts, reconstructed images have various light intensities in each part. As to this property, 1D Fourier transform holograms are more disadvantageous than 2D Fourier transform holograms. However, if we can remove this defect, 1D Fourier transform method is essentially superior to 2D method in computing time. Thus, we performed the experiment to improve the images reconstructed from 1D holograms.

In the case of the object with uniform intensity, if the total light intensity values of an object data array are different in each data array in x axis direction, the intensities of the reconstructed images must have same values in x axis direction. This causes the appearance of irregular patterns in the reconstructed images. In order to improve this energy problem, we have put the additional points for compensation at a place that cannot disturb the reconstruction of holograms. Assuming that all object data in direction of x axis are values of 1, the intensity of a single point reconstructed from the hologram of the object having a few sample data becomes larger than that having many sample points. We consequently, have performed the experiments whether we can diminish the intensity of the reconstructed single point by using compensate points in the case of the objects with a few 1D data points or not.

**Table 2. Intensity of a point reconstructed from holograms of 2 point samples with various compensation points.**

| Configuration types | Values of compensation points in the position at | | Intensity (arb.) |
|---|---|---|---|
| | 126 | 127 | |
| A | 0 | 0 | 2692(1.0) |
| B | 0 | 1 | 1969(0.731) |
| C | 1 | 1 | 1253(0.465) |
| D | 2 | 2 | 771(0.286) |

Figure 28 shows the intensity of a single sample point reconstructed from the hologram of the object having various lengths of data arrays whose values are 1. It was obtained by the numerical calculation for holograms with 2 level phase quantization. The result indicates that intensity becomes low together with increase of the sample point, as is considered before. However, for example, the intensity of each point reconstructed from the hologram of the object consisting of 10 sample points whose values are 1 does not become 1/10 for the intensity of a point image reconstructed from the hologram of a single point object.

Figure 29 shows the intensity of the complex amplitudes transformed from the object of 2 sample points with the value of 1. As can be seen in Fig.29, points with many side intensity peaks arise from reconstruction of only a few sample points. These peaks are supplied from light energy of the hologram and therefore, the amplitude of the distributed energy is different in the image field with the amplitude of input object, so that the intensity obtained are not coincident to the predicted values. Due to appearance of these side peaks, we cannot decide the numbers and the intensities of the compensate points with easy. We have investigated more optimum conditions for compensation in reconstruction with numerical experiments. The intensity of a single point that is reconstructed from the hologram of a single sample point is most intense and consequently, we must diminish its intensity approximately to the level that are shown in many sample points.

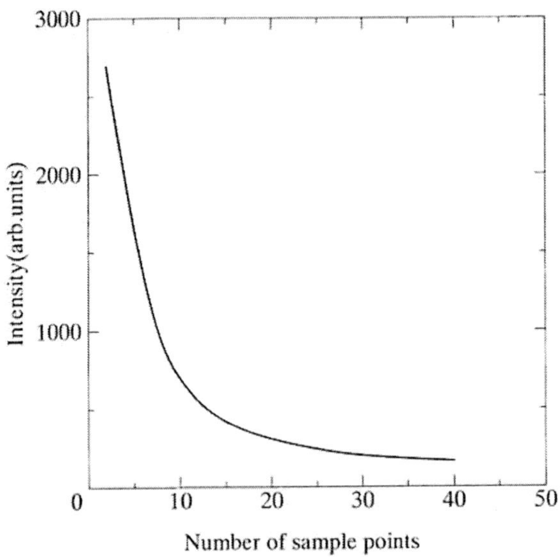


Figure 28. Intensity as a function of the numbers of sample points.

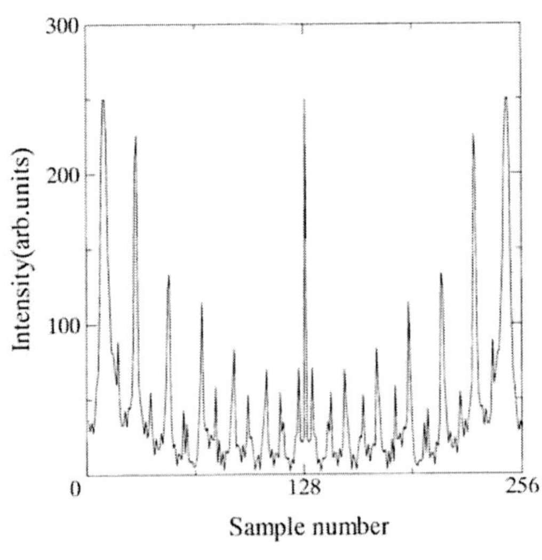


Figure 29. Intensity of image field in 1D direction reconstructed from the hologram of 2 sample points.

Table 2 shows the reconstruction intensity of the single point obtained from 2 input points having 2 compensation points. In Table 2, compensation points in type A have 0 values respectively and in type B have the values of 0 and 1 at the sample number 126 and 127. From Table 2, it is clear that intensity of a reconstructed point is decreased with increasing the number of compensation points and their values. However, from the experiment, the noise of reconstructed images is increased with increasing the number of the compensation points or the intensity values of the points. Noises were seen a little in the image obtained from the hologram made by type D configuration, but the image quality was not decreased, and the harmful noises did not occur in type C configuration. From these results, we have chosen the configurations of the compensation points from type A to type D by judging sizes of the input points. Since it is relatively difficult to relate the magnitude of the 1D input object data to the suitable configuration for the compensation points, we have chosen the configuration types through the numerical experiments. Next, we must find a rule to select the configurations. The equation for the selection is

$$X_i = (S_m - D_n) \times K$$

where $S_m$ is the maximum number of the input sample points, $D_n$ is the number of the sample points in the *i*-th row of 1D input object and K is a constant. In synthesizing hologram, compensation points must be added to the positions depending on the configuration type decided with the value of $X_i$ . In the present experiment, relations of the configuration type and the value of $X_i$ are given by

$X_i < 7.0$ type A

$7.0 \leq X_i < 8.0$ type B

$8.0 \leq X_i < 9.0$ type C

$9.0 \leq X_i$ type D

An input plane consists of 256×256 sample points and the image used as a input object has 256×128 sample points. The input plane was calculated with 1D Fourier transform from top to down in the horizontal direction. After the calculation process was finished, hologram patterns were drawn by the 2 quantized phase method[1,7,15]and then, the hologram was reconstructed numerically by computer.

Figures 30(a) and (b) are a hologram example of the triangle object without compensation points and a reconstructed image from the hologram, respectively. Figures 31 (a) and (b) are the hologram of the triangle object with the compensation points and the reconstructed image, respectively. From Fig.30(b), the sharp edge of the upper end part of the triangle is blurred by intense light, whereas the same place shown in Fig.31(b) as that in Fig.30 (b) can be more clearly seen. Furthermore, the intensity of the image shown in Fig.31(b) is more uniformly distributed than that in Fig.30(b). By utilizing these results, we have performed the experiments of making holograms of 3D objects and reconstruction.

First, we have tried making holograms without the compensation points. Figure 32 (a) is a hologram made from a 3D object having two planes which are the focal plane and the plane distant from the focal plane, and Figs.32(b) and (c) are the images reconstructed from

Fig.32(a). Figure 32(b) shows the image in the focal plane and Fig.32(c) is the image in the plane distant from the focal plane. Secondly, we have made the holograms of the object with the compensation points with the same condition as above. Figure 33 (a) is a hologram and Figs.33(b) and (c) are the images reconstructed from Fig.33(a). Figure 31(b) is the image in the focal plane and Fig.33(c) is the image in the plane distant from the focal plane. From comparison of Fig.32(b) with Fig.33(b) it is evident that the image in Fig.33(b) has comparatively uniform intensity without decreasing the image quality.

If we desire the more improved image in quality, we must employ another method for image improvement. Then, we have used the iterative method mentioned in a previous paper [31]. This method has been effective for 2D Fourier transform processes, so that we have changed the iterative method into such a method that can apply to 1D Fourier transform processes. Another methods, for instance, error diffusion methods [22-28] are not effective for the 1D Fourier transform operations, since in the case of 1D transform the error diffusion operations can be carried out in only one direction. In fact, we have tested the error diffusion method, whereas improvement has not been observed in the reconstructed images.

Figures 34(a), (b) and (c) show the hologram and images in 2 reconstruction planes, obtained from using the iterative method, after the hologram is synthesized by compensation points method. The hologram shown in Fig. 34(a) was made by 4 iterative operations. Obviously, from Fig.34(b), the image at the focal plane was sufficiently improved. The image yielded in the distant plane from focal plane was nearly identical to the first image without iterative operations. It may be caused from the reason why the Japanese letters used as the object to set the plane distant from the focal plane were not fit for a sample object because of its fine style. Generally, in the iterative method, many iterative operations do not bring about a great advance in quality of the reconstructed images.

In the 1D Fourier transformed phase type hologram, it is difficult to equalize the light intensity of a reconstructed image to that of an original input image, since a 1D data array is independent of the other arrays in 1D direction in the transformed domain. This property is disadvantageous for reconstructing accurately the intensities of objects, while it is useful that the holograms are created with shorter time than those using 2D Fourier transform. Thus, we have proposed the method for improving the images obtained from the computer-generated holograms made with 1D Fourier transform.

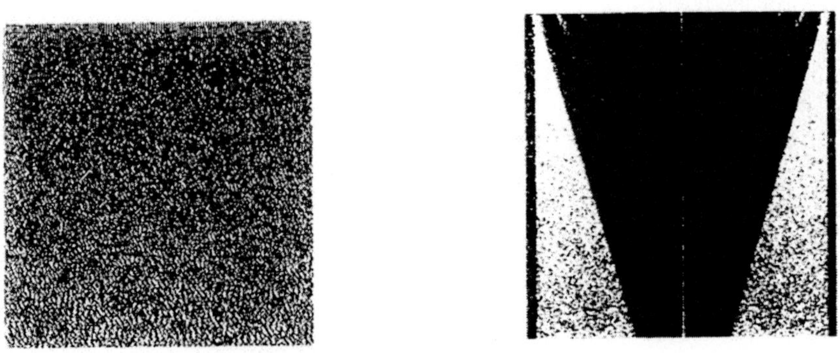

Figure 30. Hologram without compensation points and image reconstructed from the hologram. hologram and (b) reconstructed image.

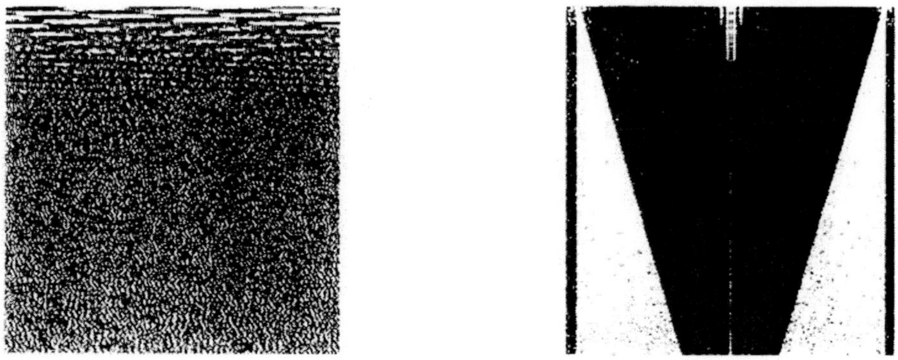

Figure 31. Hologram with compensation points and image reconstructed from the holograms. hologram and (b) reconstructed image.

a b c

Figure 32. 3D hologram without compensation points and image reconstructed from the hologram. (a) hologram. Reconstructed images (b)at the focal plane and (c) at the plane distant from the focal plane, respectively.

a b c

Figure 33. 3D hologram with compensation points and image reconstructed from the hologram. hologram. Reconstructed images (b) at the focal plane and (c) at the plane distant from the focal plane, respectively.

a b c

Figure 34. 3D hologram with compensation points and iterative operations, and image reconstructed from the hologram. (a) hologram and reconstructed images (b) at the focal plane and (c) at the plane distant from the focal plane, respectively.

For the purpose of making such holograms, we added the compensation points to input object data array in 1D direction. From the holograms obtained through this operation, we could obtain the images with uniformly distributed light intensity, and improved images were created from the holograms made with 1D Fourier transform by using the compensation points method and the iterative method.

## 7. Improvement of 1D Off-Axis Computer-Generated Holograms

The computer-generated hologram can be created from not only really existent objects but also the objects in a mathematical description. Computer-generated holograms of 2D objects belong to Fourier transform types are easily synthesized by using FFT algorithm. In the 3D object, however, computing time is increased rapidly with increasing the number of the cross sections. To decrease the increased computing time due to many iterative operations of 2D-FFT, the 3D holograms made with 1D-FFT alone have been proposed[32,35]. 1D-FFT, of course, can be exchanged for other calculation method. In this chapter, computer-generated rainbow holograms, namely, 1D Fresnel off-axis holograms are described[36]. In the experiments, creation of the holograms and improvement of reconstructed images were performed. Holograms of 3D objects were produced with the calculation of the optical transform process in the direction of x or y directions without using 1D-FFT algorithm and the improvement study of the reconstructed image was also performed by using a histogram method.

In the 3D holograms created by the method using the 1D optical transform operation, the information of an object is distributed to the direction of x- or y-axis corresponding to horizontal or vertical direction, so that the transformed complex amplitudes of the 1D arrays of the object in the direction of x- or y-axis do not affect the other arrays arranged in the same direction. For this reason, the transformed data of input objects can be divided into the several groups in the same direction and save the groups individually to the memories. If we desire to create the holograms of the images having a great many pixels, such properties of dividing and saving may be favorable in the transform operation.

We consider the arrangement for recording a 1D off-axis hologram using a reference beam, shown in Fig.35. Assuming that the reference beam R(y) is the parallel wave, the complex amplitude at point $p$ on the photographic plate can be written as

$$R(y) = r(y)\exp(i2\pi\xi y) \tag{21}$$

where $\xi=(\sin\theta)/\lambda$ and $\theta$ is the angle between the reference beam and the y-axis, and $\lambda$ is wavelength. The phase term in Eq.(21) varies depending on the coordinate of y-axis, while both the amplitude and phase of the object beam vary in the direction of x-axis and the complex amplitude can be written as

$$O(x) = o(x)\exp[i\varphi(x)] \tag{22}$$

The resultant intensity I(x,y) on the hologram plane is

$$\begin{aligned} I(x,y) &= |R(y) + O(x)|^2 \\ &= |r(y)|^2 + |o(x)|^2 + 2r(y)o(x)\cos[2\pi\xi y - \varphi(x)] \end{aligned} \tag{23}$$

When we make the off-axis types of computer-generated holograms, the first and second terms on the right-hand side of Eq.(23) may be neglected since they represent the light intensities of reference and object beams. We can therefore, use only the third term to make the hologram and the hologram pattern H(x,y) is

$$H(x,y) = 2r(y)o(x)\cos[\varphi(x) - 2\pi\xi y] + K \tag{24}$$

where K is a bias term which is needed to satisfy the condition that $H(x,y) \geq 0$.

In the calculations of the complex amplitudes on the hologram, Fourier transformation was not used from the reason that using FFT operations in the 3D objects with complicated shapes cannot decrease computing time.

Here, let coordinates of point u at the surface of the object be $(x_i, y_i, z_i)$, and let coordinates of the point u' at the hologram plane be $(x, y_i, 0)$. Then, the distance from u to u' $l_i(x)$ is defined as

$$l_i(x) = \sqrt{(x - x_i)^2 + z_i^{\,2}} \tag{25}$$

The complex amplitude is determined from the superposition of the object wavefronts, and consequently the complex amplitude distribution of the 1D array on the hologram surface in the direction of $x$, is expressed as

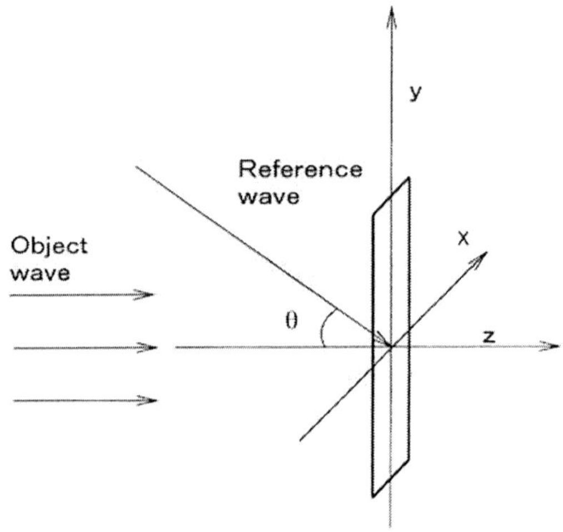


Figure 35. Schematic diagram for recording an off-axis hologram using a reference wave.

$$O(x) = \sum_{i=1}^{N} \frac{o_i}{l_i(x)} \exp\{j[kl_i(x) + \phi_i]\} \tag{26}$$

where $N$ is the number of object points in the input plane and $\phi_i$ is the constant. $k$ is the wave number and $k$=2π/λ. Complex amplitudes of the objects were computed by using Eq. (26) and the hologram patterns were displayed by a laser printer with 256 density levels.

Figure 36 is an example of a histogram obtained from an off-axis hologram generated by computer. It is clear in the figure that the amplitude intensity values of the pixels of an object are mostly concentrated to a relatively narrow intensity range. Such property has been generally observed in the holograms that were made by means of using a reference wave, even if the input objects are changed for various objects. The holograms with the properties such as Fig.36 usually have low contrast patterns. The holograms with low contrast can reconstruct the images having low light intensity, and quality of the reconstructed image is deteriorated. Thus, enhancement of the hologram pattern is needed for improvement of the reconstructed images, but the most suitable methods for the improvement of the images are unknown. Error diffusion methods[22-28], for example, may be ineffective in improving holograms since the off-axis holograms already have 256 amplitude quantization levels. Error diffusion method is essentially effective for the holograms with a few quantization.

Iterative methods[9-21] are also effective for improvement of computer-generated holograms, but in these types of holograms much computing time is required for the iteration process and then, if the iteration does not lead to the decrease of errors, there is no probability of improvement. Consequently, it may difficult to apply the both methods to our holograms. We, therefore, have employed a histogram method as another method, which is based on the assumption that if the contrast of the hologram patterns is increased, then the reconstructed images may be improved. The histogram method used in the experiment corresponds to the gray scale transformation[37]. We will discuss the histogram method next.

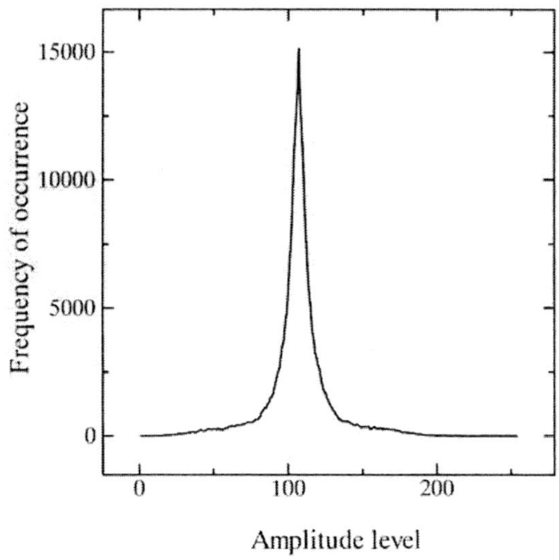


Figure 36. Example of a histogram obtained from a 1D computer-generated off-axis hologram.

Let for all points with coordinate $(x, y)$ the intensity of the image $m=f(x, y)$ be limited to the range from $t_1$ to $t_2$. That is,

$$t_1 \le m = f(x,y) \le t_2 \tag{27}$$

Assuming that the range $[t_1, t_2]$ belongs to $[d_1, d_2]$, and that

$$M = \frac{d_2 - d_1}{t_2 - t_1}(m - t_1) + d_1 \quad , \tag{28}$$

then, from the transformation shown in Eq.(28) the range stretching of the image amplitude density and the parallel shifting occur by the linear gray scale transformation[37] and the transformed density value $M$ is distributed over the $[d_1, d_2]$ density range.

If the image density values are situated mainly in the range $[t_1, t_2]$, then the transformed density value is

$$M = \begin{cases} \frac{d_2 - d_1}{t_2 - t_1}(m - t_1) + d_1 & for \quad t_1 \le m \le t_2 \\ d_1 & for \quad m < t_1 \\ d_2 & for \quad m > t_2 \end{cases} \tag{29}$$

From above transformation, the ranges $[d_1, t_1]$ and $[t_2, d_2]$ are reduced to one values such as $d_1$ and $d_2$, respectively, but if the density values mostly belonging to the range $[t_1, t_2]$, the image information is little lost by the compressing process of the ranges.

Figure 36 shows an example of a histogram of a 3D object and the transformed histogram by the gray scale transformation is shown in Fig.37. Figure 38 shows the difference of the hologram patterns obtained before and after using the histogram method. Obviously, the contrast of the transformed hologram pattern is improved by the histogram method than that of the original histogram. Parameters used in the experiment are $t_1$=70, $t_2$=150, $d_1$=0 and $d_2$=255, respectively.

We consider the hologram of a point source p on the optical axis taken with a reference beam normal to the plate. In Fig. 39, R is the radius of the holographic plate and $a$ is the local fringe spacing between the maximum intensities at the edge of the plate. The interference fringe of the point object forms the or optical density distributions with radially symmetric intensity as depicted in Fig.39. If the lattice space of the sampling lattice is a reciprocal of two times as large as a spatial frequency corresponding to the maximum spatial frequency of the original image, then using sampling theorem the same original image as is visible before sampling process can be restored completely from the sampled point data. For example, if the hologram lattice space is 1, $a$ becomes 2. We investigate a relation of the point object at point p and the amplitude distribution of the object in the hologram plane.

From Fig. 39, the following relation is obtained.

$$z_a^2 + (R-a)^2 = L^2 \tag{30}$$

$$z_a^2 + R^2 = (L+\lambda)^2 \tag{31}$$

where $z_a$ is the smallest allowed distance of the object point from the holographic plate. From Eqs.(30) and (31),

$$L = \frac{2aR - a^2 - \lambda^2}{2\lambda} \tag{32}$$

Substituting Eq.(12) into Eq.(11), $z_a$ becomes

$$z_a = \sqrt{\frac{\left(2aR - a^2 + \lambda^2\right)^2}{4\lambda^2} - R^2} \approx R\sqrt{\frac{a^2}{\lambda^2} - 1} \tag{33}$$

Next, in Fig.40 we consider the maximum distance $z_b$ between the point source and the hologram plane. From Fig.39, the following relation is obtained.

$$\left(z_b + \frac{\lambda}{2}\right)^2 = z_b^2 + R^2 \tag{34}$$

Then from Eq.(34), $z_b$ is

$$z_b = \frac{4R^2 - \lambda^2}{4\lambda} \approx \frac{R^2}{\lambda} \tag{35}$$

If $R$=10 mm and λ=0.6328 μm, the maximum distance $z_b$ is approximately 15.8 m. From this result, the maximum distance $z_b$ may be ignored in actual hologram making process.

In the first experiment, the hologram had 1600 ×1200 points area. The distance between the maximum amplitude intensities on the hologram plane $a$ was 10 μm which depends to the resolution of the film that takes pictures of hologram patterns, and consequently, the resulting hologram size reduces to 8.0 × 6.0 mm since the lattice interval is 5 μm on the hologram plane. From this hologram size, the minimum distance $z_a$ is calculated by inserting $a$=10 μm and $R$=4 mm into Eq.(33) and becomes 63.1 mm. To satisfy the condition of the minimum distance, objects were placed at the point of 64 mm from the hologram plane. On the other hand, we have not regarded the maximum distance $z_b$ since it is extremely long distance. Figure 41 shows the images reconstructed from the original hologram and the enhanced hologram using the histogram method. It is clear from the figure that the improved image can be reconstructed from the hologram with the high contrast.

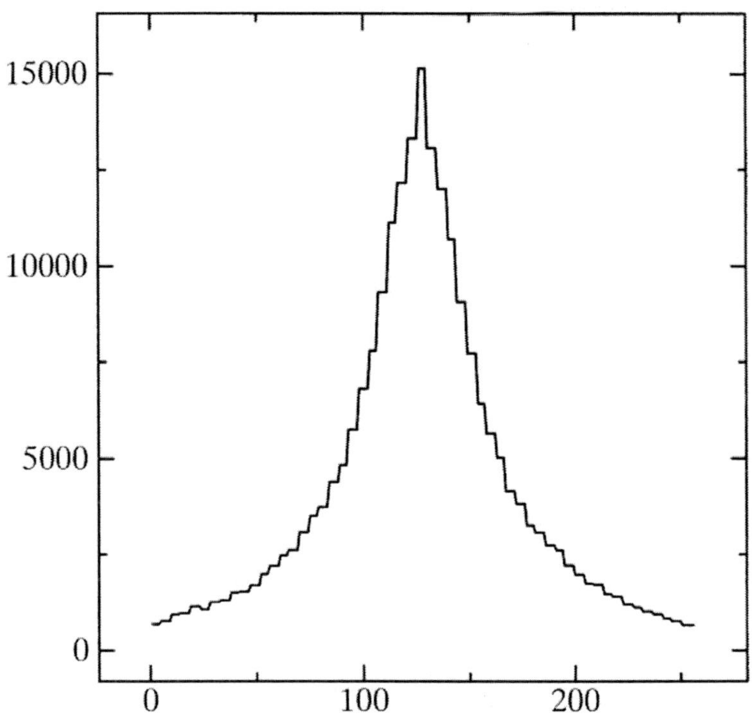


Figure 37. Transformed histogram of that shown in Fig.36 by means of the gray scale transformation.

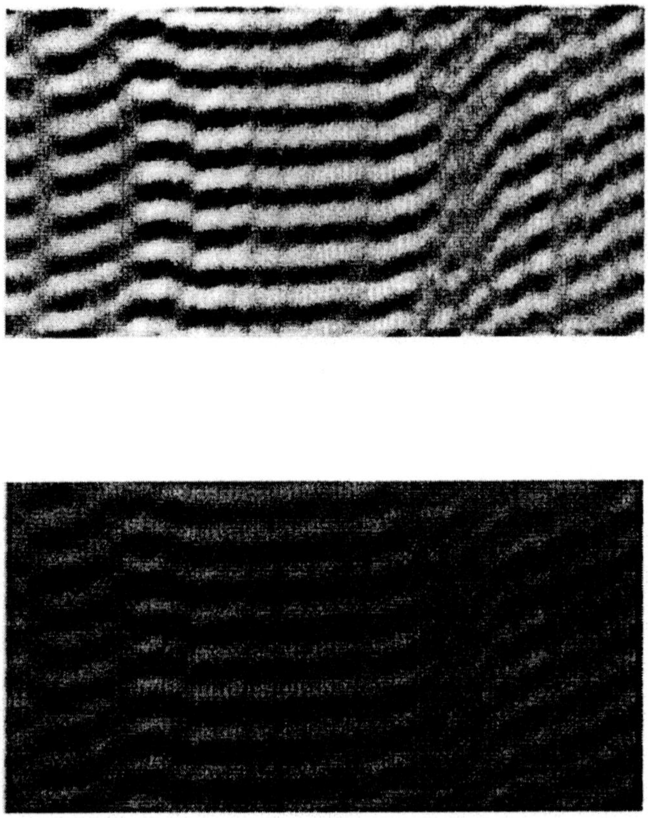

Figure 38. Parts of the hologram patterns made (a)with and (b)without a histogram method.

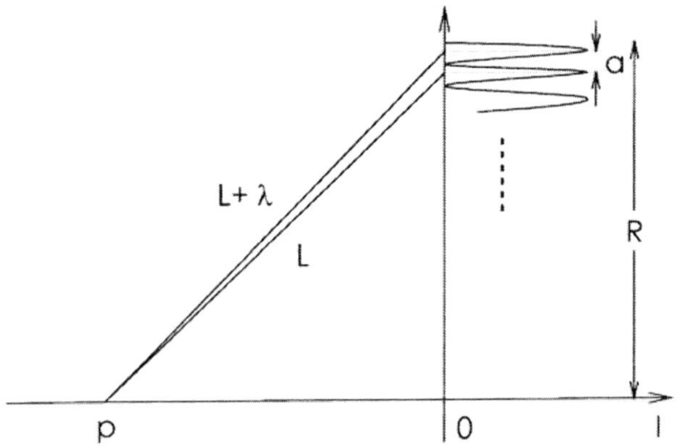


Figure 39. Calculation of the minimum distance of a point object from holographic plate. Letter I represents the amplitude intensity.

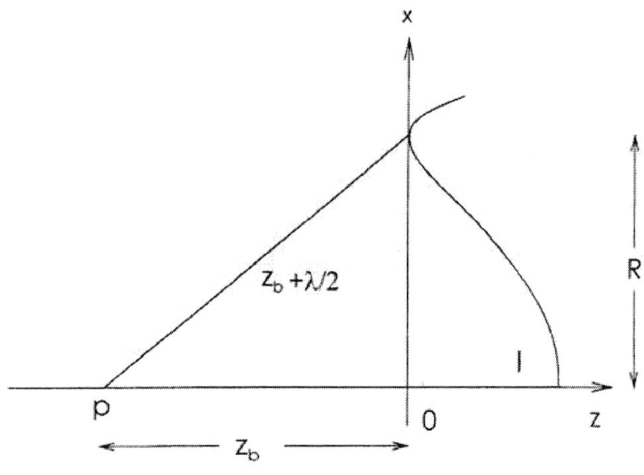


Figure 40. Calculation of the maximum distance of a point object from holographic plate. Letter I represents the amplitude intensity.

Figure 41. Image reconstruction with white light. Images reconstructed from holograms made (a)with and (b)without a histogram method.

Further, to obtain the large sizes of the images with more excellent quality, in the following experiment, holograms with 4800 × 3600 points have been produced by using a histogram method. In this case, the size of the recorded hologram is 24 × 18 mm on the photographic film and $z_a$ is estimated as 189 mm from Eq.(33). It is difficult to print out the whole hologram pattern by one process since the number of the pixels of the hologram exceeds the limitation of printing width of a laser printer. Accordingly, the hologram pattern was divided to four blocks and each was printed out separately with a laser printer. In next stage, four pieces of the hologram patterns were collected to a sheet, which was recorded to the photographic film. Figures 42 and 43 show the hologram of a 3D object (cube) and the white light reconstruction image from the hologram, respectively. Figure 43 indicates the various perspectives of the reconstructed images from different sections. We could more feel the horizontal parallax in the images from these holograms than that in the small size holograms, as is shown in Fig.43.

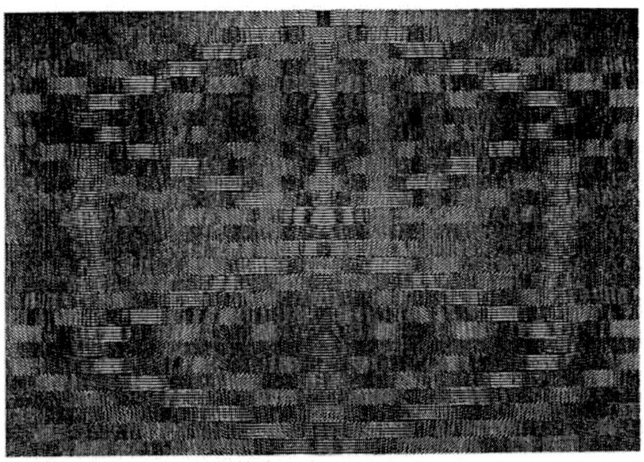

Figure 42. 1D off-axis computer-generated hologram of an object "cube", made with a histogram method.

Qualities of the images reconstructed from the Fresnel off-axis computer-generated holograms are in general, inferior to those from the optical holograms since we cannot create the same computer-generated holograms as the optical holograms. In addition, these computer-generated holograms have the relatively low contrast patterns, so that the degenerated images can be reconstructed from such computer-generated holograms. Therefore, to improve the holograms with low contrast, a histogram method was used. The holograms obtained by using the histogram method have increased in contrast and the image reconstructed from these holograms have also been improved in qualities.

# 8. Conclusion

From experiments of computer-generated holograms it became clear that 2 quantized phase holograms are effective for making hologram patterns and applying many kinds of image improving methods. In particular, iterative, error diffusion and their combination methods were suited for making holograms to improve the reconstructed images. Further, it was indicated that 3D computer-generated hologram is created by using 1D-FFT operations alone and that reconstructed images are also improved with using compensation points. In addition, it was shown that a 1D Fresnel off-axis hologram whose reconstructed image can be seen directly with eyes is improved with using histogram method.

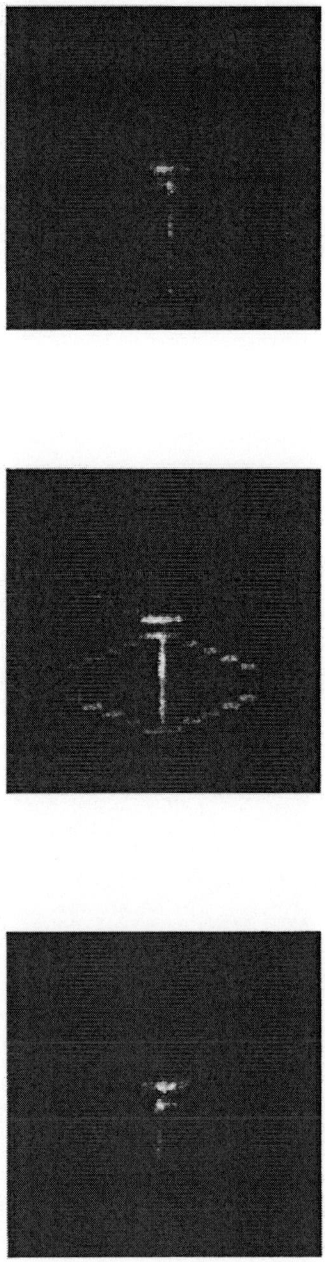

a

b

c

Figure 43. White light reconstruction images from the hologram shown in Fig.8 from different viewing positions. (a) Views from left side, (b) from the front and (c) from right side.

## REFERENCES


[1] B.R.Brown; A.W.Lohmann. *Appl Opt.* 1966, 5, 967-969.
[2] W.H. Lee. *Appl Opt.* 1970, 9, 639-643.
[3] K.Nagashima; T.Asakura. *Opt Commun.* 1976, 17, 273-276.
[4] K. Nagashima; T. Asakura. *Opt Laser Technol.* 1978, 10, 310-312.
[5] K. Nagashima; T. Asakura. *Optik* 1978, 50, 53-60.
[6] K.Nagashima; T. Asakura. *Opt Laser Technol.* 1983, 15, 133-137.
[7] L.B.Lesen; P.M. Hirsch; J.A.Jordan, Jr. *IBM J Res Develop.* 1969, 13, 150-155.
[8] D. Kermisch. *J Opt Soc Am.* 1970, 60,15-17.
[9] R.A.Gabel; B. Liu. *Appl Opt.* 1979, 9, 1180-1190.
[10] N.C.Gallagher; B.Liu. *Appl Opt.* 1973, 12, 2328-2335.
[11] J.R.Fienup. *Opt Eng.* 1980, 19, 297-305.
[12] J.R.Fienup. *Appl Opt.* 1982, 21, 2758-2769.
[13] H.Akahiri. *Appl Opt.* 1986, 25, 802-811.
[14] K.Nagashima. *Opt Laser Technol.* 1986, 48, 157-162.
[15] J.R.Fienup; C.C.Wackerman. *J.Opt Soc. Am.* 1986, A3,1897-1907.
[16] J.R.Fienup, *J.Opt. Soc.Am.* 1987, A4,118-123.
[17] F.Wyrowski; O.Bryngdahl. *J.Opt.Soc.Am.* 1988, A5, 1058-1065.
[18] M.Broja; F.Wyrowski; O.Bryngdahl. *Opt. Commun.* 1989, 69, 205-210.
[19] F.Wyrowski. *Appl. Opt.* 1989, 28, 3864-3870.
[20] B.K.jennieson; J.P.Allebach; D.W.Sweeney. *Opt. Eng.* 1989, 28, 629-637.
[21] O.K.Ersoy; J.Y.Zhuang; J.Brede. *Appl. Opt.* 1992, 31, 6894-6901.
[22] R.Hauck; O.bringdahl. *J.Opt.Soc.Am.* 1984, A1, 5-10.
[23] E.barnard. *J. Opt. Soc. Am.* 1988, A5, 1803-1817.
[24] S.Weissbach; F.Wyrowski; O.Bryngdahl. *Opt. Commun.* 1989, 72, 37-41.
[25] R.Eschbach. *Appl. Opt.* 1991, 30, 3702-3710.
[26] K.Nagashima. *Opt Laser Technol.* 1993, 25, 323-327.
[27] R.Eschbach; Z.Fan. *Appl. Opt.* 1993, 32, 3130-3136.
[28] M.P.Chang; O.K.Ersoy. *Appl. Opt.* 1993, 32, 3122-3129.
[29] T.S.Huang. *Proc. IEEE* 1971, 59, 1335-1346.
[30] J.P.Waters. *J. Opt. Soc. Am.* 1968, 58, 1284-1288.
[31] K.Nagashima. Opt.Eng. 1996, 35, 570-574.
[32] K.Nagashima. *Opt Laser Technol.* 30, 361-366, 1998.
[33] E.N.Leith; J.Upatnieks. *J. Opt. Soc. Am.* 1963, 53, 1337-1381.
[34] S.A.Benton. 1969 *Annual Meeting of Optical Society of America. FE20*, Chicago,1969.
[35] K. Nagashima. *Opt Laser Technol.* 2001, 33, 329-334.
[36] K. Nagashima. *Opt Laser Technol.* 2003, 58. 229-233.
[37] For example, E.B.Troy; E.S. Deutch; A. Rosenfeld. *IEEE Trans. Syst. Man Cybernet.* 1973, SMC-3, 91-98.

In: New Developments in Lasers and Electrc-Optics Research ISBN 1-59454-771-8
Editor: William T. Arkin, pp. 117-151 

*Chapter 4*

# THREE-DIMENSIONAL MULTI-RESOLUTION IMAGING AND MODELING

***Xiang Peng and Jindong Tian***
Institute of Optoelectronics, Shenzhen University, 518060 Shenzhen, China
Key Laboratory of Optoelectronic Systems and Devices of Education Ministry of China

## ABSTRACT

Three-Dimensional Imaging and Modeling (3DIM) has been an active and interdisciplinary research field, which covers a wide range of spectrum from range image acquisition, multi-view range images registration, multi-view range images integration, geometric representation, and manipulation of 3D image. The technology of 3DIM has been widely used for industry, medicine, culture, education and so on. In this chapter, we will focus on the technology of multi-resolution 3DIM.

The state-of-the art of techniques for multi-resolution 3D imaging will be introduced as updated methods of range imaging acquisition, which involves 3D imaging based on the strategies of a point-array encoding (incoherent illumination). In addition, a new method for generating variable frequency sinusoidal fringe patterns is presented. This method is based on acoustic-optics technique for 3D imaging (coherent illumination). The decoding and the calibration methods for 3D imaging systems with regard to different encoding schemes are described. For illustration, two typical surfaces, the free form and a step-like surface, are analyzed to demonstrate the approaches. It will be shown that the quality of reconstructed surface can be gradually improved as level-of-details, resulting in a multi-resolution acquisition. The methods of registration and integration of partial range images are also introduced in this Chapter. We will discuss some approaches for high-level range data processing, including multiple view range images registration, multiple view range images integration, and efficient geometric description of 3D graphical like subdivision. The manipulation of 3D image are also presented, which involves data reduction of 3D image and representation of 3D image with different level-of-details.

**Keywords:** multi-resolution, three-dimensional imaging, three-dimensional modeling, geometric representation, 3-D model manipulation, and level-of-details

# 1. INTRODUCTION

Optical 3-D imaging has been widely used for machine vision, industry inspection, rapid prototyping, biomedicine, etc, due to the advantages of their non-contact measurement operations, fast measurement speed, and automatic processing.[1-4] The methods to collect 3-D shape information of object surfaces were classified into two main categories: passive and active optical 3-D imaging. In the active optical 3-D imaging system, the object surface to be tested will modulate the illumination of structured light in spatial domain or time domain, resulting in an encoded light pattern[5]. The 3-D shape information encoded in the light pattern can be decoded either optically or digitally. A typical method among active techniques is laser triangulation. In the method of laser triangulation, the object height can be computed by measuring the position of imaging spot on the detector and the geometrical parameters of optical system. For those 3-D measurement methods based on triangulation, the structure light could be either laser point or laser sheet. The phase measuring profilometry[6] (PMP) is another sort of 3-D measurement technique that includes phase-shift profilometry[7], Fourier transformation profilometry[8-10] and spatial phase detection[11], etc. The laser triangulation with the point-structured light is of low efficiency and not suitable for real-time measurement because only one point can be measured each time, two-dimensional scanning mechanism must be used to realize full-field measurement. Again, the efficiency of the light-sheet structure is also not sufficient, as a one-dimensional scanning mechanism must be involved. In this regard the PMP method is featured by high-speed measurement, because a two-dimensional structured light patterns is used in this method, a full-field range image can be obtained at one time. However, the PMP method is an indirect method through measuring the phase of spatial carrier to recover the range image of object surface.

# 2. MULTI-RESOLUTION 3D IMAGING

When dealing with the surface with inherently discontinuous region, or the surface with very steep slope, phase calculation usually gets troubles such as phase ambiguity and error propagation.[12-13] To overcome these difficulties two typically reported methods are frequency-multiplex Fourier-transform [14] and temporal phase unwrapping[12]. The principle of the former one was to have multiple phase maps with various phase sensitivities in two directions, but the measurement range was also limited by the frequency of fringe patterns. The basic idea of the last one was to vary the pitch of fringes over time. A sequence of phase maps was recorder to form a 3D phase distribution. The sampling-rate along time must be high enough to ensure there were no phase ambiguities between adjacent sampling intervals. That resulted in increased computational complexity.

## 2.1 Incoherent System

In this section, a 2-D spatial point-array pattern is used to illuminate the test object as structured light, leading to a new encoding method. Compared with the PMP, the point-array encoding method does not involve phase measurement and therefore has no phase ambiguity and error propagation in phase unwrapping. The method based on point-array encoding calculates object height directly through the affine transformation[15].

The rest of this subsection is organized as follows: (1) the establishment of a mathematic and geometric model that formulizes the 3-D imaging process based on point-array encoding, (2) description of the experiment and demonstration of theoretical prediction, and (3) the presentation of experiment results and discussion.

***(1) 3-D imaging Based on Point-Array Encoding***

Figure 2.1-1 illustrates the basic geometry of the ideal camera model[16], where $(x_w, y_w, z_w)$ is the 3-D coordinates of an arbitrary object point in the world coordinate system, $(x_c, y_c, z_c)$ is the coordinates of an object point in the 3-D camera coordinate system, whose origin is located at point $O_c$. $O_c$ is the center of entrance pupil of the camera lens. The $z$-axis of camera coordinate system is the same as the optical axis. $(X_u, Y_u)$ is the 2-D image coordinate system centered at $O_u$ (intersection of the optical axis $z$ and the front image plane) and parallel to $X_c$ and $Y_c$ axes and $f$ is the distance between front image plane and the optical center. $(x_u, y_u)$ is the image coordinate of $(x_c, y_c, z_c)$ if a perfect pinhole camera model is used. The coordinate $(u, v)$ used in the computer is the number of pixels for the discrete image in the frame memory. The overall transform from $(x_w, y_w, z_w)$ to $(u, v)$ is described in Fig. 2.1-2, three steps are necessary since only ideal (undistorted) camera model is addressed.

The imaging geometry of a point-array is shown in Fig. 2.1-3, in which $PO_w$ is the optical axis of a projector lens. It crosses the optical axis of a camera lens, $O_cO_w$, at point $O_w$ on a reference plane $R$. $PO_w$ is perpendicular to the reference plane. The point $P$ is the center of the exit pupil of the projector lens. $O_w$ denotes the origin of the object coordinate system. The points $P$, $O_c$ and $O_w$ form a triangulation system. A structure light transmitter, say, a digital projector projects a point-array pattern on the object surface $S$, and the projected line $PP_1$, $PP_2$ and $PP_3$ cross the surface $S$ at points $B_1$, $B_2$, $B_3$ respectively. Taking $PO_c$ as the baseline, a number of triangles can be formed with the centers of point-array projected on the surface $S$. Thus, the object height located at these points can be uniquely specified. The camera lens, with the entrance pupil at $O_c$, images the point-array patterns to the front image plane $X_uO_uY_u$ following by the perspective projection. The transformations for any point in the 3-D world coordinate system can be depicted as follows:

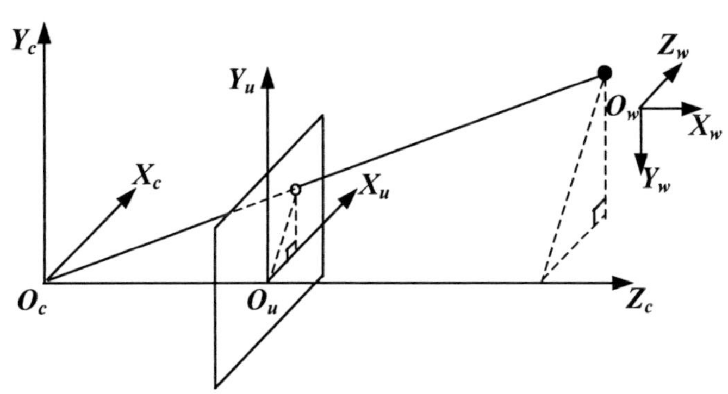


Figure 2.1-1. Basic geometry of camera model

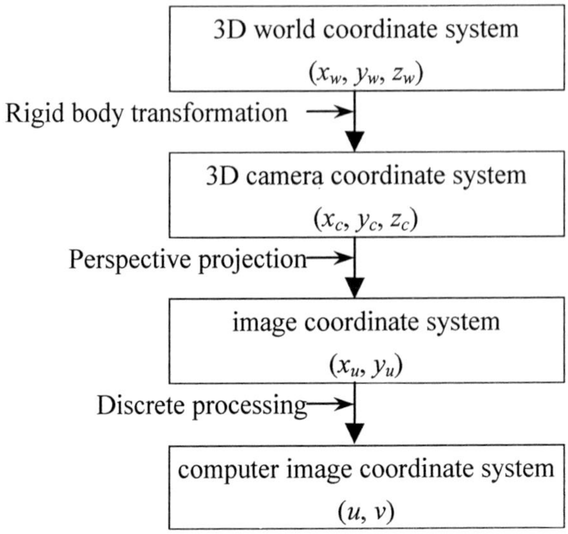


Figure 2.1-2. transform from $(x_w, y_w, z_w)$ to $(u, v)$

Step 1: Transformation from the world coordinate system $(x_w, y_w, z_w)$ to the camera coordinate system $(x_c, y_c, z_c)$. In homogeneous coordinates, the transformation is expressed as[15]

$$\begin{bmatrix} x_c \\ y_c \\ z_c \\ 1 \end{bmatrix} = \begin{bmatrix} R & T \\ 0_3^T & 1 \end{bmatrix} \begin{bmatrix} x_w \\ y_w \\ z_w \\ 1 \end{bmatrix} \tag{2.1-1}$$

where $R$ is the rotation matrix and $T$ is the translation vector. As shown in Fig. 2.1-3, the world coordinate system rotates through an angle $\alpha$ about the $y_w$ axis and translates $L$ along the opposite direction of $z_c$ axis. Therefore, the matrixes $R$ and $T$ is

$$R = \begin{bmatrix} \cos\alpha & 0 & \sin\alpha \\ 0 & 1 & 0 \\ -\sin\alpha & 0 & \cos\alpha \end{bmatrix}, \quad T = \begin{bmatrix} 0 \\ 0 \\ L \end{bmatrix} \tag{2.1-2}$$

Step 2: Transformation from 3-D coordinate system $(x_c, y_c, z_c)$ to ideal image coordinate $(x_u, y_u)$ using perspective projection

$$x_u = f\frac{x_c}{z_c}, \quad y_u = f\frac{y_c}{z_c}$$

where $f$ is the effective focal length. In homogeneous coordinates, we have

$$\begin{bmatrix} x_u \\ y_u \\ 1 \end{bmatrix} = \frac{1}{z_c}\begin{bmatrix} f & 0 & 0 & 0 \\ 0 & f & 0 & 0 \\ 0 & 0 & 1 & 0 \end{bmatrix}\begin{bmatrix} x_c \\ y_c \\ z_c \\ 1 \end{bmatrix} \tag{2.1-3}$$

Step 3: Transformation from ideal image coordinate ($x_u$, $y_u$) to computer image coordinate ($u$, $v$)

$$u = \frac{x_u}{dx} + u_0, \qquad v = \frac{y_u}{dy} + v_0$$

where $dx$ and $dy$ are the center to center distance between adjacent sensor elements in $X$ and $Y$ directions respectively, $u_0$ and $v_0$ are row and column numbers of the center of computer frame memory. In homogeneous coordinates, we obtain

$$\begin{bmatrix} u \\ v \\ 1 \end{bmatrix} = \begin{bmatrix} 1/dx & 0 & u_0 \\ 0 & 1/dy & v_0 \\ 0 & 0 & 1 \end{bmatrix}\begin{bmatrix} x_u \\ y_u \\ 1 \end{bmatrix} \tag{2.1-4}$$

The overall operation can be formulated by following matrix

$$\begin{bmatrix} u \\ v \\ 1 \end{bmatrix} = \frac{1}{z_c}\begin{bmatrix} 1/dx & 0 & u_0 \\ 0 & 1/dy & v_0 \\ 0 & 0 & 1 \end{bmatrix}\begin{bmatrix} f & 0 & 0 & 0 \\ 0 & f & 0 & 0 \\ 0 & 0 & 1 & 0 \end{bmatrix}\begin{bmatrix} \cos\alpha & 0 & \sin\alpha & 0 \\ 0 & 1 & 0 & 0 \\ -\sin\alpha & 0 & \cos\alpha & L \\ 0 & 0 & 0 & 1 \end{bmatrix}\begin{bmatrix} x_w \\ y_w \\ z_w \\ 1 \end{bmatrix} \tag{2.1-5}$$

For an arbitrary point projected on the object surface, we get its $u$ coordinate in the computer image coordinate system as

$$u_b = \frac{x_w \cos\alpha + z_w \sin\alpha}{L - x_w \sin\alpha + z_w \cos\alpha} \cdot \frac{f}{dx} + u_0 \tag{2.1-6}$$

It is significant that the object height $z_w$ is contained in Eq. 2.1-6

The coordinate of corresponding point projected from reference plane ($z_w$ =0) can therefore be easily derived from Eq. (2.1-6).

$$u_r = \frac{x_w \cos\alpha}{L - x_w \sin\alpha} \cdot \frac{f}{dx} + u_0 \tag{2.1-7}$$

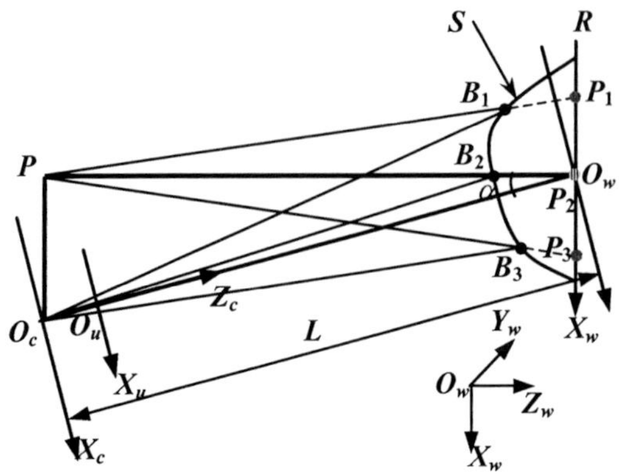


Figure 2.1-3. The imaging geometry of a point-array

The both imaging points projected from reference plane and the object surface, respectively, form a pair of point-array. According to Eq. (2.1-6) and (2.1-7), the position difference of a pair of points along $U$ axis in the computer image coordinate is

$$\Delta u = u_b - u_r = \frac{f}{dx}\left(\frac{x_w \cos\alpha + z_w \sin\alpha}{L - x_w \sin\alpha + z_w \cos\alpha} - \frac{x_w \cos\alpha}{L - x_w \sin\alpha}\right) \tag{2.1-8}$$

Since $u_0$ is canceled in Eq. (2.1-8), it is unnecessary to know the row numbers of the center of computer frame memory. That is helpful for improving the measurement precision.

Figure 2.1-3 also shows the projection geometry of the point-array. The coordinates of the intersections of the projection lines and the object surface can decide the spatial position of the point-array in the 3-D world coordinate system. For example, the projection geometry of an arbitrary projection line $PP_r$ is shown in Fig. 2.1-4.

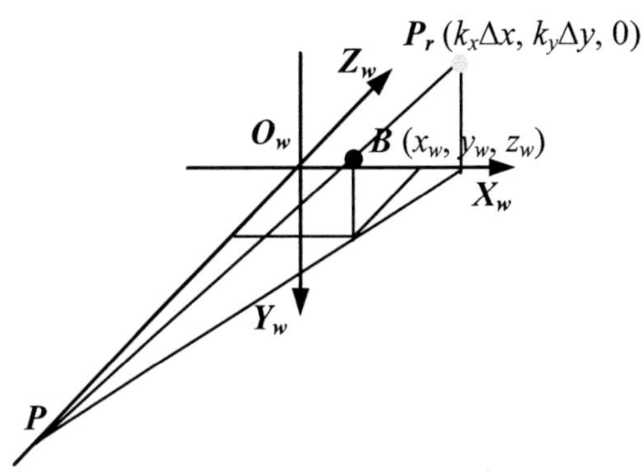


Figure 2.1-4. The projection geometry of an arbitrary projection line $PP_r$

$PP_r$ crosses the object surface at $B$, and crosses the reference plane at $P_r$. Let $(x_w, y_w, z_w)$ be the 3-D world coordinates of $B$, then $(k_x\Delta x, k_y\Delta y, 0)$ is the 3-D world coordinates of $P_r$. We find that

$$\frac{z_w}{L'} = \frac{k_x \Delta x - x_w}{k_x \Delta x} \Rightarrow x_w = \frac{L' - z_w}{L'} k_x \Delta x$$
$$\frac{z_w}{L'} = \frac{k_y \Delta y - y_w}{k_y \Delta y} \Rightarrow y_w = \frac{L' - z_w}{L'} k_y \Delta y \qquad (2.1\text{-}9)$$

where $L'$ is the distance between the points $P$ and $O_w$. $\Delta x$ and $\Delta y$ denote the center-to-center distance between adjacent points in $X_w$ and $Y_w$ direction respectively. $k_x = -m, \ldots, -1, 0, 1, \ldots m$, $k_y = -n, \ldots, -1, 0, 1, \ldots n$, $m$ and $n$ express the number of points along lateral and longitudinal direction, respectively. $k_x$ and $k_y$ multiply $\Delta x$ and $\Delta y$ to express the $(x_w, y_w)$ coordinates of the point-array projected on reference plane in the 3-D world coordinates system.

Substituting Eq. (2.1-9) into Eq. (2.1-8) and letting $C$ represent the terms that are irrelative with $z_w$

$$C = \frac{\Delta u dx}{f} + \frac{k_x \Delta x \cos\alpha}{L - k_x \Delta x \sin\alpha} \qquad (2.1\text{-}10)$$

Substituting Eq. (2.1-10) into Eq. (2.1-8) yields Eq. (2.1-11)

$$z_w = L' \frac{k_x \Delta x \cos\alpha - LC + C k_x \Delta x \sin\alpha}{C\left(k_x \Delta x \sin\alpha + L' \cos\alpha\right) + k_x \Delta x \cos\alpha - L' \sin\alpha} \qquad (2.1\text{-}11)$$

The $z_w$ coordinate of the intersection point represents the object height of this point. Equation (2.1-11) is the mathematic model that correlates the lateral displacement of point-array to the object height. $\Delta x$, $k_x$ and other parameters are all known. Firstly, we can measure the spatial position of the both images of the point-array projected on reference plane and the object surface respectively. Secondly, computes the lateral displacement for each pair of points in the point-array. Finally, calculates the object height at every discrete point to form the range image of the object according to Eq. (2.1-11).

Discrete operation losses spatial resolution, the point-array with high density is advantageous for improving spatial resolution, but if the density is very high, the image spots may be crossed each other in the image plane. That will bring the difficulty for determining the center of the image spots of point-array. In our system, the structured light illumination device is a digital projector and it allows easy alteration in spatial density of point-array. In order to achieve high spatial resolution, the projected point-array can be shifted digitally along both lateral and longitudinal directions with one-pixel minimum step to fill up the lost information caused by the discrete processing. Consequently, we will obtain an image sequence of shifted point-array. Through each image of this sequence and the point-array image of reference, the object height can be computed by Eq. (2.1-11), the related $(x_w, y_w)$ coordinates can be calculated by Eq. (2.1-9). In this way the range image of test object is therefore reconstructed. If merging all calculated range images of the sequence according to the $(x_w, y_w)$ coordinates of each discrete point, we can obtain a range image with high spatial resolution. How many shifts are usually used in the real measurement system is determined by the topology complexity of test object, more shifts are necessary for complex object. For example, four-step shifts were usually used for the free-form surface (the mannequin face). As a result, more measurement time including acquisition time and additional computation cost is required for those object surfaces with complex geometries. The pattern movement was unnecessary for the surface with simple geometry by complex topology such as the step-like object surface. During processing the point-array shift, the other matter has to be dealt with is to ensure the center-to-center distance between adjacent points of the camera is higher than that of projector in the field of view. In this case, the camera can detect the point movement with enough spatial resolution.

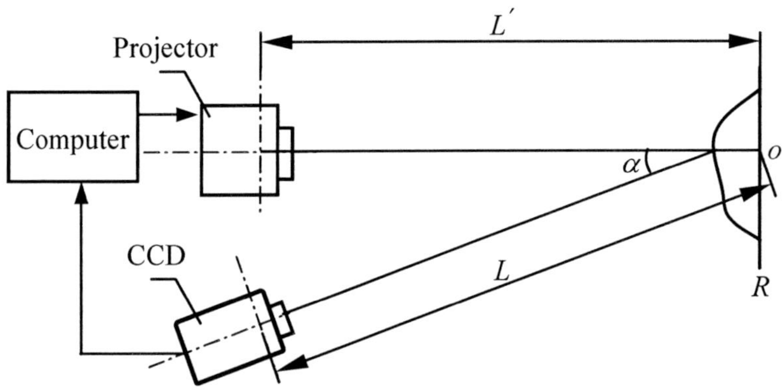


Figure 2.1-5. Schematic diagram of the experimental setup

***(2) Demonstration of Theoretical Prediction***

The schematic diagram of the experimental setup is shown in Fig. 2.1-5. A digital projector (Benq 2115) is used to project the point-array patterns. A JVC TK-C1481BEC color video camera with 25mm focal length lens is used to acquire point-array where the topographic information of object surfaces had been encoded. A frame grabber OK-USB20A with USB 2.0 interface was used to capture and save sets of position-shift point-array images for further processing. The USB interface frame grabber is more convenient because it is able to connected with desktop or laptop computer directly.

In the experiment, overall processes are the following steps:

Step 1: the spatial period and the position of the point-array patterns are initialized by the computer in advance, and then the patterns are projected on the reference plane sequentially according its spatial position. The point-array images are captured and stored in the computer.

Step 2: the same point-array patterns are projected again by the same sequence on the object surface. The modulated or encoded point-array images by the object surface are captured and stored in the computer too.

Step 3: locate the center of each image spot of all point-array images and compute the lateral displacement of every related pair of points.

Step 4: for each image spot calculate the $z_w$ coordinate by Eq. (2.1-11) and ($x_w$, $y_w$) coordinates by Eq. (2.1-9) to form a sequence of range images.

Step 5: merge all range images according to the sequence of ($x_w$, $y_w$) coordinates.

***(3) Experimental Results and Discussion***

Figure 2.1-6 shows the measurement result of a mannequin face. Figure 2.1-6(a) is an image of point-array modulated by the face. Figure 2.1-6(b) shows the range image of selected region in Fig. 2.1-6(a). A regularly arranged point-array with 64×64 points was used in the experiment. The size of the pattern is 512×512 pixels. The pattern is shifted along both lateral and longitudinal direction with two-pixel step to fill up the lost information caused by the discrete processing. The center-to-center distance between adjacent points of projector and camera are 8 pixels and 14 pixels, respectively. When the point-array moves in projector, the camera has enough spatial resolution to detect the movement.

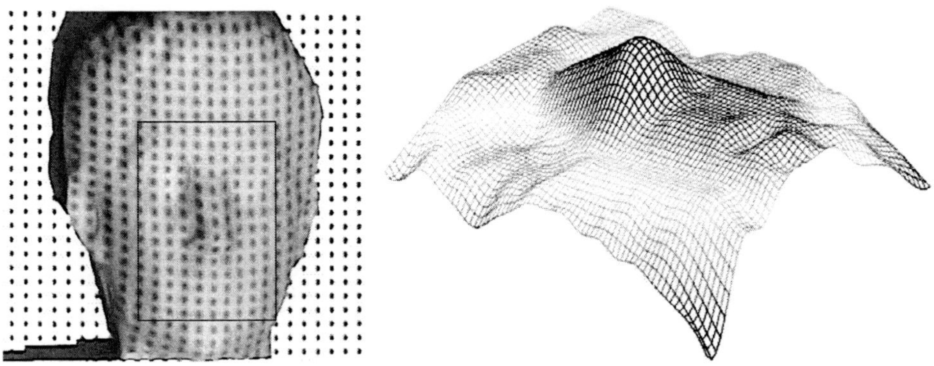

(a) (b)

Figure 2.1-6. Experimental results: (a) Modulated point-array, (b) reconstructed 3D surface

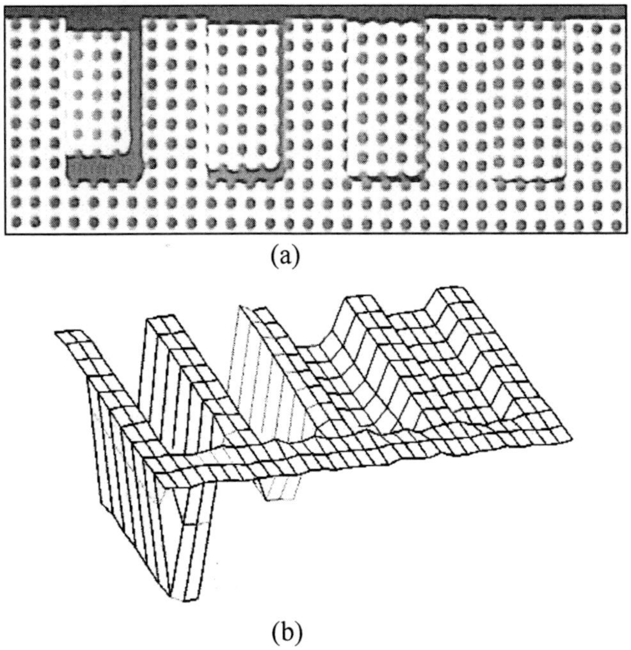

(a)

(b)

Figure 2.1-7. Experimental results: (a) modulated point-array, (b) reconstructed surface

Another kind of object surface used to demonstrate our approach is a step-like object surface. This is typical subclass of surfaces with very deep slopes. Figure 2.1-7(a) is an image of the object illuminated by point-array structured light. Figure 2.1-7 (b) shows the range image of Fig. 2.1-7(a). Only one 64×64 point-array pattern without movement is used in this experiment. As we have already known that, for fringe projection techniques, spatial phase unwrapping methods usually get troubles in reconstructing such kind of object surface with discontinuous region or the surface with very deep slope, just like the object surface shown in Fig. 2.1-7(a). For the point-array encoding method, the point-pair is independent each other. Though these points vanished in the shadow regions cannot form a pair of points, they do not propagate the errors to neighboring point. As illustrated in this experiment, a range image can still be reconstructed from presented approach.

In summary, a point-array encoding method for 3-D imaging is proposed. In order to decode the range image of the test object, we have established a mathematic model to formulize the process of 3-D digital imaging. According to this model, we can reconstruct the range image of object surface by measuring the lateral displacement of each point-pair. In comparison with conventional triangulation methods, our approach is more convenient as it does not require any scanning mechanism. Compared with PMP methods, it does not exist the problems of phase ambiguity and error propagation as encountered in spatial phase unwrapping. The use of a digital projector allows for flexible change in spatial density of point-array, making realizable to implement 3-D imaging with various spatial resolution. In addition, the projected point-array can be digitally shifted along both lateral and longitudinal direction step by step to fill up the lost measurement points caused by discrete processing to achieve high spatial resolution. With this method, we have successfully measured two representative classes of objects surface. Experimental results have shown the effectiveness of this approach and have revealed that the method can be applicable for a broad range of objects, especially for objects with large height discontinuities and/or surface isolation.

## 2.2 Coherent System

Projecting a periodic pattern upon an object and observing its deformation from a different view angle has been a common principle for various 3-D shape measurement techniques, including Fourier-transform profilometry (FTP). Phase-mapping based profilometry involves in two basic operations: (1) *phase encoding*. In this process the spatial phase of fringe pattern is modulated by height variation of object surface as the pattern is projected upon the test object. We referred to this process as phase encoding because the topographic information about the object surface has been encoded into deformed fringe pattern through the phase modulation, (2) *phase decoding*. This process can be divided into another two sub-operations: phase evaluation and phase unwrapping. The process of phase evaluation can be performed by either phase-shifting techniques or by Fourier transform based techniques. However, the operation of phase unwrapping plays more significant role in the process of phase decoding because it would become a difficult task if the object surface has complicated geometric shape or topology. For example, if object surface has a portion with steep slope or there are holes or breakout on the surface, then most of conventional phase unwrapping techniques will not work well. This limits the applicable scope of phase-mapping based profilometry.

To overcome the discontinuities, the basic idea is that changing the measurement system's sensitivity results in fringe or projected strip density changes. This means that the integer order of the fringes sweep through the discontinuity. It can be viewed in both spatial and temporal domains and results in various different methods. In order to extend the scope of phase-mapping-based techniques, allowing it applicable to broader class of object surfaces temporal phase unwrapping techniques[12] have been proposed to deal with those objects that have discontinuous height steps and/or surface isolations hinder the unique assignment of fringe orders for Moiré technique or unique phase unwrapping for FTP. In additional to temporal phase unwrapping technique, spatial frequency-multiplexing techniques[14] and hybrid encoding method[17] have also been also proposed to make phase unwrapping technique adapted the to those object with complex surface geometry and topology.

Except for the efforts made in developing new techniques for phase unwrapping, a lot of efforts have been made in developing new hardware suitable for profiling those objects with complex surface geometry and topology. For example, the video projector has been used to generate variable sensitivity fringe patterns for structured light illumination[18-20]. Recently, a so-called accordion fringe interferometry using an acoustic-optic modulator (AO-AFI) has been proposed to reach the video-rate image acquisition and 3-D arbitrary shape sensing[21]. As a structured-light active triangulation technique, AO-AFI overcomes many of weaknesses typical of active triangulation techniques. AO-AFI provides absolute range measurements that are (1) computational inexpensive, (2) at video rate, (3) full field, and (4) unambiguous. However, in the AO-AFI scheme, amplitude-modulated (AM) laser power must be used to freeze the traveling interference pattern at a particular phase, resulting in the complexity in building up the system. The effect is analogous to a stroboscope only lighting up a moving fringe pattern when it aligns to the desired spatial phase. In this subsection, we present a new 3-D imaging modality that is constructed with two acousto-optic deflectors (AODs) in order to simplify the system setup of AO-AFI. We refer to this new scheme as dual acousto-optic profilometry (DAOP)[22].

***(1) AO Deflection***

An acousto-optic cell utilizes the effect of Bragg diffraction of the laser beam incident on a volume grating. When an ultrasonic wave propagates through the crystal material, different regions of expansion and compression are created inside the Bragg cell, causing changes in density. The index of refraction is periodically modulated, and the medium becomes equivalent to a moving grating, which can be described by the following equation

$$\Delta n(x,t) = \Delta n \sin(\omega_s t - k_s x) \tag{2.2-1}$$

where $x$ is position in the Bragg cell along the vertical axis; $\omega_s$ is the sound angular frequency and $k_s$ is the sound wave vector given by $\omega_s / v$ where $v$ is the acoustic velocity.

Diffraction of light by ultrasonic wave is used to change the direction of the laser beam. The electromagnetic wave with angular frequency $\omega$ and wave-vector $k$ can also be considered to consist of photons (light particles) with energy and momentum. If one regards acousto-optic interaction as a photon-photon interaction, then due to the conservation of energy, one has[23]

$$f_i = f_d - f_s \qquad (2.2\text{-}2)$$

and

$$\vec{k}_i = \vec{k}_d - \vec{k}_s \qquad (2.2\text{-}3)$$

For the isotropic case when $n_i = n_d$, it can be shown that

$$\sin\theta_B = \sin\theta_i = \sin\theta_d = \frac{\lambda_l}{2\lambda_s} = \frac{\lambda_l}{2v_s} f_s \qquad (2.2\text{-}4)$$

where $\lambda_l = \lambda_0 / n$ is the wavelength of the light in the material itself and $\theta_B$ is the Bragg angle. For uniaxial crystal, for diffraction in a plane perpendicular to the optical axis, we have $n = n_o \quad or \quad n_e$. Under the assumptions of $n_i = n_d = n$ and small $\theta_i$ and $\theta_d$, we can determine the deflection angle, $\theta_D$, by[23]

$$\theta_D = \frac{\lambda_0 f_s}{n v_s} - \theta_B \qquad (2.2\text{-}5)$$

By differentiating (2.2-4) we obtain the spread in the Bragg angle, $\Delta\theta_B$,

$$\Delta\theta_B = \frac{\lambda_l}{2v_s \cos\theta_B} \Delta f_s \qquad (2.2\text{-}6)$$

This implies that changing the frequency of acoustic wave results in the change to the Bragg angle.

***(2) DAOD-Based Projector***

A fringe projector plays a central role in Moiré technique and the Fourier-transform profilometry (FTP). Using the effect of acousto-optic deflection, we construct a fringe projector with dual acousto-optic deflectors. A schematic diagram of the dual-acousto-optic deflector (DAOD) based fringe projector is shown in Fig. 2.2-1.

The DAOD is composed of following units: a coherent light source, a beam splitter (BS), two acousto-optic deflectors (AODs), two optic wedges, a beam synthesizer, a lens, a microscopic objective, the RF signals, and the direct digital synthesizer (DDS).

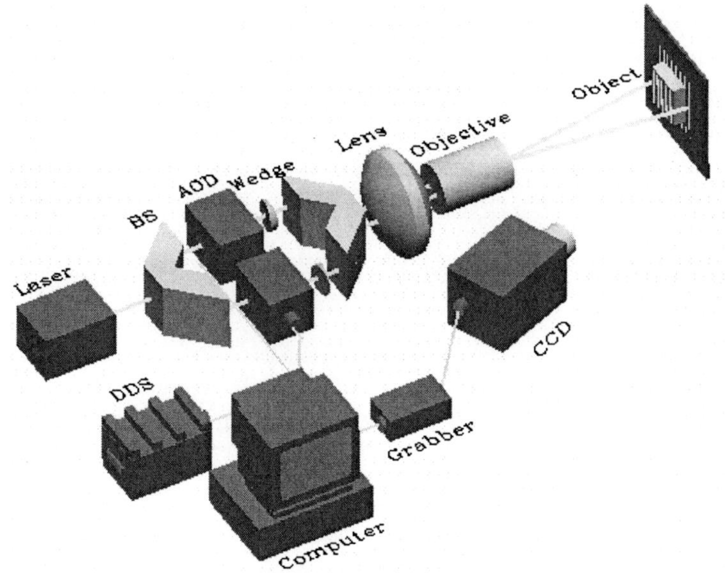


Figure 2.2-1. Schematic diagram of DAOD-based profilometer

The working principle of DAOD can be described as follows: an incident light is divided into two parallel light beams by a beam splitter and input to acousto-optic deflectors (AODs), respectively. By adjusting two optic wedges, we are able to compensate the deviations of the propagation directions of two first-order diffraction beams from AODs. Then the two output beams are recombined and converged by the beam synthesizer and lens to form an interference region in a micro scale region. Furthermore, this micro interference region is expanded by a microscopic objective and thus projected upon the test object surface. The amplified RF signals used to drive the AODs are generated and controlled by DDS. By changing the time frequency of RF signal, we are able to generate spatial gratings with varying frequencies according to Eq. (2.2-6). In comparison with other fringe generators, the proposed DAOD projector has following advantages: (1) static optical fringes with sinusoidal profile can be obtained, (2) the sensitivity of the fringe pattern can be precisely changed and controlled, (3) any moving mechanical part are not needed due to the adaptation of acousto-optic interaction, (5) perturbation immunity can be reached because of inherent symmetric property and equal optical path of the configuration, (6) quick variation of optical sensitivity due to the use of electronic circuit.

***(3) DAOD-Based 3D Imaging Modality***

A deformed fringe pattern is captured by a CCD camera and digitized via a frame grabber as also shown in Fig. 2.2-1. Through the desktop computer, a time-sequence control can be realized to change the sensitivity of projecting fringes via changing the time frequency of the RF driven signal. The deflection or the propagation direction of the first-order diffraction beam can be adjusted by changing the time frequency of the RF signal according to the Eq. (2.2-6). The spatial frequency of the fringe patterns can be altered through changing the bandwidth of the sound wave in a magnitude of the millisecond so that the speed for generating time-sequence fringe patterns with variable sensitivities can be much faster than video-frame rate.

***(4) Experiment and Results***

The experiment setup is composed of a 'transmitter' and a 'receiver'. The main components of the transmitter consist of the light source, beam splitter, AODs, DDS, beam synthesizer, optic-wedge, converging lens, and microscopic objective. The laser source is a single longitudinal mode laser with wavelength $532nm$ and $21mW$ output power (SUTTECH G-SLM-020), the AODs used in the experiment is AO deflector made by A.A OPTO-ELECTONIQUE designed for $532nm$ laser. The central frequency is $80MHz$ and the bandwidth is $\pm 25MHz$ so that the AO laser frequency shift is $80 \pm 25MHz$. The material of acousto-optic birefringent crystal is tellurium oxide (TeO2) and the Bragg (incident) angle is close to auto-collimation that is approximately perpendicular to input face. The AO efficiency is larger than 70% for RF power being smaller than $1W$. The active aperture of the AOD is $4.5mm$ while the laser beam diameter is smaller than $4mm$. The driver for AOD is direct digital synthesizer (DDS) OEM version plus amplifier. The communication time of the DDS is less than $40ns$ and the rise time/fall time is less than $10ns$. The desktop computer controls the DDS though a parallel port and $15bits$ TTL logic. The receiver is composed of an imaging lens and a CCD sensor as well as a frame grabber. The control software is programmed to generate a time-sequence to synchronize the spatial frequency variation and image acquisition. A time-varying sequential deformed fringe patterns with different optical sensitivities are thus obtained for later processing. A modified spatial-temporal phase-unwrapping algorithm[13] is applied to extract the phase maps with different resolution levels and then synthesized using a recursive scheme.

In order to demonstrate that our approach is valid for those object surfaces with steep slope, we select a step-like object as test object. The deformed fringe patterns with 25, 47, and 95 fringes are shown in Fig. 2.2-2 (a)-(c), respectively, and corresponding unwrapped phase maps obtained from spatial-temporal phase unwrapping algorithm are presented as Fig. 2.2-2 (d)-(f), respectively. Fig. 2.2-2 (g) shows reconstructed object surface.

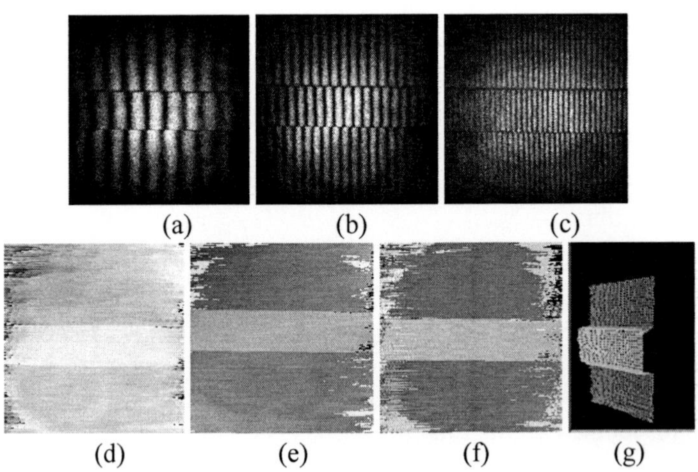


Figure 2.2-2. Experiment results. Each of them has been explained in the text

In conclusion, we have introduced a new 3-D imaging modality that is constructed with two acousto-optic deflectors (AODs). We refer to this new scheme as dual acousto-optic profilometry (DAOP). The DAOP has several unique features such as (1) static optical fringes with sinusoidal profile can be obtained, (2) the sensitivity of the fringe pattern can be precisely changed and controlled, (3) any moving mechanical part are not needed due to the adaptation of acousto-optic interaction, (5) perturbation immunity can be reached because of inherent symmetric property and equal optical path of the configuration, (6) quick variation of optical sensitivity due to the use of electronic circuit. The control software is designed for the DAOP operation and programmed to generate a time-sequence to synchronize the spatial frequency variation and image acquisition. A time-varying sequential deformed fringe patterns with different optical sensitivities are thus obtained and a modified spatial-temporal phase-unwrapping algorithm is applied to extract the phase maps with different resolution levels and then synthesized using a recursive scheme. An object surface with steep slope has been used to validate the approach presented in this section.

# 3. MULTI-VIEW RANGE IMAGE REGISTRATION

In practice there is a limitation on the field of view (FOV) and depth of view (DOV) to any optical digitizer for a single-view acquisition. Therefore, we need to capture multiple 2.5D range images of a complete object surface from different views. In this case, each partial range image is represented in its local coordinate frame, namely, view-dependent coordinate frame. Thus, it is necessary to transform these partial range images into a common coordinate frame in order to register them together. We briefly introduce a visualized registration scheme[3] that can provide a reliable estimate for the fine registration. In conjunction with iterative closest points algorithm (ICP), one is able to achieve an optimum registration of multiple range images[24].

## 3.1 Initial Estimation

We have proved a Lemma[3] saying that for two sets of sampling points corresponding to the identical part of the object surface measured from different range views, if (1) They are overlapped when projected, in a perspective manner, onto a screen plane; and (2) They are still overlapped on the same plane when rotated an arbitrary angle, then they are said being overlapped in three-dimensional space. Based on this Lemma, we constructed a visualized matching scheme of initial estimate for range images registration. The implementation of visualized registration for a pair of range images is straightforward. However, it is necessary for the pair of range image data to have overlay subset when using visualized matching scheme. The inputs are a pair of range images, with which we fix one projection of range-image and adjust interactively another to make them overlapped on the projection plane. As long as we have confirmed this pair of projected range images being overlapped on the projection plane through several different viewing-angle observations, we are able to infer that these two range images are overlapped in the 3-D space as stated by visualized matching theorem[3]. Meanwhile, we have recorded the parameters of relative rotations and translations between the two viewing coordinate frames. These parameters are composed of initial entries of transformation matrix for the ICP method. In this way, we demonstrate a procedure to obtain such an initial estimate, which is based on the visualized operations of two range images. Both range images correspond to the measurement of the same part of the object surface and are captured from different viewing points.

Fig. 3-1 (a)-(b) show the registration of a pair of range images using visualized matching scheme. This procedure allows a reliable initial estimate for further fine registration. We should point out that it would be sometimes difficult to identify the overlapping subset between the two range-images utilizing only mesh projection since a large number of points involved. In this case, discriminating overlapping subset with the help of rendered range image or texture map will be necessary.

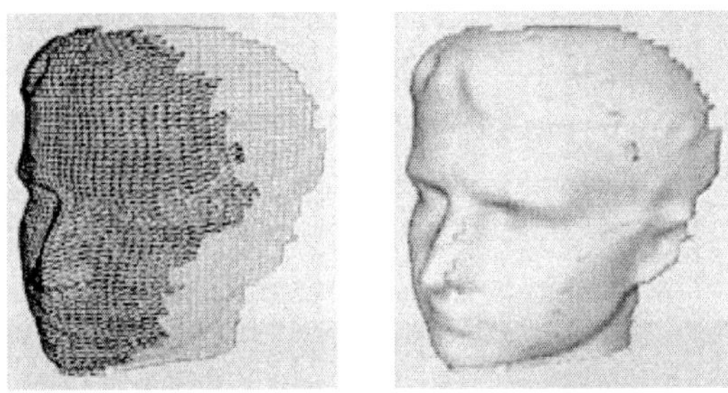

(a) (b)

Figure 3-1. Registration results using visualized matching scheme: (a) with the projection of mesh, (b) with rendered range images

## 3.2 Fine Matching

The goal of fine registration is to accurately transform sets of surface measurements into a common coordinate system. Besl and Mckay[24] introduced the Iterative Closest Point (ICP) algorithm to register two sets of points on a free-from surface. Modifications to original ICP algorithm have been made to improve the rate of convergence and register partially overlapping sets of points[25-26]. ICP is a general-purpose, representation independent method for the accurate and computationally efficient registration method of 3-D shapes, including free-from curves and surfaces. Extensions of this algorithm are now widely used for registration of multiple sets of surface data. The original ICP algorithm operates as follows[24]

Given point set P and surface Q where P is a subset of Q:

(1) Nearest point search: For each point p in P find the closest point q on Q.
(2) Compute registration: Evaluate the rigid transform T that minimizes the sum of squared distances between pairs of closest points (p,q).
(3) Transform: Apply the rigid transform T to all points in set P.
(4) Iterative: Repeat step 1 to 3 until convergence.

This approach will converge to the nearest local minimum of the sum of squared distances between closest points. A good initial estimate of the transformation between point sets is required to ensure convergence to the correct registration. Incorrect registration may occur if the error in the initial transformation is too large or if the surface does not contain sufficient shape information for convergence. With this good estimate obtained by visualized matching scheme, we are able to achieve a precise registration together with the ICP algorithm[24].

# 4. MULTI-VIEW RANGE IMAGE INTEGRATION

In previous section, we have introduced a visualized registration procedure that can provide a reliable estimate for the fine registration of multiple range images in a unified coordinate system. In this section we shall present a new approach for merging multiple range images to build up a complete geometric model of the objects in 3-D space. This approach is based on the segmentation of a pair of range images using the concepts of "double stitching lines", as well as a constrained local triangulation process.

The goal of geometric fusion is to integrate registered sets of surface measurements into a single 3-D surface model. The generic problem of surface reconstruction is to estimate a manifold surface that approximates the unknown object surface from a set of sampled points. This makes no prior assumptions about the surface shape and topology. Correct approximation of the object surface requires that the model has the same topology and is an accurate approximation of the surface geometry. In the general problem of the reconstruction of a manifold surface model from a set of unorganized points no information is available about measurement connectivity. Surface reconstruction from multiple 2.5D range images is a sub-problem that can also be addressed as a set of unorganized points. However, reconstruction from 2.5D range image is a more constrained problem as the 2D image structure can be used to estimate the local surface topology.

Reconstruction of an integrated 3-D surface from multiple 2.5D range images has received considerable interest. Recently, two approaches have been proposed for fusion of multiple overlapping surface measurements into a single surface model: mesh integration[27-28], and volumetric fusion[29-30]. In the following subsection, we briefly introduce an alternative method for merging multiple range images. This method is based on a concept of "double-stitching lines" and may be categorized as one of the methods in mesh integration[31].

## 4.1 Segmentation of Range Images

When looking at the object surface from a view angle, we intuitively infer that the confidences of measured points are not identical over the field of view (FOV). For every point in space (except the center of projection), a single ray passes through that point in a particular direction. The directions evaluated throughout space define a vector field. Then we could define the weight of the point by dotting the normal of the element with the projection direction at the center of the element. Thus, we will use two criteria to segment the overlapped region of a pair of range images: (1) the weights of sampled points within the overlapped region of the pair of range images, and (2) the distances from sampled point in one range image to the surface patches of another range image.

Suppose a pair of range images is measured from different view angles and they have an overlapped data set corresponding to identical part of the object surface. We denote two range images as $R_1(V(i))$ and $R_2(V(i))$, respectively, where $V(i)$ represents the sampled point set of two range images. Denoting the overlapping region (OR) as $O(V(i))$, we have

$$O(V(i)) = R_1(V(i)) \bigcap R_2(V(i)) \tag{4-1}$$

where the symbol $\bigcap$ stands for the intersection between sets $R_1(V(i))$ and $R_2(V(i))$. Further we denote the subset of $R_1(V(i))$ within the OR as $O_1(V(i))$, and the subset of $R_2(V(i))$ within the OR as $O_2(V(i))$. The non-overlapping regions (NOR) of the $R_1(V(i))$ and $R_2(V(i))$ can therefore be denoted with following set representations

$$U_1(V(i)) = R_1(V(i)) \bigcap \overline{R_2(V(i))} \tag{4-2}$$

$$U_2(V(i)) = R_2(V(i)) \bigcap \overline{R_1(V(i))} \tag{4-3}$$

For the sampled points of each range images within the OR, we define a reservation-region (RR), $S_k(V(i))$, and a deletion-region (DR), $D_k(V(i))$, respectively, where $k = 1,2$. In the overlapping region, if the weight of a sampled point in one range image is larger than its counterpart of another range image then we assign this sampled point to the RR otherwise assign it to the DR. According to the definitions of RR and DR, we can further segment the regions of $O_1(V(i)), O_2(V(i))$ with $S_k(V(i))$ and $D_k(V(i))$:

$$S_1(V(i)) = \{V(i) \in O_1(V(i)) \, for W_1(i) \geq W_2(i)\} \tag{4-4}$$

$$D_1(V(i)) = \{V(i) \in O_1(V(i)) \, for W_1(i) < W_2(i)\} \tag{4-5}$$

$$S_2(V(i)) = \{V(i) \in O_2(V(i)) \, for W_2(i) \geq W_1(i)\} \tag{4-6}$$

$$D_2(V(i)) = \{V(i) \in O_2(V(i)) \, for W_2(i) < W_1(i)\} \tag{4-7}$$

where $W_1(i)$ and $W_2(i)$ are the weights of sampled points corresponding to each range image view, respectively. Since $S_1(V(i)) \in O_1(V(i))$ and $D_1(V(i)) \in O_1(V(i))$, it is obvious that following relations are valid according to the segmentation mentioned above:

$$S_1(V(i)) \bigcup D_1(V(i)) = O_1(V(i)) \tag{4-8}$$

$$S_1(V(i)) \bigcap D_1(V(i)) = \Phi \tag{4-9}$$

where the symbol $\bigcup$ stands for union of sets $S_1(V(i))$ and $D_1(V(i))$, and $\Phi$ stands for the null set.

## 4.2 Definition of Double Stitching Lines

**Definition 1:** $\forall \vec{p}(x,y,z) \in S_1$, we define the connection of sampled points adjacent to $S_1(V(i)) \bigcap D_1(V(i)) = \Phi$ as the Primary Stitching Line (PSL) as illustrated in Fig. 4-1. Similar to the definition of the PSL, we can also define another stitching line.

**Definition 2:** $\forall \vec{p}(x,y,z) \in S_2$, we define the connection of sampled points adjacent to $S_2(V(i)) \bigcap D_2(V(i)) = \Phi$ as the Secondary Stitching Line (SSL) (see Fig. 4-1). Again, we have

$$S_2(V(i)) \bigcup D_2(V(i)) = O_2(V(i)) \tag{4-10}$$

$$S_2(V(i)) \bigcap D_2(V(i)) = \Phi \tag{4-11}$$

where $S_2(V(i)) \in O_2(V(i))$ and $D_2(V(i)) \in O_2(V(i))$.

Together with thc definitions of the PSL and SSL, we are able to represent the non-redundant set of the object surface, depicted by two viewing-angle range images, as

$$U_1(V(i)) \bigcup U_2(V(i)) \bigcup S_1(V(i)) \bigcup S_2(V(i)) \tag{4-12}$$

The integration method proposed here is therefore try to look for the sampled points located at $S_k(V(i))$ and $D_k(V(i))$, where $k = 1,2$, and reconstruct the geometric model of the object surface by the use of the PSL and SSL.

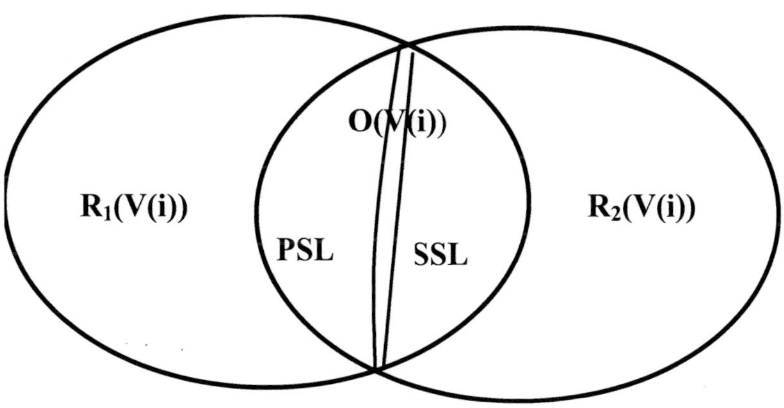


Figure 4-1. Projection illustration of the PSL and SSL of a pair of range image views

### 4.3 Integration algorithm based on PSL and SSL

The whole implementation procedure of merging the range data of different views can be divided into three phases: the construction of the primary stitching line (PSL), the construction of the secondary stitching line (SSL), and constrained local triangulation of the gap between the PSL and SSL.

***(a) Construction of PSL***

This procedure involves three steps: (1) Convert the sampled points and mesh in the view frame of the $R_2(V(i))$ to the view frame of the $R_1(V(i))$ through the affine transformations. (2) Calculate the distance from each sampled point in the $R_1(V(i))$ to corresponding surface patch in the $R_2'(V(i))$, where the superscript indicates the geometric transformation result. If this distance less than a given threshold and the weight of the sampled point in $R_1(V(i))$ is larger than its corresponding point in the $R_2'(V(i))$, then this point is marked as $S_1(V(i))$ otherwise marked as $D_1(V(i))$. Traverse all points in $R_1(V(i))$ till the process is completed. (3) Find out those points adjacent to the boundary of $S_1(V(i))$ and $D_1(V(i))$ but within the $S_1(V(i))$, then connect them sequentially to form the PSL. With the aid of PSL, we are also able to deal with the sampled points in the $R_2'(V(i))$ to find out the SSL.

***(b) Construction of SSL***

Similarly, we need to identify the $S_2(V(i))$ and $D_2(V(i))$ in order to build up the secondary stitching line. This procedure is addressed as follows:

(1) Transforming the triangular mesh of $R_1(V(i))$ and the sampled points of $R_2(V(i))$ into an intermediate frame that is perpendicular to the bisection of two view directions. In this way, we are able to simplify the judgment of point-to-mesh correspondence, and to keep the mappings from $R_2(V(i))$ to $R_1(V(i))$ injective.

(2) Check for each sampled point in the $R_2(V(i))$ to see if it falls into a reserved triangle $T_1(i)$ within the $S_1(T(i))$. If yes, mark it being within the $S_2(V(i))$ otherwise mark it within the $D_2(V(i))$.

(3) Mark the triangle patches according to the identification of the marked points in the $S_2(V(i))$. If three vertices of a triangle patch are marked as $S_2(V(i))$, then the patch is marked being within the $S_2(T(i))$.

Building up the SSL within the $S_2(V(i))$ according to the connectivity of $S_2(T(i))$. The SSL is a boundary of the triangles within the reservation region of the $R_2(V(i))$.

***(c) Local Triangulation (Stitching the Gap)***

After forementioned processing, we have obtained the triangular meshes corresponding to the reservation regions, namely $S_1(T(i))$ or $S_2(V(i))$, of the pair range images. The remaining problem is to perform a constrained local triangulation for stitching the gap between the PSL and SSL, leading a non-redundant geometric representation or a single geometric model. This process can be addressed as follows:

(4) Read in the sampled points on the PSL and SSL and register their positions. Two lists are then applied to store these points corresponding to each stitching line.

(5) Triangulate the sampled points. Pick up the first point $p_1$ on the PSL and take it as the first vertex, A, of the first triangle patch. Then pick up the first point $q_1$ on the SSL as the second vertex B. Take A and B as endpoints of a line segment that is referred as a base edge of the triangle patch. Further, pick up probed point from the PSL or from the SSL, either point is adjacent to $p_1$ or to $q_1$ and is denoted as $p_2$ or $q_2$. In term of the vertex, we mark $p_2$ and $q_2$ as C and D, respectively. Now we compare the angles of C and D subtending to the base edge AB, and the larger one determines the second edge of the triangle patch. For example, if the angle formed by □ACB is larger than one formed by □BDA, then CB will be the second edge, AB the third and C is set to be the third vertex of the triangle patch. Otherwise, D will be the third vertex and DA will be the second edge, DB the third. If the angles are equal, then compare the differences of the distance formed by C and D to endpoints A and B. The smaller one will be chosen as the third vertex, else if the differences of the distance are also identical, then select $p_2$ as the third vertex C to form the first triangle patch.

(6) Check the handled point on the PSL and SSL to see whether it is the final point or not. If yes, then terminate the process. Else if repeat the step (2) till all sampled points on the PSL and SSL are processed completely.

With the process above, we are able to get a non-redundant geometric fusion of a pair of range images. For the case of more than two range images, we shall add additional range image to fused range image in order to form a new pair of range images and repeat the process (1) to (3). This process will be repeated till all the range images are merged completely.

## 4.4 Experimental Results

Fig. 4-2 (a)-(f) show experiment results for the integration of a pair of range images by the use of our approach. Two range images captured from different views are illustrated in (a). The primary stitching line PSL and the secondary stitching line SSL used to segment the range images are shown as (b) and (c). The reservation regions associated with each view are presented in (d). Finally, the stitching results before and after constrained local triangulations are illustrated as (e) and (f), respectively.

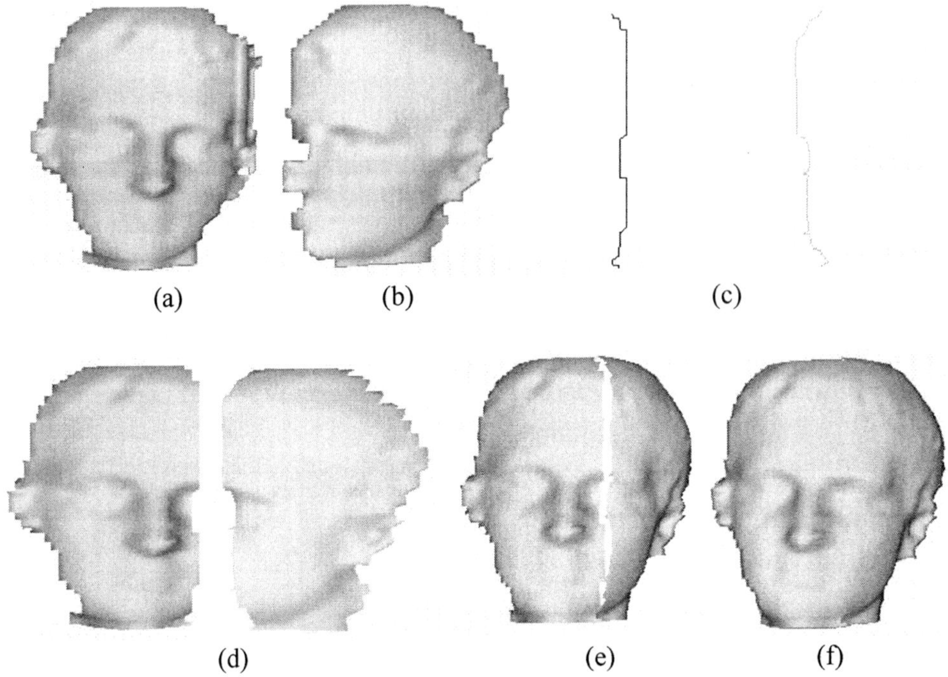

Figure 4-2. Experiment results of a pair of range images obtained from stitching-based integration approach: (a) Two range images measured form different views; (b) PSL; (c) SSL; (d) Reservation regions segmented by the PSL and SSL; (e) Before stitching; (f) After stitching

It should be pointed out that only valid points are involved in above figures. In other words, we have removed those invalid points already by selecting the ROI and preprocessing.

## 5. Simplification of 3D Model

Once a 3D image is obtained, the treatment of a significant quantity of data is required to represent a captured 3D image with high resolution. To make efficient storage and transmission of 3D digital content under the environment of network, it is essential to consider efficient treatment of large amount of data, namely the manipulation of 3D digital image.

A 3D image usually involves a huge amount of data consisting of polygonal surface models. Consequently, there has been considerable interest in automatic reduction of dense polygonal models into faithful approximations with much less number of polygons. Recently, a variety of approaches have been proposed for achieving such a goal[32-34]. Those approaches involved different strategies such as vertex clustering, vertex decimation, region merging, and edge collapse. However, in many cases those mentioned approaches produced plausible results at low level of details. It was usually difficult to preserve the sharp features and thus not only the visual appearance of simplified geometric model looked unfavorable, but also the accuracy of later refinement significantly will be severely affected.

In this section, we propose a modified simplification algorithm in order to solve the problem. Our approach can be categorized as a quadric-error-metrics (QEM) -based simplification scheme. The criterion QEM was defined as the summation of square distance of a vertex to its associated planes and it could be mathematically described as a 4×4 symmetric matrix[34]. This matrix was used to compute the cost of an edge collapse operation and the optimal position of the constituent vertex. However, previously reported QEM-based algorithms lost the sharp features at very low level of details. To solve this problem, Kho et al [35] presented a user-guided approach for mesh simplification based on QEM. Their method allowed users to selectively control the important parts in different surface regions and preserve sharp features through the imposition of geometric constraints. This approach could produce good simplified models with important features, but it was not automatic. Subsequently Kim[36] introduced an automatic algorithm that employed edge collapse operator and exploited discrete curvature norm to measure geometric deviation. This method seemed to be good at preserving visually important detail of a model at a low level of detail, but it was with low computational efficiency.

## 5.1 Method

Because the quadric error metric (QEM) was fundamental criterion of previous QEM-based simplification scheme, we may possibly modify this criterion to allow for achieving preservation of sharp features. Our approach is driven by edge collapse operation[37] and employs the sharp degree of vertex. The motivation of our approach is to achieve automatic simplification and increase the computational efficiency of the algorithm.

### *(a) Edge Collapse Operation*

For an edge collapse operation[32] (Fig.5-1), we denote $(v_i, v_j) \rightarrow \bar{v}$ to imply an operation that we move the vertices $v_i$ and $v_j$ to a new position $\bar{v}$; connect all their incident edges to $v_i$, and delete the vertex $v_j$; those edges or faces to be degenerated are removed. The effect of an edge collapse is highly localized.

### *(b) Derivation of New Error Metric*

First, we briefly describe the concept of the quadric error metric: Given a vertex $v$ and a set of planes associated with it $planes(v)$, Garland[32] defined the error metrics of the vertex $v$ as the sum of squared distances to these planes:

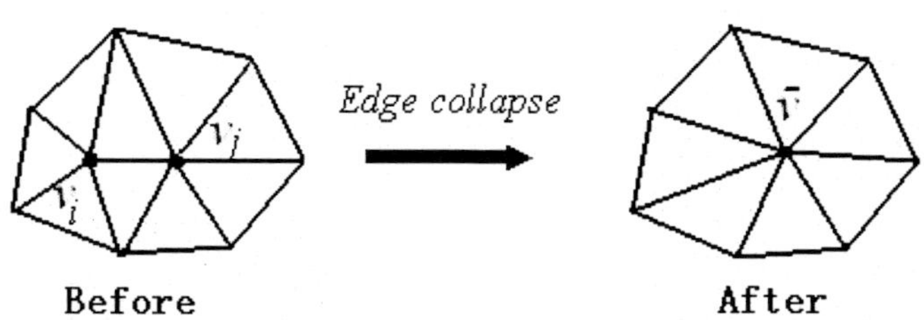


Figure 5-1. Edge Collapse

$$\Delta(v) = \sum_{p\in planes(v)} d_p^2(v) = \sum_{p\in planes(v)} v^{\mathrm{T}}(K_p)v = v^{\mathrm{T}}\Big(\sum_{p\in planes(v)} K_p\Big)v \tag{5-1}$$

where $v=[v_x\ v_y\ v_z\ 1]^{\mathrm{T}}$; $p=[a\ b\ c\ d]^{\mathrm{T}}$ represents the plane defined by the equation $ax+by+cz+d=0$ where $a^2+b^2+c^2=1$; $K_p = pp^{\mathrm{T}} = \begin{bmatrix} a^2 & ab & ac & ad \\ ab & b^2 & bc & bd \\ ac & bc & c^2 & cd \\ ad & bd & cd & d^2 \end{bmatrix}$.

The quadric error matrix $Q$ of vertex $v$ is defined: $Q(v)=\sum_{p\in planes(v)} K_p$. Once edge collapse $(v_i, v_j)\to \bar{v}$ is finished, the new vertex $\bar{v}$ accumulates the planes associated with the edge end points by $Q_{\bar{v}} = Q_i + Q_j$.

To keep more important features of the original model, we now modify the error metric by introducing a new concept of *sharp degree*, including sharp edge and sharp degree of vertex:

**Sharp edge:** given a threshold $\theta$ (e.g. 40 degrees), we define the sharp edge as an edge whose dihedral angle is large than $\theta$, and the edges on the boundary are sharp edges.

**Sharp degree of vertex:** We define the sharp degree of a vertex is the total number of the sharp edges associated with the vertex, so the maximum of a vertex's sharp degree is the vertex's degree; the minimum of a vertex's sharp degree is zero.

With above definitions, we add the sharp degree to the vertex structure:

```
Structure Vertex
{ float pos[3];
 int sharp;}
```

where pos[3] presents the position of a vertex; *sharp* presents the sharp degree of vertex.

Now we add a penalty term $c_{sharp}\cdot sharp$ to Eq. (5-1) to obtain following new error metric

$$\Delta'(v) = \Delta(v) + c_{sharp}\cdot sharp = v^{\mathrm{T}}\Big(\sum_{p\in planes(v)} K_p\Big)v + c_{sharp}\cdot sharp \tag{5-2}$$

where $c_{sharp}$ is a penalty coefficient, its value is set by user. Therefore, the error value of the vertex with non-zero sharp degree is larger than that before. And the greater the sharp degree is, the larger the error value will be. Because the edge collapses are operated by the order of the error values, the edge whose end points are with non-zero sharp degrees may be collapsed later. In addition, this additional penalty term can keep the linearity when computing new vertex.

Consider $\sum_{p\in planes(v)} K_p$ being a 4×4 symmetric matrix in QEM, we construct a similar matrix in order to unify the error metrics with QEM:

Let $K_{sharp} = \begin{bmatrix} 0 & 0 & 0 & 0 \\ 0 & 0 & 0 & 0 \\ 0 & 0 & 0 & 0 \\ 0 & 0 & 0 & \sqrt{c_{sharp} \cdot sharp} \end{bmatrix}$, then Eq. (5-2) becomes:

$$\Delta'(v) = v^{\mathrm{T}}(\sum_{p\in planes(v)} K_p + K_{sharp})v$$

The new quadric error matrix of vertex $v$ is thus defined: $Q'(v) = \sum_{p\in planes(v)} K_p + K_{sharp}$, and it keeps the form of a 4×4 symmetric matrix unchanged.

When the edge-collapse operated, $\Delta(v)$ is increased. In order to achieve the penalty to the degree of sharp features, we need amplify the coefficient $c_{sharp}$ as the edge-collapse processes. Besides, there are too many edges involved in a model, if altering $c_{sharp}$ one time after each edge-collapse operation, we shall subject to time-consuming computation and require large extra memory space to store the results. In practical implementation we take sum of the terms ($c_{sharp} \cdot sharp$) for two end-points of an edge as the penalty term of a new vertex in order to reduce the cost of computation and storage. In this way, once edge collapse $(v_i, v_j) \to \bar{v}$ is completed, the quadric error matrix of a new vertex $\bar{v}$ is: $Q'_{\bar{v}} = Q'_i + Q'_j$.

***(c) Computation of New Vertices***

In order to perform the collapse $(v_i, v_j) \to \bar{v}$, we must also choose a position for $\bar{v}$. Since the error function $\Delta'$ is quadratic, finding its minimum is a linear problem. Thus, we find $\bar{v}$ by solving

$$\partial\Delta'/\partial x = \partial\Delta'/\partial y = \partial\Delta'/\partial z = 0.$$

For $\partial(C_{sharp} \cdot sharp)/\partial x = \partial(C_{sharp} \cdot sharp)/\partial y = \partial(C_{sharp} \cdot sharp)/\partial z = 0$, the result is similar to QEM algorithm:

$$\bar{v} = \begin{bmatrix} q_{11} & q_{12} & q_{13} & q_{14} \\ q_{12} & q_{22} & q_{23} & q_{24} \\ q_{13} & q_{23} & q_{33} & q_{34} \\ 0 & 0 & 0 & 1 \end{bmatrix}^{-1} \begin{bmatrix} 0 \\ 0 \\ 0 \\ 1 \end{bmatrix}$$

where $$\begin{bmatrix} q_{11} & q_{12} & q_{13} & q_{14} \\ q_{12} & q_{22} & q_{23} & q_{24} \\ q_{13} & q_{23} & q_{33} & q_{34} \\ q_{14} & q_{24} & q_{34} & q_{44} \end{bmatrix} = Q'$$ .

## 5.2 Experimental Results

In order to demonstrate proposed method, we apply this improved simplification technique to a practical example of 3D image. Fig. 5-2 (a) shows the mesh representation of original 3D image (graphic model), in which there were 6730 vertices and 13140 triangle facets. The simplified model with our approach under 50% data reduction rate is represented in Fig. 5-2 (b), where there were 3395 vertices and 6570 facets. Fig. 5-2 (c) presents the simplified model with our approach under 20% data reduction rate with only 1414 vertices and 2628 facets. From Fig. 5-2 we can observe that those sharp features in original 3D image such as nose, eye, mouth, ear, can be preserved even if only a few vertices or facets are involved. This can be helpful for later reconstruction with high accuracy. Furthermore, we also observe that the triangle facets are reasonably distributed as those dense meshes are distributed in high-curvature regions and sparse meshes in flat regions. This is certainly helpful for saving storage of memory.

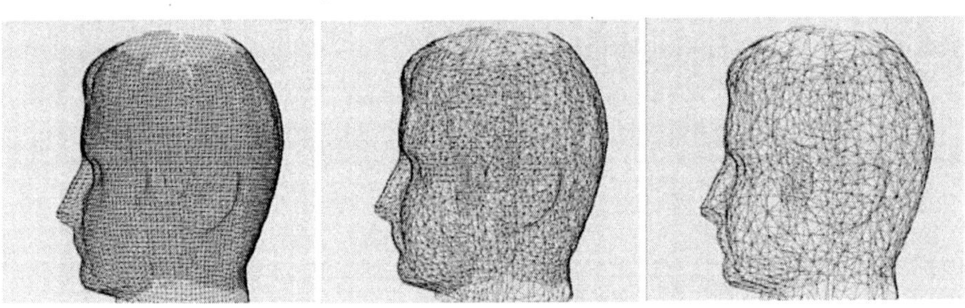

(a) (b) (c)

Figure 5-2. Experiment result with our approach: (a) original 3D image represented in a mesh model, in which there were 6730 vertices and 13140 triangle facets. (b) The simplified model with our approach under 50% data reduction rate, in which 3395 vertices and 6570 facets were included. (c) illustrated the simplified model with only 1414 vertices and 2628 facets were included.

# 6. Refinement of Simplified 3D Model

3D image refinement is a reverse process to that of simplification. In the context of geometric representation of 3D image, subdivision schemes are becoming more attractive because of their arbitrary topology, uniformity of representation, numerical stability and code simplicity[38]. Up to now, a number of subdivision methods have been proposed, such as Catmull-Clark subdivision[39] and $\sqrt{2}$-subdivision[40] based on quadrilateral meshes, and Loop subdivision[41], Butterfly subdivision[42] and $\sqrt{3}$-subdivision[43] based on triangular meshes. However, the shortcomings in existed methods include: (1) the number of meshes increases exponentially during the subdivision process. This implies that a subdivision process requires a great deal of memories to store the information of vertices and edges as well as facets. (2) Previous subdivision methods emphasized on the smoothness too much whereas the sharp features would be missed. However it is an important to hold the sharp features for the reconstruction with high accuracy.

In this section, we present an improved $\sqrt{3}$-subdivision scheme to reduce computational cost while preserve sharp features at the same time. To reduce consumed memory, we only choose those meshes facing to the viewpoint as preferable areas to be further subdivided in the pre-process. Then in each step of adaptive subdivision, we choose the normal angle of adjacent faces as a criterion to judge whether or not the approximation is sufficient. Furthermore, the selected meshes are divided into two different classes: those could be subdivided and those could not. Finally, we modify previous subdivision masks to rebuild sharp features. Sharp edges are automatically tagged according to the given threshold $\theta$ and subdivision masks are designed based on different sharp degrees.

## 6.1 Brief Review of $\sqrt{3}$ Subdivision

In contrast to 1-to-4 split scheme like Loop subdivision and Catmull-Clark subdivision, the $\sqrt{3}$-subdivision[43] performed 1-to-3 split so that the number of triangles may increase slower. In this sense, the $\sqrt{3}$ subdivision possessed finer gradation for hierarchy levels. Moreover, it enabled a simple implementation of adaptive refinement strategies without inconsistent intermediate states. In $\sqrt{3}$-subdivision scheme, a new vertex was inserted at the center of each triangle patch and it was connected to the surrounding vertices, then original edges were flip and connected with two old vertices. Fig. 6-1 illustrates the 1:3 split and the results with $\sqrt{3}$-subdivision. It can be observed from Fig. 6-1 that all newly inserted vertices have exactly valences six and the old vertices keep their valences unchanged after a subdivision operation. Performing this scheme twice results in 1-to-9 face-split, and thus the method was named $\sqrt{3}$-subdivision. For a triangle mesh with arbitrary topology□ the new face vertex of triangle $\Delta(v_1, v_2, v_3)$ had its centroid: $v = \frac{1}{3}(v_1 + v_2 + v_3)$□the new vertex is $S(v) = (1 - n\alpha_n)v + \alpha_n \sum_{i=0}^{n-1} v_i$, where $\alpha_n = \frac{4 - 2\cos(\frac{2\pi}{n})}{9n}$□ $v$ was a vertex with valence $n$ and $v_0, \ldots, v_{n-1}$ were its adjacent neighbors in the unrefined mesh[43].

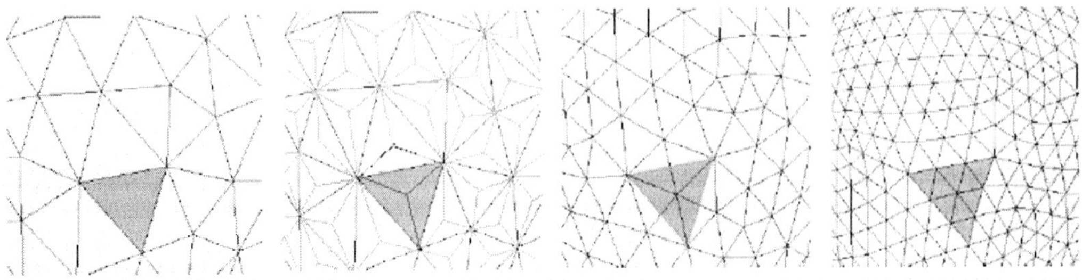

(a) Given mesh (b) 1:3 split (c) $\sqrt{3}$ subdivision (d) 2 times $\sqrt{3}$ subdivision

Figure 6-1. $\sqrt{3}$ subdivision

## 6.2 Improved Scheme to $\sqrt{3}$ Subdivision

To model sharp features, Hoppe, et al[44] proposed an algorithm based on Loop subdivision. They manually tag edges that need to generate sharp features, and modify the masks to construct crease, corner and dart. The motivation of our approach is to automatically tag sharp edges by computing the normal angle of the adjacent faces. Furthermore we modify previous $\sqrt{3}$ subdivision masks to obtain an improved scheme. The procedure details are described as follow:

***(a) Features Tagging***

Features tagging is achieved by considering the concept of sharp degree introduced in section 3B. A smooth vertex is the one where sharp degree $s = 0$; a dart vertex has $s = 1$; crease vertex has $s = 2$; and corner vertex has $s > 2$.

***(b) Topology Rules***

Based on the specific face-split and edge-flip characteristics of $\sqrt{3}$-subdivision, we modify previous topology rules. First an indicator, $index$, indicates the times of subdivision and a flag, $f.flag$, denotes whether the face $f$ has sharp edge, if it does, then $f.flag = 1$, else $f.flag = 0$. Initially we let every original triangle be $f.flag = 0$. Thus the topology rule is consist of two successive steps:

- Face splitting. If $f.flag = 0$, new vertices are inserted on the centroid of the facet else if $f.flag = 1$ and $f.flag \neq 0$, new edge points are inserted on the sharp edges of the facet. Then the new mesh is obtained by connecting new vertices with old ones, as shown in the facet splitting step of Fig. 6-2;
- Edge flipping. Considering the new mesh, we find that every triangle $f$ consists of two new edges and one old edge. If the old edge of $f$ is not sharp, we tag $f.flag = 0$ and flip this edge, else we tag $f.flag = 1$, as shown in the facet-flipping step of Fig. 6-2.

Fig. 6-2 shows this sharp-feature retained $\sqrt{3}$ subdivision process, where double lines denote sharp edges. We find that performing this scheme twice results in a triadic edge and a 1-to-9 face-split. In this way, our approach inherits the topology characteristic of previous $\sqrt{3}$ subdivision. When performing edge flipping in the second refinement, we flip all the old edges of the meshes. That is because the edges called old ones are derived at the first refinement, and they don't contain any sharp feature. We thus unify the edge flipping of the second refinement to the proposed operation. All the odd refinements are essentially the same as the first one, while all the even ones are the same as the second.

***(c) Geometric Rules***

To compute the vertex position in the next subdivision step by accounting for the sharp degree of each vertex, we design the different weight masks to hold sharp features. The Fig. 6-3 illustrates the geometric rules of vertices and new facet points, where the new position of each vertex should be computed between the facet splitting and edge flipping.

Two new edge vertices are inserted into the sharp edge, and their computing formulas are empirically given by

$$v_4 = \frac{1}{2}v_1 + \frac{1}{4}v_2 + \frac{1}{8}v_3 + \frac{1}{8}v \,, \quad v_5 = \frac{1}{4}v_1 + \frac{1}{2}v_2 + \frac{1}{8}v_3 + \frac{1}{8}v \qquad (6\text{-}1)$$

The formulas of the edge vertices on the boundary are

$$v_4 = \frac{2}{3}v_1 + \frac{1}{3}v_2 \,, \quad v_5 = \frac{1}{3}v_1 + \frac{2}{3}v_2 \qquad (6\text{-}2)$$

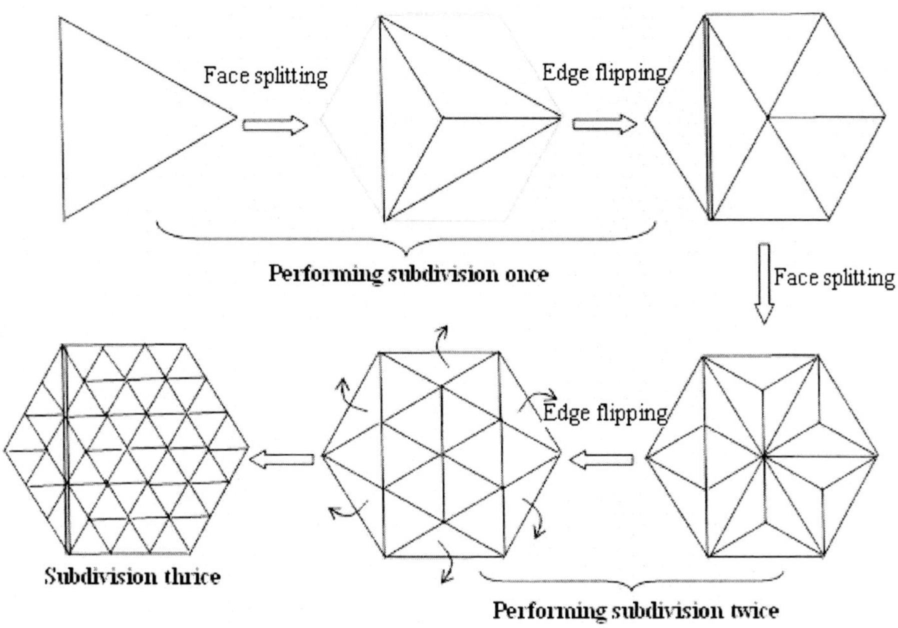


Figure 6-2. Sketch map of $\sqrt{3}$ subdivision holding sharp features

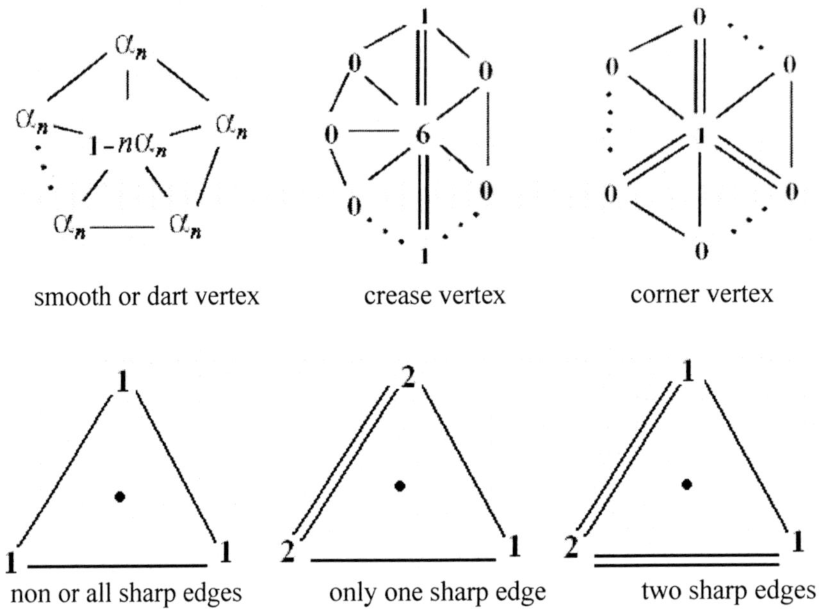


Figure 6-3. Vertex and face vertex subdivision masks, double lines denote sharp

## 6.3 Local Feature-Retained $\sqrt{3}$-Subdivision Algorithm

Now we introduce a view-dependent local adaptive algorithm based on the feature-retained $\sqrt{3}$-subdivision. This is referred to as a local feature-retained $\sqrt{3}$-subdivision algorithm since we apply adaptive subdivision only to those facets facing to viewpoint, so that we are able to reduce required memory in subdivision process while keep sharp features in original 3D image. The implementation of local feature-retained $\sqrt{3}$-subdivision algorithm is composed of two steps: extraction of facets facing to the viewpoint and adaptive subdivision. In the first step, we extract region of interest (ROI) following three successive sub-steps: defining vision frustum and selecting front-facing triangles as well as maintaining silhouette areas[45]. We also prescribe that the further iterations of refinement are only applied to the front-facing and silhouette items located within the vision frustum. In addition, we choose the normal angle of adjacent facets as a criterion to judge whether or not the approximation is satisfied[46]. Given a threshold $\theta$, we compute the normal angles of the facets and three adjacent facets separately, and decide whether each facet need to be further subdivided. If at least two angles are larger than $\theta$, then perform the features-retained $\sqrt{3}$ subdivision algorithm; else detect whether the other angle is larger than the sharp threshold $\theta$, and decide how to perform the successive refinement accordingly. Fig. 6-4 shows the topology rules for our first and second refinement.

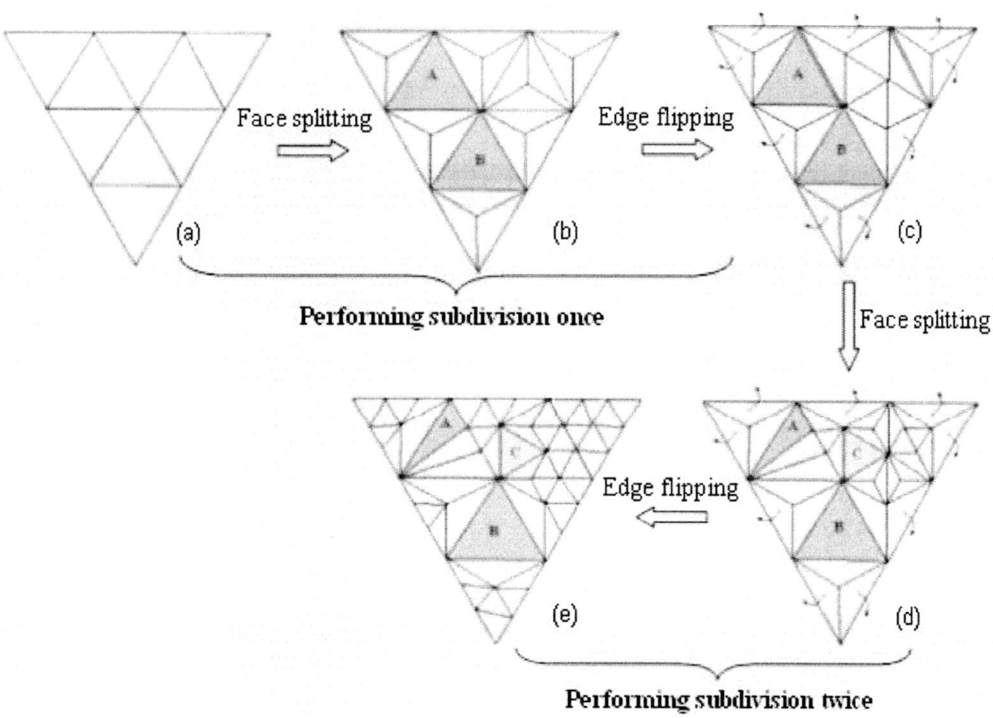


Figure 6-4. Local feature-retained $\sqrt{3}$ subdivision process

## 6.4 Practical Examples

To demonstrate the effectiveness of proposed method, we implemented it under the environment of Microsoft VC++6.0. Fig. 6-5 shows the simplified 3D image model, in which only 1414 vertices and 2628 facets were included. The refinement results using feature-retained local subdivision scheme are given in Fig. 6-5 (b) where subdivision has been performed once and with 2400 facets, and Fig. 6-5 (c) in which two times subdivision have been performed and with 7200 facets.

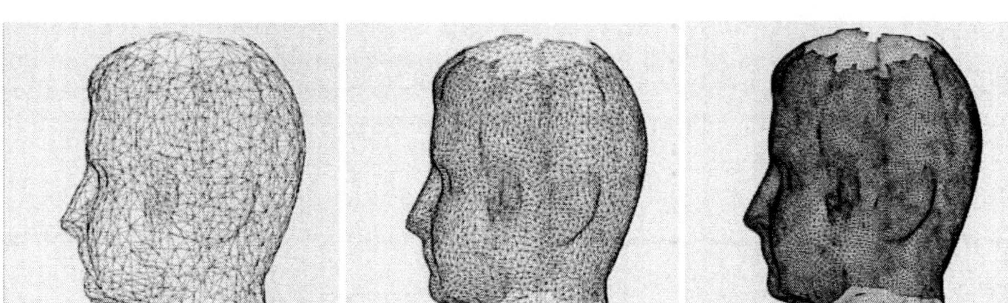

(a) (b) (c)

Figure 6-5. 3D image refinement with local feature-retained $\sqrt{3}$ -subdivision scheme

# 7. Conclusion

In this chapter, we have presented the core techniques involved in multi-resolution three-dimensional imaging and modeling (3DIM). These core techniques include multi-resolution 3D imaging of object surface with complex geometry and topology. In particular, we have discussed the sate-of-the-arts 3D imaging techniques for those object surfaces having discontinuities or a scene with isolated portions. In this case multi-resolution techniques are particularly useful. An incoherent 3D imaging technique based on point-array structured light illumination and a coherent 3D imaging modality based on dual-acousto-optic deflectors are introduced, respectively.

A range image acquired by a 3D optical digitizer is of 2.5-dimension. In order to obtain a complete object surface in three-dimensional space, one must take range images from different viewing points around the test object. In this regard it is necessary to register partial range images at local coordinate frames to unified coordinate system. A registration scheme with a visualized initial matching together with ICP algorithm for fine alignment is presented, leading to an efficient registration to multi-view range images. Furthermore we also discussed a methodology of multi-view range images integration with which a stitching-based technique has been used to segment the overlapped portion of two adjacent range images and then followed by a local triangulation procedure in order to merge partial range images.

Finally we have talked about the 3D image manipulation, including the simplification and the refinement of the raw data of a 3D image. This seems equivalent to the discussion of compression and decompression of a 3D image. Such a topic would be valuable in the application, for example, of storage or transmission under the environment of network. The experimental results for each part of the chapter have demonstrated our theoretical analysis. We wish that the work reported here would be helpful to those who are working in this field.


## Acknowledgement

The financial support from National Science Foundation of China (No. 60275012 and No. 60472107), Nature Science Foundation of Guangdong Province (No. 031804 and No. 04300864), Natural Science Research Grant of Higher Education of Gaungdong province (No. 04Z010) and Science and Technology Bureau of Shenzhen (No. 200341 and No. 200426) are greatly acknowledged.

The author (X. Peng) would like also to thank his former and current students Dr. Shaoming Zhu, Dr. Zonghua Zhang, Weiqiang Shi, Changqing Liu, Peng Zhang, Zhaoliang Yang, Pengdong Gao, Yabin Ding, Xiaoli Liu, Ying Liu, Linbin Wei, and Wenjie Qiu for their contributions to the research projects mentioned above. Special thanks go to Dr. Zeyi Liu, a Professor of Mathematics, for his long-term collaboration with the author in this interesting research field.


The authors' e-mail addresses: xpeng@szu.edu.cn, jindt@szu.edu.cn

## REFERENCES


[1] F. Chen, G. M. Brown, M. Song, "Overview of three-dimensional shape measurement using optical methods", *Opt. Eng.*, 39, 10–22 (2000).

[2] Hans J. Tiziani, "Optical 3D-shape, surface, and material analysis", in *Second International Conference on Experimental Mechanics*, F. S. Chau and C. Quan, eds., *Proc. SPIE* 4317, 204-210 (2001).

[3] X. Peng, Z. Zhang, and Hans J. Tiziani, "3-D imaging and modeling – Part I: acquisition and registration", *Optik*, 113, 448-452 (2002).

[4] S. Luis, L. Esteban, S. Javier, G. Victor and S. Manuel, "Profilometry by fringe projection", *Opt. Eng.* 42, 3307–3314 (2003).

[5] Jorge L.C. Sanz, *Advances in Machine Vision* (Springe-Verlag, Berlin, 1989).

[6] B. Fiona, P. Paul, C. James, "A theoretical comparison of three fringe analysis methods for determining the three-dimensional shape of an object in the presence of noise", *Optics and Lasers in Engineering*, 39, 35-50 (2003).

[7] J. H. Bruning, D. R. Herriott, J. E. Gallagher, D. P. Rosenfeld, A.D. White and D. J. Brangaccio, "Digital wavefront measuring interferometer for testing optical surfaces and lenses", *Appl. Opt.* 13, 2693-2703 (1974).

[8] M. Takeda, K. Mutoh, "Fourire transform profilometry for the automatic measurement of 3-D object shapes", *Appl. Opt.* 22, 3977-3982 (1983).

[9] X. Su, W. Chen, "Fourier transform profilometry: a review", *Opt. Laser Eng.* 35, 263-284 (2001).

[10] R. Vander, S. G. Lipson, and I. Leizerson, "Fourier fringe analysis with improved spatial resolution," *Appl. Opt.* 42, 6830–6837 (2003).

[11] S. Toyooka, Y. Iwaasa, "Automatic profilometry of 3-D diffuse objects by spatial phase detection", *Appl. Opt.* 25, 1630-3 (1986).

[12] H.O. Saldner, J.M. Huntley, "Temporal phase unwrapping application to surface profiling of discontinuous objects", *Appl. Opt.* 36, 2770-2775 (1997).

[13] X. Peng, Z. Yang, H. Niu, "Multi-resolution reconstruction of 3-D image with modified temporal unwrapping algorithm", *Optical Communication*, 224, 35-44 (2003).

[14] M. Takeda, Q. Gu, M. Kinoshita, H. Takai, and Y. Takahashi, "Frequency-multiplex Fourier-transform profilometry: a single-shot three-dimensional shape measurement of object with large height discontinuities and/or surface isolation", *Appl. Opt.*, 36, 5347-5354 (1997).

[15] J. Tian and X. Peng, "3-D digital imaging based on shifted point-array encoding," *Appl. Opt.* 44, 5491-5496 (2005).

[16] R. Jain, R. Kasturi, B. G. Schunck, *Machine Vision* (McGraw Hill, New York, 1995)

[17] J. Tian and X. Peng: Three-dimensional vision from multi-sensing mechanism, *Appl. Opt.*, 45, 3003-3008 (2006).

[18] M. K. Kalms, W. Juptner, and W. Osten, "Automatic adaptation of projected fringe patterns using a programmable LCD projector," *Proc. SPIE*, 3100, pp.156-165 (1997).

[19] H. Saldner and J. M. Huntley, " Profilometry by temporal phase unwrapping and spatial light modulator based projector," *Opt. Eng.* 36, pp. 610-615 (1997).

[20] P. S. Huang, Q. Hu, F. Jin, and F. P. Chiang, "Color-enhanced digital fringe projection for high-speed 3D surface contouring," *Opt. Eng.* 38, pp. 1065-1071 (1999).

[21] M. S. Mermeistein, D. L. Feldkhum, and L. G. Shirley, "Video-rate surface profiling with acousto-optic accordion fringe interferometry," *Opt. Eng.* 39, pp. 106-113 (2000).

[22] X. Peng, J. Tian, P. Zhang, L. Wei, W. Qiu, E. Li, D. Zhang, "3D vision with dual-acousto-optic deflections encoding," *Opt. Lett.*, 30, 1965-1967 (2005).

[23] P. K. Das, *Optical Signal Processing*, Springer-Verlag, Berlin, 1991, pp. 150-152.

[24] P. J. Besl, N. D. Mckay: A method for registration of 3-D shapes. *IEEE Trans PAMI* 14, 239-256 (1992).

[25] Z. Zhang: Iterative point matching for registration of free-from curves and surfaces. *Int. J. Computer vision* 13, 111-152 (1994).

[26] R. Bergevin, M. Soucy, H. Gagnon, D. L aurendeau: Towards a general multi-view registration technique. *IEEE Trans. PAMI* 18, 540-547 (1996).

[27] M. Soucy, D. Laurendeau: A general surface approach to the integration of a set of range views *IEEE Trans. PAIM* 14, 344-358 (1995).

[28] G. Turk, M. Levoy: Zippered polygon meshes from range images. *Proc. SIGGRAPH'94*, 311-318 (1994).

[29] B. Curless, M. Levoy: A volumetric method for building complex models from range images *Proc. SIGGRAPH'96*, 302-312 (1996).

[30] Hilton, A. J. Stoddard, J. Illingworth: Implicit surface-based geometric fusion *Computer Vision and Image Understanding.* 69, 273-291 (1998).

[31] X. Peng, Z. Zhang, and Hans J. Tiziani: 3-D imaging and modeling – Part II: integration and duplication. *Optik*, 113 453-459 (2002).

[32] M. Garland, P. S. Heckbert, "Surface simplification using quadric error metrics," In: *Proceedings of the Computer Graphics*, 31, 209-216 (1997).

[33] Kalvin, R. Taylor, "Superfaces: Ploygonal mesh simplification with bounded error," *IEEE Computer Graphics and Applications*, 16, 64-77 (1996).

[34] H. Hoppe, "Progressive meshes," In *Computer Graphics Proceedings (Proc. SIGGRAPH 96)*, ACM Press. 99–108 (1996).

[35] Y. Kho, M. Garland, "User-guided simplification. Computer Graphics," *Proceedings of ACM Symposium on Interactive 3D Graphics*, (2003).

[36] S..J. Kim, C..H. Kim, D.Levin, Surface simplification using a discrete curvature norm, *Computers and Graphics*, 26, 657-663 (2002).

[37] X. Peng, P. D. Gao, Z. Y. Liu and H. B. Niu, "Data reduction and extrapolation for 3-D digital image," *OPTIK*, 114, 226-236 (2003).

[38] D. Zorin, "Overview of subdivison schemes," *Computer Graphics Proceedings, Annual Conference Series, ACM SIGGRAPH.* New York: ACM Press, 65-84 (2000).

[39] E. Catmull, J. Clark, "Recursively generated B-Spline surfaces on arbitrary topological meshes," *Computer Aided Design*, 10, 350-355 (1978).

[40] Guiqing Li, Weiyin Ma, Hujun Bao, "$\sqrt{2}$ Subdivision for quadrilateral meshes," *Visual Computer*, 20, 180-198 (2004).

[41] C. Loop, Smooth subdivision surfaces based on triangles, *Thesis*, University of Utah, Department of Mathematics, (1987).

[42] N. Dyn, D. Levin, J. A. Gregory, "A butterfly subdivision scheme for surface interpolatory with tension control," *ACM Trans Graph*, 9, 160-169 (1990).

[43] Leif Kobbelt, "$\sqrt{3}$ -Subdivision," SIGGRAPH 2000, *Proceedings of the 27th Annual Conference on Computer Graphics*, New Orleans, Louisiana, USA, 103-112 (2000).

[44] H. Hoppe, T. DeRose, T. Duchamp, *et al.*, "Piecewise smooth surface construction," *Computer Graphics Proceedings, Annual Conference Series, ACM SIGGRAPH*, Orlando, Florida, ACM Press, 295~302 (1994).

[45] Pierre Alliez, Nathalie Laurent, Henri Sanson, *et al.*, "Efficient view-dependent refinement of 3D meshes using $\sqrt{3}$-subdivision," *Visual Computer*, 18, 205-221 (2003).

[46] Zheng Xu, K. Kondo, "Local subdivision process with Doo-Sabin subdivision surfaces," *International Conference on Shape Modeling and Applications*, Banff, Canada, May 17- 22, (SMI'02), 7-12 (2002).

In: New Developments in Lasers and Electrc-Optics Research ISBN 1-59454-771-8
Editor: William T. Arkin, pp. 153-171 

*Chapter 5*

# IN-LINE HOLOGRAM RECONSTRUCTION BY USING ITERATIVE ALGORITHMS

***Yan Zhang***
Department of Physics, Capital Normal University
Xisanhuan Beilu 105, 100037 Beijing, China
Tel. and Fax: 86-10-68902178
E-mail: yzhang@mail.cnu.edu.cn

## ABSTRACT

Digital holography is a quickly developing technology for high speed imaging. In-line holography has been extensively investigated due to it can effectively utilize the space-bandwidth product of digital recording instruments. However, since the reference wave and the object wave are overlapped during the hologram is recorded; the directly inverse Fresnel transform can not present the original object well. The reconstructed image is blurred by its conjugated image and the zero-order diffraction of the hologram. This shortcoming limits the application of the in-line holography. Therefore, an effective reconstruction algorithm is important for generalizing its application.

In this chapter, some approaches based on the phase retrieval algorithm for in-line hologram reconstruction are reported. Firstly, the Yang-Gu algorithm and the GS algorithm are used to restructure pure absorption and pure phase object from their in-line holograms, respectively. The differences between these two algorithms are analyzed. Then the GS algorithm is extended to reconstruct whole optical field from double or multiple holograms. At last, a new approach for reconstructing object from a hologram series is presented. Experimental results show that all these methods can reconstruct original object well.

**Keywords:** digital holography, in-line holography, reconstruction, iterative algorithm

# 1. INTRODUCTION

Holography is a powerful approach for storing and reconstructing both the amplitude and phase information of a wave front [1]. In the conventional approach the hologram is recorded on the photosensitive materials. The reconstruction is performed by illuminating the developed materials with the reference wave and the wet processing is necessary; therefore, the traditional holography is time consuming. Using electrical imaging instruments instead of the photosensitive materials, a new kind of technology called digital holography (DH) is proposed. In this technology, the hologram is digitalized and fed into a computer. The reconstruction is achieved in the computer. The DH has many advantages: It can provide both the amplitude and phase information of a wave front simultaneously. The reconstruction does not require the wet processing, therefore, the DH can get all information of an object in real time. Using a high-speed recorder, the DH can also record the dynamic varieties of an object, thus it can provide four-dimensional imaging. At present, the DH has already used in many fields such as deformation analysis [3], microscopy [4], particle measurement [5], object contouring [6], and so on.

However, the size of the object to be imaged is limited by the CCD's size and resolution. In-line holography uses one wave which severs as both the illuminating light and reference light simultaneously, therefore, it can effectively utilize the spatial band product of the CCD. Due to no beam splitter or mirror is required; this technology is quite suitable for short wavelength light imaging such as X-ray imaging. The directly inverse Fresnel transform of the in-line hologram will present the real image, the zero order term, and the conjugation image together. Since the reference light and the object light are overlapped during the hologram is recorded, the real image and its conjugation image can not be separated. Many researchers tried to solve this problem. Lai et. al. use off-axis reconstruction of the in-line hologram to avoid ghost image [7], but this method is limited to small object due to losing low frequency components. Phase-shifting method [8] can obtain the phase and amplitude information of a wave front by processing three or more holograms, however, the corresponding optical setup is quite complex and the phase shifter for recording each hologram must be controlled strictly.

In this chapter, some reconstruction methods based on the iterative phase retrieval algorithm are introduced. The YG algorithm [9] and the GS algorithm [10, 11] are used to reconstruct pure absorption object or pure phase object from their in-line holograms, respectively. The differences between these two algorithms in in-line hologram reconstruction are analyzed. Then the GS algorithm is extended to reconstruct whole optical field form double or multiple holograms. At last, a new approach for reconstructing object from a hologram series is presented.

# 2. FORMULAS USED IN RECONSTRUCTION

The in-line digital hologram recording system is schematically shown in Fig. 1. The object is illuminated by a plane wave. A part of light passes the object without any changes; other part suffers an intensity or phase modulation. Assuming that the distribution function of the object is $a(x', y')$, the intensity distribution recorded at the CCD surface will be:

$$I(x, y) = |A(x, y)|^2 \quad , \tag{1}$$

with

$$A(x, y) = \frac{1}{\lambda d} \exp(i\frac{2\pi}{\lambda} d) \int a(x', y') \exp\{\frac{i\pi}{\lambda d}[(x - x')^2 + (y - y')^2]\} dx'dy' \quad , \tag{2}$$

where $\lambda$ is the wavelength of the incident light and $d$ is the distance between the object and the CCD surface. The reconstruction problem can be described as: Known the intensity distribution $I(x, y)$ of the hologram, how can one retrieval the distribution $a(x', y')$. Usually, there is no unique resolution for this problem. However, if some information about the object is known, for example, the object is a real or pure phase one; the reconstruction problem can be solved by using the phase retrieval algorithm.

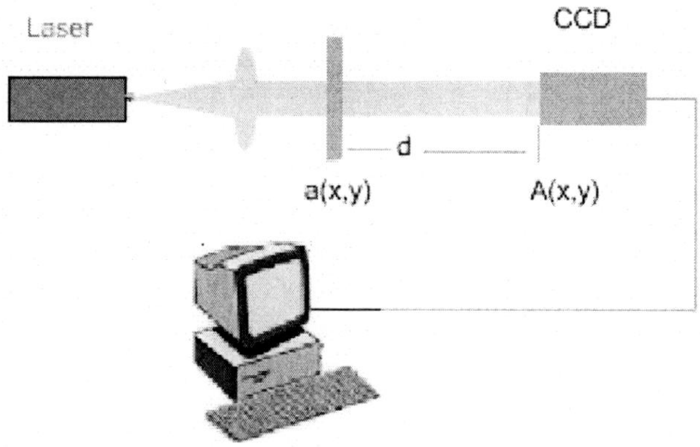


Figure 1. Digital hologram recording system.

Only one-dimensional case is considered here for the sake of brevity; all the formulas obtained here can be expanded to the two-dimensional case directly. In the matrix form, Eq. (2) can be rewritten as:

$$A(x) = \mathbf{G}\ a(x') \quad , \tag{3}$$

where **G** expresses the Fresnel integral operator. If the system is a diffractive loss or nonparaxial one, **G** will be a nonuniatry matrix and $\mathbf{G}^*\mathbf{G} = \mathbf{A} \neq \mathbf{I}$, where the superscript * represents the Hermitian conjugation and $\mathbf{I}$ is the identity transform. For reconstructing the object in a computer, the continuous wave functions must be sampled. Assume that the numbers of sampling pixels in the object and hologram planes are $N_1$ and $N_2$, respectively, then the distributions $a(x')$ and $A(x)$ can be expressed by vectors and the operator **G** corresponds to a matrix:

$$a_l = \rho_{1l}\exp(i\varphi_{1l}), \quad l = 1, \ 2, \ 3, \ ..., \ N_1, \tag{4a}$$

$$A_m = \rho_{2m}\exp(i\varphi_{2m}), \tag{4b}$$

$$A_m = \sum_{l=1}^{N_1} \mathbf{G}_{ml} a_l \quad m = 1, \ 2, \ 3, \ ..., \ N_2. \tag{4c}$$

where $\rho_1$, $\varphi_1$, $\rho_2$, and $\varphi_2$ are amplitude and phase distribution of the $a(x')$ and $A(x)$, respectively. In order to describe the closeness of $\mathbf{G}a(x')$ to $A(x)$, a distance measure is defined as

$$D(\rho_1, \varphi_1, \rho_2, \varphi_2) = \|A - \mathbf{G}a\| = \left[\sum_{m=1}^{N_2} |A_m - (\mathbf{G}a)_m|^2\right]^{1/2}. \tag{5}$$

If $D = 0$, then $A(x)$ will exactly equal to $\mathbf{G}a(x')$ Thus the reconstruction problem can be formulated as the search for the extreme of the function $D^2$ with respect to the function arguments $\rho_1$, $\varphi_1$, $\rho_2$ and $\varphi_2$, i.e.

$$\begin{aligned} &\Delta_{\rho_1} D^2 = 0, \quad \Delta_{\varphi_1} D^2 = 0, \\ &\Delta_{\rho_2} D^2 = 0, \quad \Delta_{\varphi_2} D^2 = 0, \end{aligned} \tag{6}$$

where $\Delta_\xi D^2$ represents the functional variation of $D^2$ with respect to the function argument $\xi$ [11].

It is easy to obtain the following equations by using matrix algebra

$$\rho_{1k} = \frac{1}{\mathbf{A}_{kk}} Abs\left[\sum_{j=1}^{N_2} \mathbf{G}^*_{jk} \rho_{2j} \exp(i\phi_{2j}) - \sum_{j \neq k} \mathbf{A}_{kj} \rho_{1j} \exp(i\phi_{1j})\right], \tag{7a}$$

$$\varphi_{1k} = Arg\left[\sum_{j=1}^{N_2} \mathbf{G}^*_{jk} \rho_{2j} \exp(i\phi_{2j}) - \sum_{j \neq k} \mathbf{A}_{kj} \rho_{1j} \exp(i\phi_{1j})\right], \tag{7b}$$

$$\rho_{2k} = Abs\left[\sum_{j=1}^{N_1} \mathbf{G}_{kj} \rho_{1j} \exp(i\varphi_{1j})\right], \tag{7c}$$

$$\varphi_{2k} = Arg\left[\sum_{j=1}^{N_1} \mathbf{G}_{kj}\rho_{1j}\exp(i\varphi_{1j})\right]. \tag{7d}$$

If **G** is a unitary operator, Eqs. (7) will reduce to those of the GS algorithm

$$\rho_1 = Abs(\mathbf{G}^* A), \tag{8a}$$

$$\varphi_1 = Arg(\mathbf{G}^* A), \tag{8b}$$

$$\rho_2 = Abs(\mathbf{G}a), \tag{8c}$$

$$\varphi_2 = Arg(\mathbf{G}a). \tag{8d}$$

Generally, Eqs. (7) and (8) have no analytical solution and should be solved numerically. The corresponding algorithms are called YG iteration algorithm and GS iteration algorithm, respectively. Eqs. (7a) and (7d) are used for the YG algorithm and Eqs. (8a) and (8d) are used for the GS algorithm to reconstruct pure absorption object. Eqs. (7b) and (7d) are used for the YG algorithm and Eqs. (8b) and (8d) are used for the GS algorithm to reconstruct pure phase object. The accuracy of the solution is evaluated by the sum-squared error (SSE), defined as:

$$SSE = \frac{\sum\left[\rho_2 - \left\|\mathbf{G}\rho_1^n \exp(i\varphi_1)\right\|\right]^2}{\sum \rho_2^2}, \tag{9}$$

with $\rho_1^n$ denotes the $n$th iteration for $\rho_1$.

## 3. Differences between the YG and GS Algorithms

In this section, some numerical simulations are shown to compare the YG algorithm and the GS algorithm for pure absorption object reconstruction. The model object is presented in Fig. 2. It is a pure absorption object. The parameters of the simulating system are selected as: The wavelength of the illuminating light is $\lambda = 0.532\mu m$, the sampling ranges in the object and hologram planes are $a_1 = a_2 = 6.7 \times 1024\mu m$, and the numbers of sampling pixels in the object and hologram planes are $N_1 = N_2 = 1024$. It is found that only for $d = 8.64 \times 10^4 \mu m$ the corresponding parameters satisfy the relation $N_1 \lambda d / a_1 a_2 = 1$, thus the operator $\mathbf{G}$ is unitary [12]. The operator $\mathbf{G}$ will become more nonunitary with distance $d$ increasing.

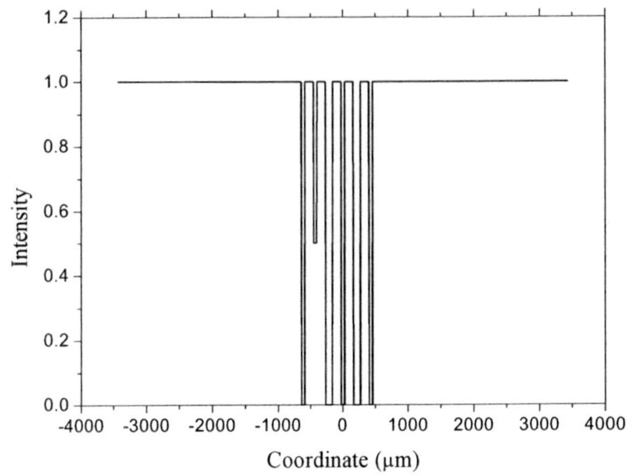


Figure 2. One dimensional object used in simulations.

In order to give a qualitative description, the ratio $B/C$ is used to measure the degree of unitarity of the transform operator $\mathbf{G}$. The quantity $C$ and $B$ is defined as the average value of the diagonal elements and of the off-diagonal elements of the matrix $\mathbf{A}$

$$C = \frac{1}{N_1}\sum_{i=1}^{N_1}\left|A_{ii}\right| \tag{10}$$

$$B = \frac{1}{N_1(N_1-1)}\sum_{i=1}^{N_1}\sum_{j\neq i}^{N_1}\left|A_{ij}\right| \tag{11}$$

If the transform is unitary, one has $B=0$ and $C=1$. When $\mathbf{G}$ is nonunitary, one has $B\neq 0$, the lager the ratio $B/C$, the stronger the nonunitary of the transform. The variation of $B/C$ with recording distance $d$ is shown in Fig. 3. It can be seen that only when the recording distance is near to $8.64\times10^4\,\mu m$, the ratio $B/C$ close to zero, thus the transfer is unitary. When $d$ is lager than $8.64\times10^4\,\mu m$, the ratio $B/C$ increases monotonically with $d$ expect for the existence of a few links.

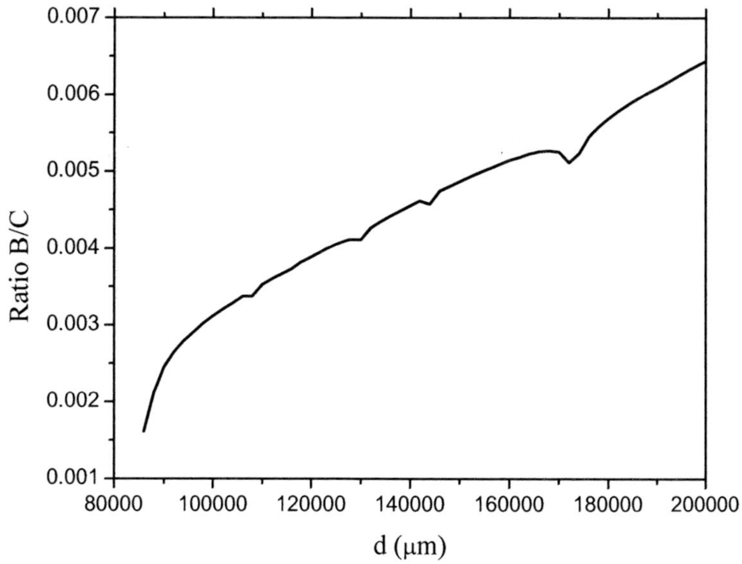


Figure 3. Variation of the ratio $B/C$ with recording distance $d$.

Figure 4 describes the relationship between the SSE and the recording distance $d$ for the YG algorithm and the GS algorithm, respectively. It can be found that the SSE for the GS algorithm increases quickly with $d$ increasing and the corresponding reconstruction quality becomes worse. However, the SSE for the YG algorithm can keep a small value until the distance $d$ is bigger than $1.7\times10^5\mu m$. When the distance is selected as $d=8.64\times10^4\mu m$, the corresponding value of $B/C$ is $9.48\times10^{-5}$ and the system is nearly unitary, both the GS and YG algorithms can provide good reconstruction results. The retrieved results obtained are shown in Fig. (5a) by using the GS algorithm and in Fig. (5b) by using the YG algorithm. It can be found that the original object has been recovered quite well expect for small oscillations. The corresponding SSEs are $1.12\times10^{-6}$ for the GS algorithm and $1.08\times10^{-6}$ for the YG algorithm. This result are as expected due to two algorithms are identical when the system is unitary. For $d=9.50\times10^4\mu m$, the corresponding $B/C$ is $2.85\times10^{-3}$ and the system is nonunitary. The reconstructed results using two algorithms are shown in Fig. 6. The result obtained by using the GS algorithm in Fig. (6a) has stronger oscillations than that obtained by using the YG algorithm in Fig. (6b). The corresponding SSEs are $1.28\times10^{-3}$ for the GS algorithm and $3.40\times10^{-4}$ for the YG algorithm after 200 iterations.

Experiments have also been carried out to test these two algorithms [13]. The setup shown in Fig. 1 is used to record the holograms. The working wavelength is again $\lambda=0.532\mu m$, and the used CCD has $1300\times1030$ pixels with pixel size $6.7\mu m\times6.7\mu m$. Only $1024\times1024$ pixels are used for reconstruction. The used object is a cross made with two hairs. The hologram recorded at $d=8.64\times10^4\mu m$ is shown in Fig. 7(a). Fig. 7(b) and 7(c) present the recovered images obtained by the GS algorithm and the

YG algorithm after 5 iterations, respectively. Although the illuminating light is not so uniform and the hologram has been blurred by the disturbed specks, both two algorithms can recover the original object well. The corresponding SSEs are $7.52\times10^{-4}$ for the GS algorithm and $7.51\times10^{-4}$ for the YG algorithm. The reconstructed results obtained in a nonunitary system are also presented for a clear comparison. Fig. 8(a) shows the recorded hologram and Fig. 8(b) and 8(c) present the reconstructed object obtained by the GS algorithm and the YG algorithm after 5 iterations, respectively. It can be seen that Fig. 8(b) has stronger oscillation in the background. The corresponding SSEs are $8.96\times10^{-3}$ and $6.21\times10^{-3}$ for the YG algorithm and GS algorithm, respectively.

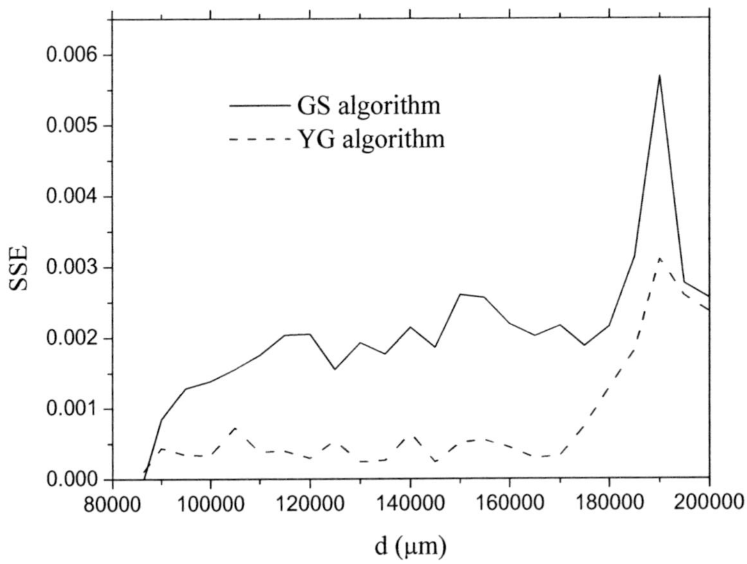


Figure 4. Relationship between the SSE and the distance $d$.

The reconstruction of a pure phase object has also been carried out experimentally [14]. The sample used is the logo of the *Institut für Technische Optik* shown in Fig. 9, it has a step height of $\sim 330\text{nm}$ at the location of underlying glass substrate, and the corresponding phase difference is $\sim 1.8\text{rad}$. The widths of the lines are $40 \sim 60\mu\text{m}$ depending on the difference parts in the logo. The distance between the object and CCD is $d = 8.64\times10^{4}\,\mu m$. Eqs. 8(b) and 8(d) are used to reconstruct the original object. Fig. 10 presents the reconstructed phase-contrast image obtained by the modified GS algorithm after 500 iterations. The corresponding SSE is 0.1. This phase contrast reveals the topography of the sample. The height distribution is proportional to the reconstructed phase distribution. However, the values of recovered phase are limited in the internal of $[-\pi,\pi]$, the height difference greater than $\dfrac{\lambda}{2(n-1)}$ ($n$ is the refractive index of the sample) will give rise to indeterminacy, but can be solved by used of phase-unwrapping methods.

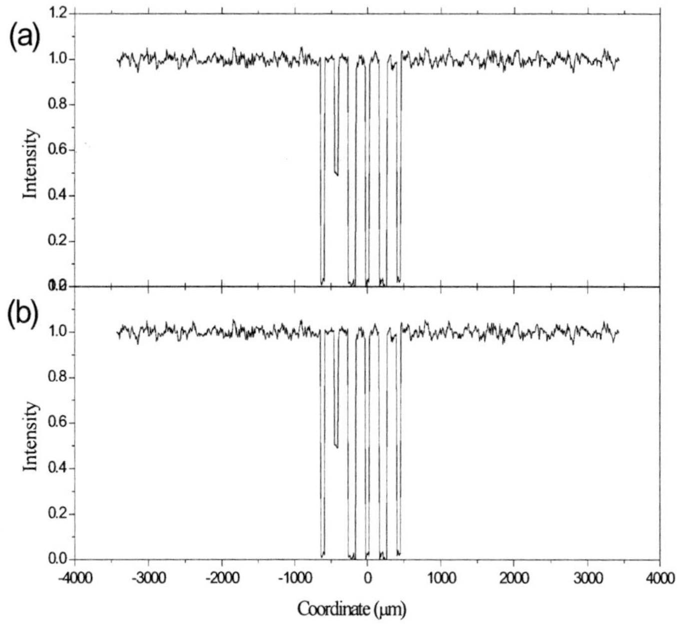


Figure 5. Reconstructed image obtained (a) by the GS algorithm and (b) by the YG algorithm after 200 iterations in an unitary system.

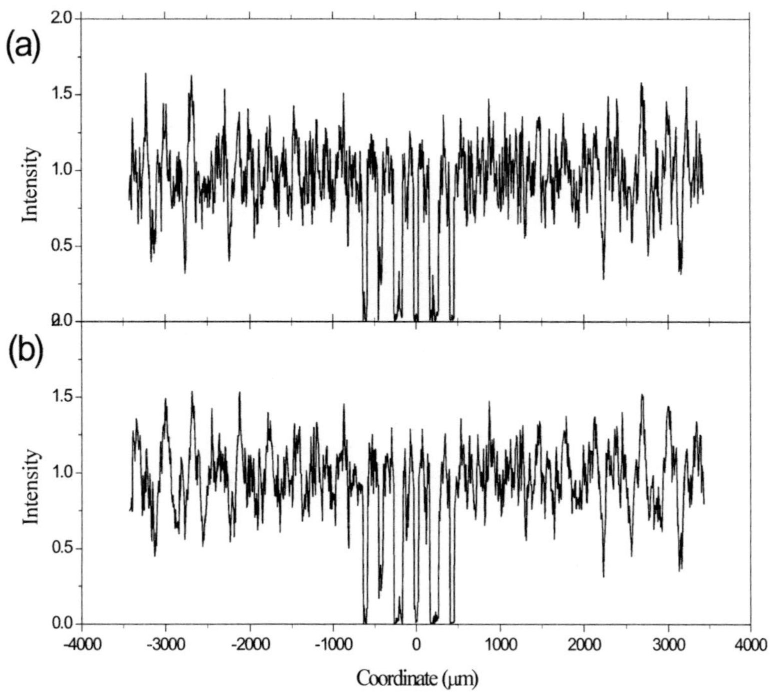


Figure 6. Reconstructed image obtained (a) by the GS algorithm and (b) by the YG algorithm after 200 iterations in a nonunitary system.

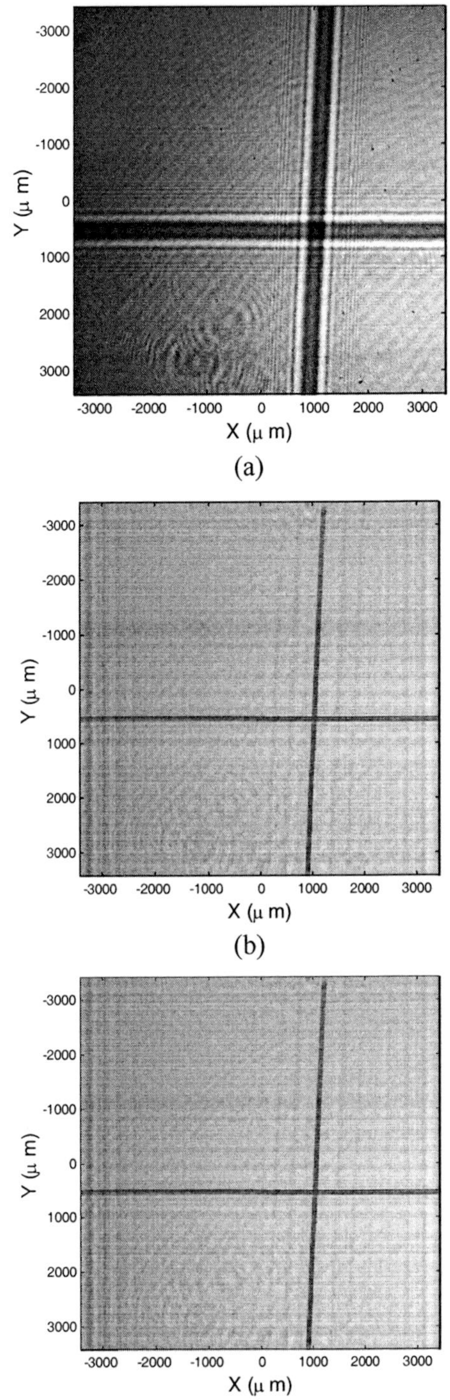


Figure 7. The reconstruction of real digital hologram recorded at $d = 8.64 \times 10^4\ \mu m$. (a) Digitalized hologram, (b) result obtained by the GS algorithm after 5 iterations, and (c) result obtained by the YG algorithm after 5 iterations.

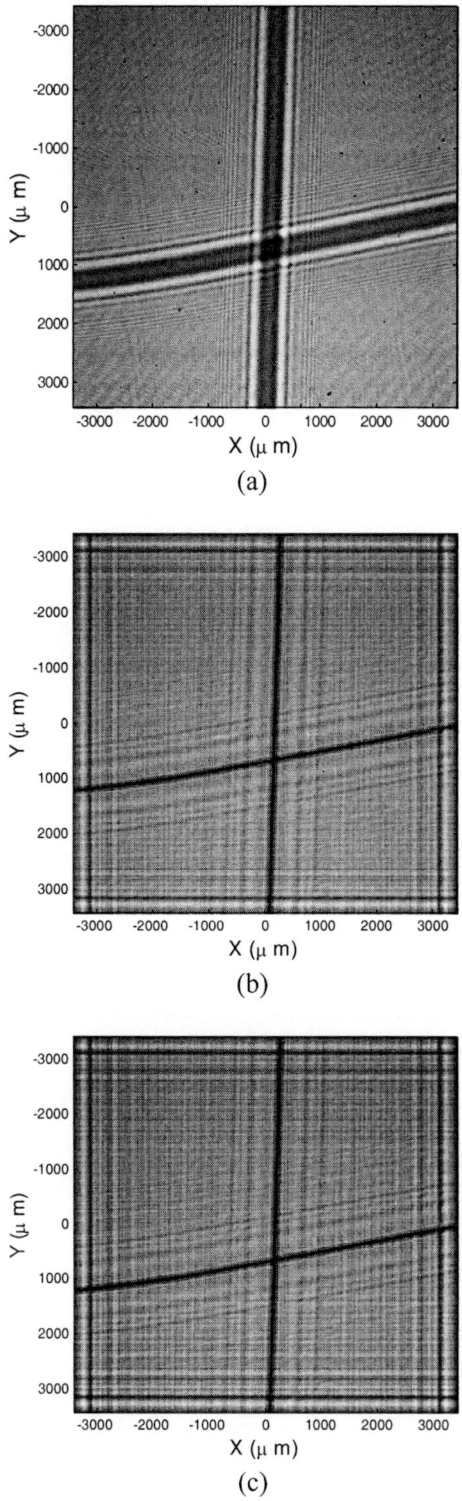


Figure 8. Same as Fig. 6 expect for $d = 9.50 \times 10^4 \, \mu m$.

Figure 9. Pure phase object used in experiment.

Figure 10. Reconstructed pure phase object.

## 4. Object Reconstruction from Double or Multi Holograms

The object can also be reconstructed without know any prior information [15]. As shown in Fig. 11, two holograms $H_0$ and $H_1$ are recorded, then this two holograms are used to recover the phase distribution on the $H_1$ plane. Thus the recovered phase information and known amplitude distribution can be used to reconstruct original object. Furthermore, the whole wave field between the object plane and the CCD plane can also be obtained by using known wavefield. The object used in the experiment is also the logo of the *Institut für Technische Optik,* as shown in Fig. 9. Parts of three recorded holograms at distance $d = 7.40cm,\ 8.15cm,\ 8.90cm$ are shown in Fig. 12. The holograms (a) and (b) are used to retrieval the phase information. After 500 iterations, the SSE arrives 0.2956. The recovered amplitude and phase information at $d = 8.90cm$ , i.e. the object plane, are shown in Fig. 13.

The variations of the amplitude and phase of the wave field with the propagation distance can also been achieved by reconstructing the complex wave field at difference locations. This character can be used to analyze the propagation of light in the phase-distorting media. However, it can be found from Fig. 13 that the recovered amplitude distribution at $d = 8.90cm$ is not an ideal plane wave, and the high frequency components have been filtered out. This is caused by the limitation of the size and resolution of the used CCD. Furthermore, the noise in holograms and errors in distance measurement also bring mistakes in the reconstruction. This algorithm is sensitive to the value of the initial phase, but this drawback can be overcome by using multi holograms reconstruction algorithm.

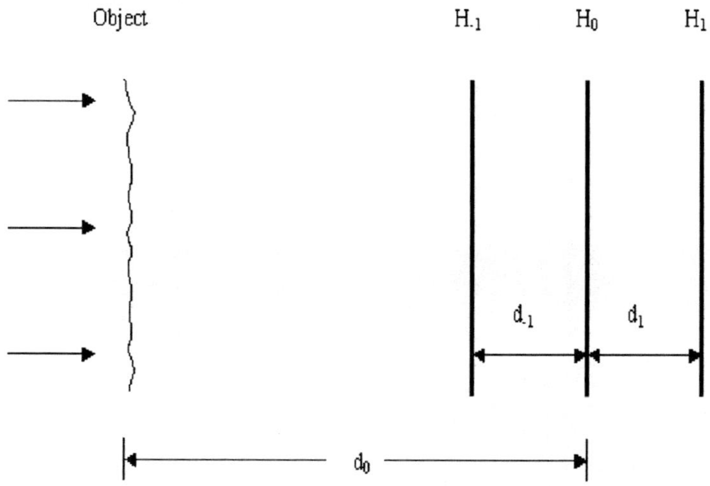


Figure 11. Schematic diagram of the iterative method for whole wave field reconstruction.

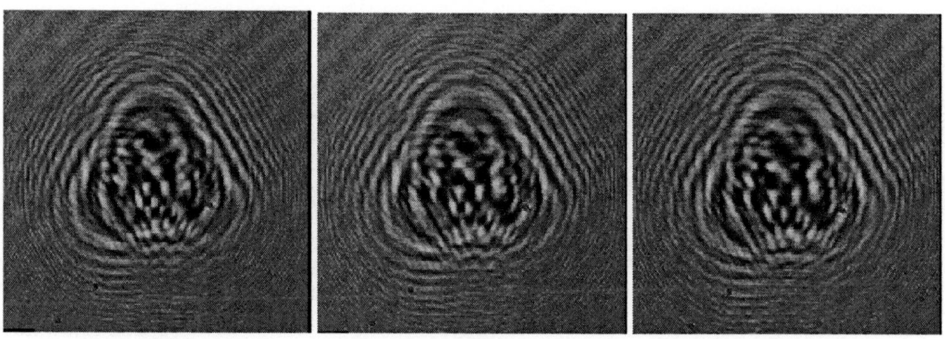

(a) (b) (c)

Figure 12. Holograms recorded at different distance. (a) $d = 7.40cm$ (b) $d = 8.15cm$ and (c) $d = 8.90cm$.

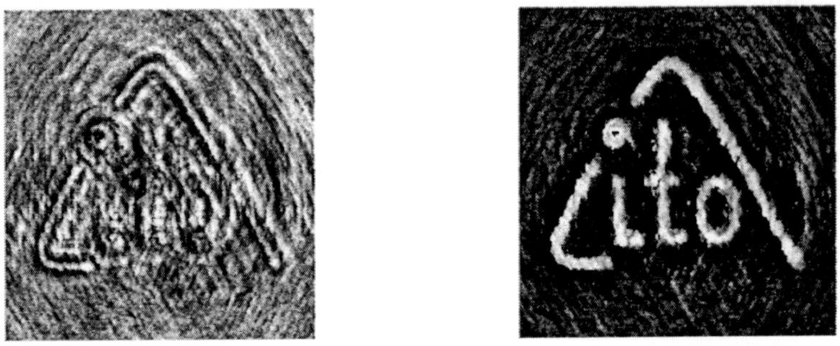

(a) (b)

Figure 13. (a) Intensity and (b) phase distributions of the reconstructed wave field at $d = 8.90cm$, obtained by using double holograms.

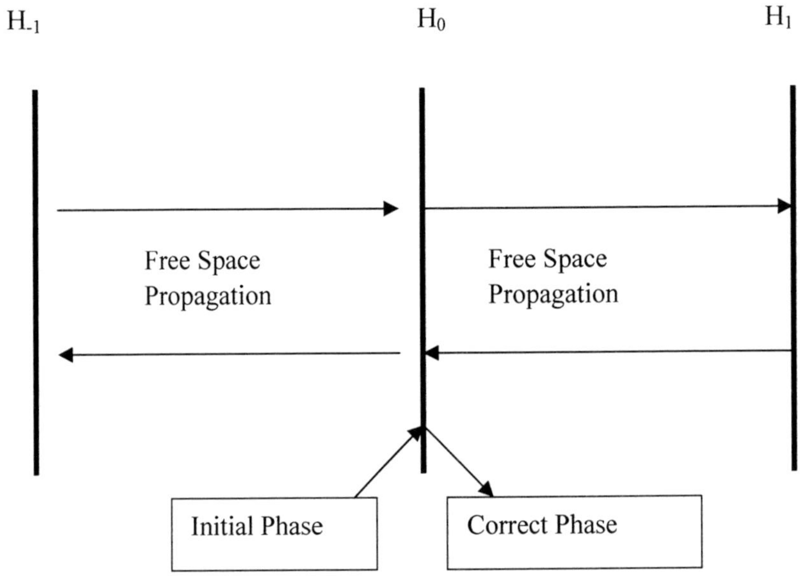


Figure 14. Schematic diagram of the iterative method of phase retrieval by using three holograms.

Since the algorithm based on two holograms is under constrained and susceptible to uniqueness concerns, more information are required to recover the phase information, thus more constraints are available and a deterministic solution is expected. As an example, three holograms are used to reconstruct the original object. As shown in Fig. 14, the iterative process begins from the central ($H_0$) plane, the amplitude obtained form $H_0$ and an arbitrary initial phase are employed to construct a wave function. This wave function propagates to the plane $H_1$, the phase distribution is kept unchanged and the amplitude is replaced by the amplitude information obtained from $H_1$, the new wave function propagates back to the plane $H_0$. Again, the amplitude is replaced by the amplitude information calculated from $H_0$ and the phase information is leaved unchanged. The new composed wave field continually propagates back to the plane $H_{-1}$, the amplitude is again corrected to the value obtained form $H_{-1}$. Then the corrected wave propagates to the central plane. The SSE is checked in every cycle and the iteration will be stopped if the SSE is sufficient small or the SSE no longer decreases.

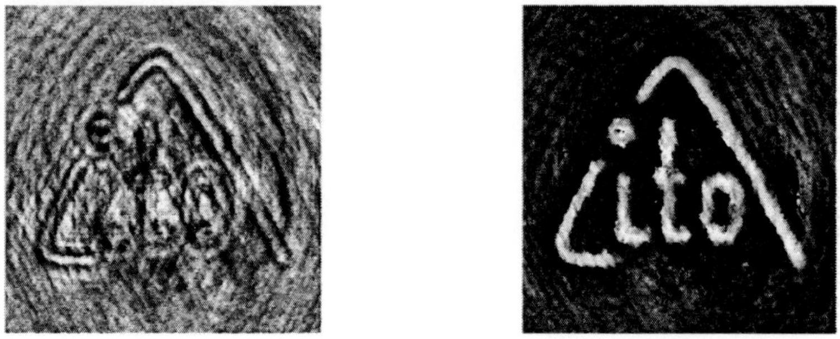

(a) (b)

Figure 15. (a) Intensity and (b) phase distributions of the reconstructed wave field at $d = 8.90cm$, obtained by using three holograms.

A demonstration reconstruction is preformed by using three holograms shown in Fig. 12. After 75 iterations, the SSE arrives at its minimal value of 15.96. Using the recovered phase distribution and the intensity distribution on the plane of $H_1$, the complex wave field at $d' = 8.90cm$ is reconstructed and presented in Fig. 15. Although three holograms are used, the SSE has not been reduced and reconstruction quality has not been improved comparing with the case by using double holograms. The noise in holograms and errors in distance measurements could be the main reasons. These errors or noise will be transferred into the recovered phase distribution in the iterative processes, thus the quality of the whole optical field reconstructed based on this phase distribution will be effected.

## 5. OBJECT RECONSTRUCTION FROM A HOLOGRAM SERIES

The original object can also be reconstructed from a series of holograms [16]. As shown in Fig.16, a series of holograms $H_n (n = 1,2,3,...,N)$ are recorded. The distance between the first hologram and the object is $d$ and the interval between two adjacent holograms is fixed as $\Delta d$. A propagation of the hologram $H_1$ is performed from $d$ to $d + \Delta d$ for obtaining complex amplitude distribution of $A_2$, the phase distribution of $A_2$ is kept unchanged and the square root of the intensity distribution of hologram $H_2$ is used as the amplitude distribution for composing a new complex wave field $A'_2$. Then this new composed distribution $A'_2$ is performed a propagation from $d + \Delta d$ to $d + 2\Delta d$. The same procedure is repeated until all holograms are processed. Thus the wave field $A_n$ can be obtained with high accuracy, and the original object can be reconstructed using the inverse propagation with a distance of $d + n\Delta d$. This approach is very similar to the algorithms for retrieving the phase information from two intensity distributions [10-14], in which the phase information is revised again and again by using the correct intensity distributions. This approach uses multiple holograms instead of two intensity distribution, and more information can be used to retrieve the phase information.

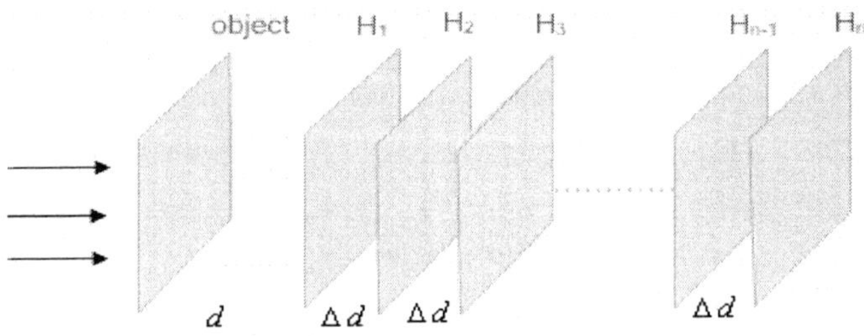


Figure 13. Schematic diagram for reconstruction from a hologram series.

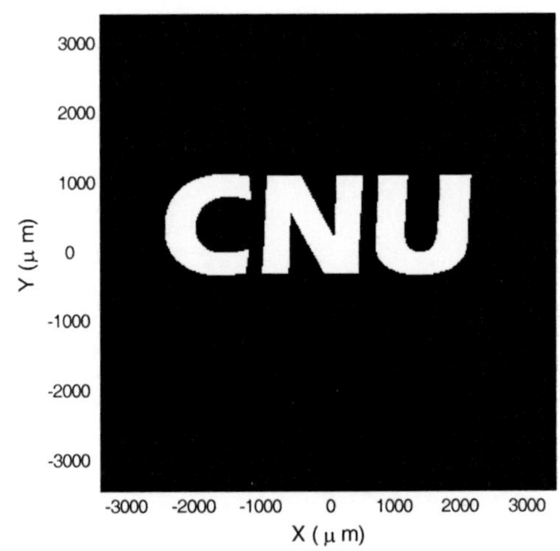


Figure 14. Original image used for simulation.

The object used in numerical simulations is shown in Fig. 14, its size is 6.8 x 6.8 mm$^2$. The parameters for the simulation are selected as follows: The wavelength of the illuminating light is also $\lambda = 0.532\mu m$; the area of the CCD window is $6.7 \times 1024\mu m \times 6.7 \times 1024\mu m$, the number of the pixels is $1024 \times 1024$, the distance from the object to the first hologram plane is $80\,mm$ and $\Delta d$ =$100\,mm$. The relationship between values of $c$ and $n$ is shown in Fig. 15. It can be found that the value of $c$ decreases with $n$ increasing till $n = 27$, and then it increases with $n$. This phenomenon can be understood as follows: When the number of used holograms increase, more and more information is used to retrieve the phase of the wave field; therefore, the recovered wave fields will be more and more close to the real value. However, if the CCD is located too far from the object, it can not collect enough information due to the size limitation of the CCD, and thus the quality of the reconstructed wave filed will not be improved even more holograms are used.

In our simulations, we used 27 holograms to reconstruct the original object; the reconstructed image is shown in Fig. 16. It can be see that the recovered object is nearly equal to the original one; all the details have been retrieved. For comparison purposes, the reconstructed image obtained by using the inverse Fresnel transform of the hologram recorded at distance $(80+27\times100)\ mm$ from the object is shown in Fig. 17, it is hard to recognize the original object due to the appearance of the ghost image. For a complex object, more holograms are required to reconstruct high quality image.

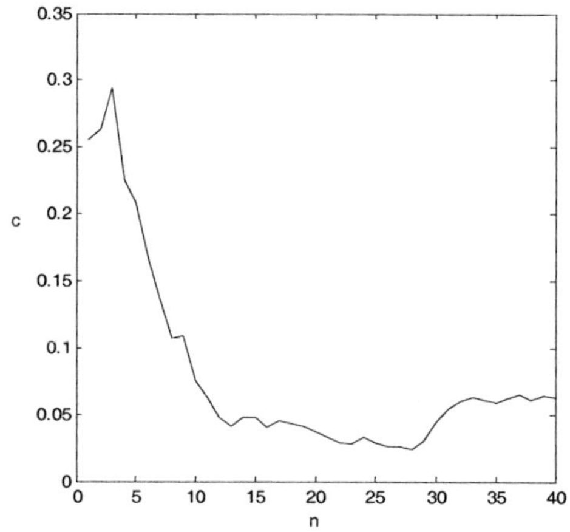


Figure 15. Relationship between the values $c$ and n.

Figure 16. Image reconstructed by using our method.

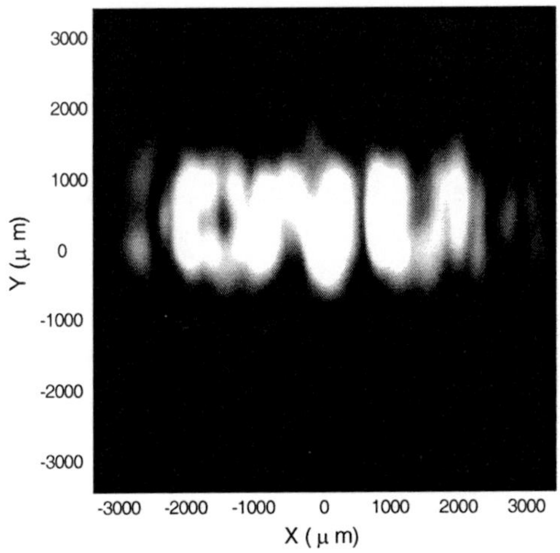


Figure 17. Image reconstructed by using the inverse Fresnel transform directly.

## 6. Conclusion

Several approaches based on iterative phase retrieval method are described in the presentation to reconstruct the full wave front form single, double, multi, and a series of in-line hologram(s). By processing these in-line holograms, the phase information in a hologram plane can be retrieved, thus the complete wave field can be achieved, and the original object can be reconstructed using the inverse propagation. Since that no reference light is required in these approach, the corresponding system is relatively simple. The signal-noise-ratio can be relatively improved. Both the computer simulations and experiment results demonstrate the validly of these approaches, the images reconstructed form experimentally recorded holograms exhibit good agreements with the corresponding original objects.

## Acknowledgments

The author would like to express his acknowledgements to Prof. H. J. Tiziani, Prof. W. Osten, and Dr. G. Pedrini for fruitful discussions and valuable suggestions.

## References

[1] D. Gabor, "A new microscopic principle," *Nature* (London) **161** 777-778, 1948.

[2] J. W. Goodman and R. W. Lawrence, "Digital image formation form electronically detected holograms," *Appl. Phys. Lett.*, **11** 77-79, 1967.

[3] S. Schedin, G. Pedrini, H. Tiziani, A. K. Aggrarawl, and M. E. Gusev, "Highly sensitive pulse digital holography for built-in defect analysis with laser excitation," *Appl. Opt.*, **40** 100-107, 2001.

[4] U. Takaki and H. Ohzu, "Fast numerical reconstruction technique for high-resolutions hybrid holographic microscopy," *Appl. Opt.*, **38** 2204-2211, 1999.

[5] S. Murata, N. Yasuda, "Potential of digital holography in particle measurement," *Opics & Laser Techlonogy*, **32** pp. 567-574, 2000.

[6] C. Wagner, W. Osten, S. Seebacher, "Directly shape measurement by digital wavefront reconstruction and multiwavelength contouring," *Opt. Eng.*, **39** pp. 79-85, 2000.

[7] S. Lai, B. Kemper, G. V. Belly, "Off-axis reconstructions of in-line holograms for twin-image elimination," *Opt. Commun.* **169** 37-43, 1999.

[8] T. Zhang, I. Yamaguchi, "Three-dimensional microscopy with phase-shifting digital holography," *Opt. Lett.* **23** 1221-1223, 1998.

[9] R. W. Gerchberg and W. O. Saxton, "Practical algorithm for the determination of phase from image and diffraction plane pictures," *Optik* (Stuttgart), **35** 237-246, 1972.

[10] J. R. Fienup, "Phase retrieval algorithms: a comparison," *Appl. Opt.*, **21** 2758–2768, 1982.

[11] G. Z. Yang, B. Z. Dong, B. Y. Gu, J. Zhuang, and O. K. Ersoy, "Gerchberg–Saxton and Yang–Gu algorithms for phase retrieval in a nonunitary transform system: a comparison," *Appl. Opt.*, **33** 209-218, 1994.

[12] Y. Zhang, B. Z. Dong, B. Y. Gu, and G. Z. Yang, "Beam shaping in the fractional Fourier domain," *J. Opt. Soc. Am. A* 15, 1114-1120 (1998).

[13] Y. Zhang, G. Pedrini, W. Osten, and H. J. Tiziani, "Image reconstruction for in-line holography using the Yang-Gu algorithm," *Appl. Opt.* **42** 6452-6457, 2003.

[14] Yan Zhang, Giancarlo Pedrini, Wolfgang Osten, Hans J. Tiziani, "Phase retrieval microcopy for quantitative phase-contrast imaging", *Optik*, **115** 94-97 (2004).

[15] Y. Zhang, G. Pedrini, W. Osten, and H. J. Tiziani, "Whole optical wave field reconstruction from double or multi in-line holograms by phase retrieval algorithm," *Opt. Express*, **11** 3234-3241, 2003.

[16] De-Xiang Zheng, Yan Zhang, Jing-Ling Shen, Cun-Lin Zhang, "Wave field reconstruction from a hologram sequence", *Optics Commun.*, **249** 73-77. (2005).

In: New Developments in Lasers and Electrc-Optics Research ISBN 1-59454-771-8
Editor: William T. Arkin, pp. 173-222 

***Chapter 6***

# SPECKLE TECHNIQUES FOR OPTICAL METROLOGY

***Nandigana Krishna Mohan***
Applied Optics Laboratory, Department of Physics
Indian Institute of Technology Madras
Chennai - 600 036, INDIA

## ABSTRACT

The self-interference of a large number of random coherent waves scattered from a diffuse object surface or propagated through a medium of random refractive index fluctuations results in a granular structure known as speckle pattern. This chapter illustrates a concise development on various speckle techniques with latest developments and refinements in terms of optical systems, fast and reliable image processing systems and modern fringe analysis methods for optical metrology. Particular emphasis is provided for the measurement of mechanical behavior of MEMS structures and non-destructive evaluation.

## INTRODUCTION

The speckle phenomenon came into prominence after the advent of laser. When an optically rough surface is illuminated with a coherent beam, a high contrast granular structure, known as *speckle pattern* is formed in space. Speckle pattern is a 3-D interference pattern formed by the interference of secondary, de-phased wavelets scattered by an optical rough surface or transmitted through a scattered medium that imposes random phases of the wave. It is assumed that the scattering regions are statistically independent and uniformly distributed between $-\pi$ and $\pi$. The pioneering contributions of Leendertz[1] and Archbold[2] in 1970 demonstrating that the speckles in the pattern undergo both positional and intensity changes when the object is deformed, triggered the origin of speckle work. Since then, the developments in speckle based techniques has gone from strength to strength backed by spates of activities in different research laboratories and industries around the world. The randomly coded pattern that carries the information about the object deformation provided to develop a wide range of techniques, which can be classified into three broad categories:

speckle photography, speckle interferometry and speckle shear interferometry. Speckle photography includes all those techniques where positional changes of the speckle are monitored, whereas speckle interferometry includes methods that are based on the measurement of phase changes and hence intensity changes. If instead of phase change, we measure its gradient, the technique falls into the category of speckle shear interferometry. All these techniques can be performed using digital/electronic detection using a CCD and imaging processing system. The attraction to speckle techniques lies their applicability for optical metrology; (i) static and dynamic deformation and its gradient measurements, (ii) the shape of the three-dimensional objects, (iii) surface roughness and (iv) non-destructive evaluation (NDE) of engineering and biomedical specimens. The range of the objects that can be evaluated using the speckle techniques is of the order of few hundred microns such as Micro Electro Mechanical Systems (MEMS) to few meters such as space craft engineering structures. The prominent advantages of the speckle techniques with the real-time operation systems (photorefractive crystals, CCD and image processing systems) are (i) non-contact measurement, (ii) real-time whole field operation, (iii) data storage and retrieval for analysis, (iv) quantitative evaluation using phase shifting, (v) variable measuring sensitivity and also (vi) possible to reach remote areas.

The goal of this chapter is to present a concise development on various speckle techniques starting from speckle photography with conventional recording and filtering techniques and proceed further with the digital methods (*digital speckle photography*) and real-time evaluation methods (*photorefractive speckle photography*). The chapter also covers the topics on speckle interferometry, speckle shear interferometry (*shearography*) and Electronic/Digital speckle pattern interferometry (ESPI/DSPI) with latest developments and refinements in terms of optical systems, fast and reliable image processing systems, modern fringe analysis methods. Particular emphasis is given in this chapter for the measurement of mechanical behavior of MEMS structures, applications in medical diagnostics and non-destructive testing using these techniques. The chapter also touches the current advancement in speckle techniques for optical metrology with latest references in this field.

## 1. Speckle Photography

The speckle pattern formed in space is referred as objective speckle pattern. For most of the metrology applications, the subjective speckle pattern, wherein object is imaged by a lens to establish one to one correspondence between the points on the object. The technique involving the recording of objective/subjective(imaging geometry) speckle patterns is referred as speckle photography and it is one of the simplest technique to set-up[2-4] in which a laser-lit diffuse surface is recorded at two different states of stress. The speckle photography is exploited for metrology by using a either (i) photographic plate, (ii) CCD detector, or (iii) photorefractive material as a recording medium. The first two procedures need post-processing, while the third one can be performed in real-time.

## 1.1. Speckle Photography using a Photographic Plate

An imaging geometry arrangement for recording speckle pattern on a photographic plate is shown in Fig.1(a). An expanded laser beam illuminates the object that scatters light, and two exposures are made on a single photographic plate placed at the image plane, and the object is given an in-plane displacement between the exposures. The recorded and developed photographic plate consists of speckled image but mutually translated due to displacement. These two traditionally registered speckle patterns are identical, except that one is displaced with respect to the other by an amount which depends on the magnitude of the displacement of the object and imaging geometry. A magnified speckle pattern is demonstrated in Fig.1(b). The recorded speckle patterns after development can be analyzed by means of two classical filtering techniques. The first method, a point-by-point approach, consists of interrogating the speckle pattern by means of a narrow laser beam to generate a diffraction halo in the far field at each data point. The halo is modulated by a system of Young's fringes. The fringe spacing and orientation are governed by the displacement vector. The fringe orientation is always perpendicular to the direction of displacement, while the fringe spacing is inversely proportional to the magnitude of displacement. The Young's fringes are observed when the two identical speckles are separated by a distance equal to or more than the average speckle size. The second method based on spatial filtering or Fourier filtering of the speckle pattern provides a direct whole field visualization of the distribution of interest. This approach also offers a degree of latitude in sensitivity selection during the reconstruction process. In short point-wise filtering yields absolute value of displacement, while the Fourier filtering gives fringes representing incremental change[5-8]. The advantages of speckle photography such as its relatively lower sensitivity to external vibrations and a reasonably low initial cost of developing a speckle photography based system advocate strongly in favour of a widespread use of the method in non laboratory environments. On the other hand, some important drawbacks that have prevented the widespread use of speckle photography reside in its inherent requirement of two step process: exposing two speckle patterns pertaining to initial and displaced states of the object, respectively, on a photographic plate, and, adopting the filtering methods to extract the information. The disadvantage related with the photographic plate has been addressed by the use of a CCD detector for recording the speckle pattern and quantitative analysis by means of digital processing (*digital speckle photography*).

## 1.2. Speckle Photography with Video/Digital Recording

Digital speckle photography (DSP)[9-16] also called as speckle correlation (SC), electronic speckle photography (ESP) or digital image correlation (DIC) is the digital/electronic version of speckle photography for the evaluation of displacement fields. The relative low complexity in the hardware for a basic DSP system where the photographic plate is replaced with a CCD camera is shown in Fig.2. The registration of the speckle pattern is made by video/digital recording of the object and the evaluation process is performed on a PC. The analysis is much faster since no film processing or reconstruction (filtering) is needed. The method is based on a cross-correlation analysis of two speckle patterns one stored before and another after the object is deformed. As an example, the experimental analysis for in-plane rotation of a specimen is shown in Fig.3. The technique is recently exploited for measuring the shear magnitude between two sheared images in a shearography arrangement, and the results are shown in Fig.4 [15]. The correlation analysis can be seen in Fig.4(c), where the arrows represent how much and in which direction the images are shifted. With the developments provided by the computer industry and with the increased resolution and stability of modern CCD detectors, DSP systems are getting more and more attractive for material testing. An overview of the state-of-the-art in digital speckle photography is given by Mikael Sjödahl in Ref.17.

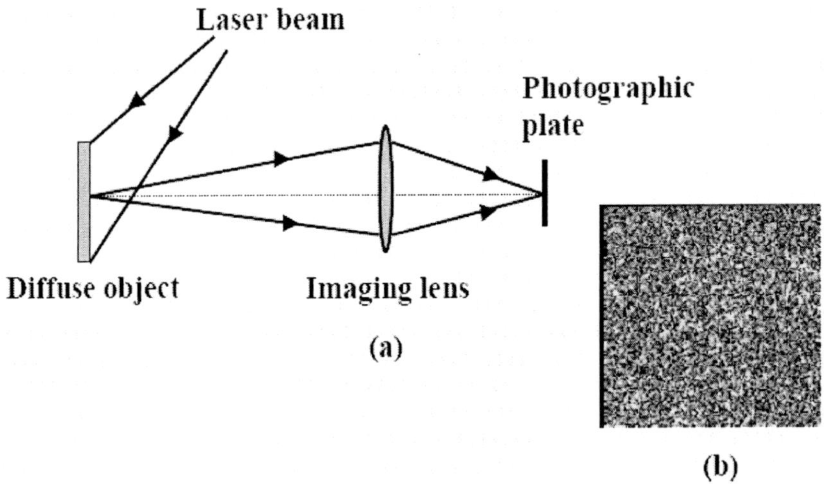


Figure 1. (a) Schematic for recording a speckle pattern on a photographic plate using a imaging system, (b) Enlarged speckle pattern at the recording plane.

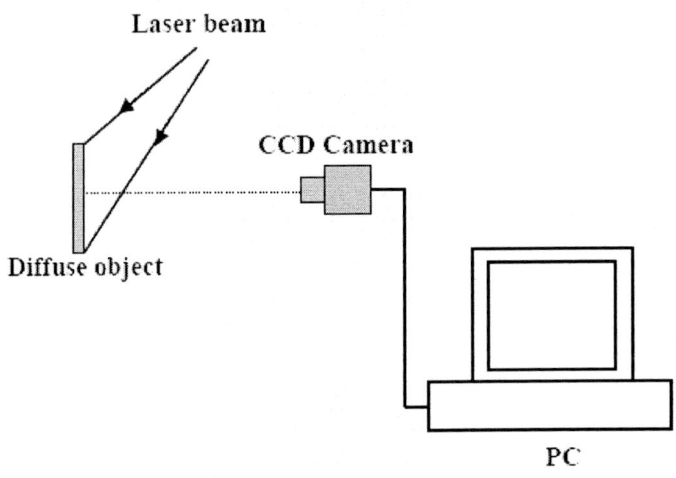


Figure 2. Schematic of a digital speckle photography (DSP) system.

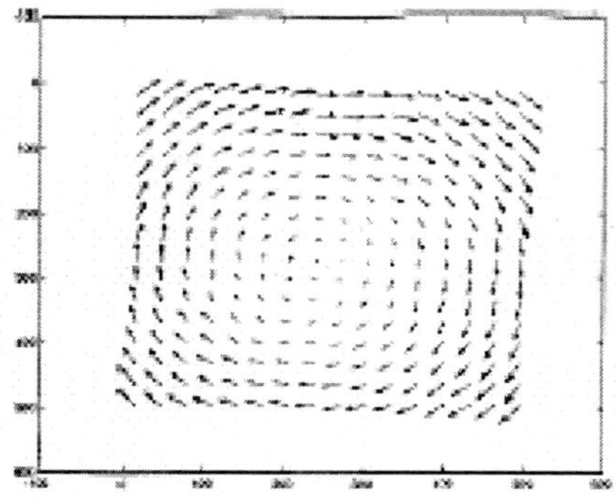

Figure 3. An arrow plot representing in-plane rotation of an object.

Figure 4. DSP technique to determine the magnitude and direction of the shear in a shearography arrangement. (a) and (b) are the speckle images obtained when one or the other of the two arms in the Michelson interferometer is blocked and (c) represents the speckle correlation result. In the present analysis, an average horizontal shear of 4.89 mm.

## 1.3. Speckle Photography with Photorefractive Materials

The emergence of photorefractive materials with their unique properties of higher resolution, low intensity operation and the capability of real-time response, has been shown to hold promise in several photonic applications such as optical data storage, information processing, coherent image amplification, optical logic operations, dynamic holography, self-pumped phase conjugation, neural networks and speckle metrology [18]. Speckle photography first demonstrated for real-time displacement, tilt and vibration analysis with a BSO crystal by Tiziani et.al.[19]. An Argon Ion laser is used to create the dynamic grating of the speckle pattern inside the crystal and a low power He-Ne laser reads the grating to obtain the fringe pattern. Liu et al. [20] proposed a two-beam coupling novelty filtering technique in a slow response crystal such as $BaTiO_3$. In this real-time filtering technique, a shifted speckle pattern is compared with the un-shifted original speckle pattern and the difference between these two is shown as a speckle correlation fringe pattern. With two-beam coupling, the crystal can be used as a novelty filter in two configurations, i.e., when the phase shift between the dynamic grating and the interference pattern is $\pm\pi/2$. The optical configuration with $+\pi/2$

gives a bright speckle background due to the amplification of the object beam, whereas -π/2, yields a dark speckle background due to de-amplification of the object beam[21-23]. However the optical characteristics are different for the two configurations.

The schematic optical arrangement using a $BaTiO_3$ crystal as recording medium for real-time visualization of speckle correlation fringes is shown in Fig.5. The object is illuminated with a narrow laser beam and the scattered speckle pattern generated from the object is imaged onto a from the image is imaged onto a $BaTiO_3$ crystal (5x5x5 $mm^3$) by the lens L. An unexpanded plane wave (pump beam) making an angle with respect to the object beam is added to the crystal. The speckle patterns are observed and recorded by using a ground glass screen and a CCD camera.

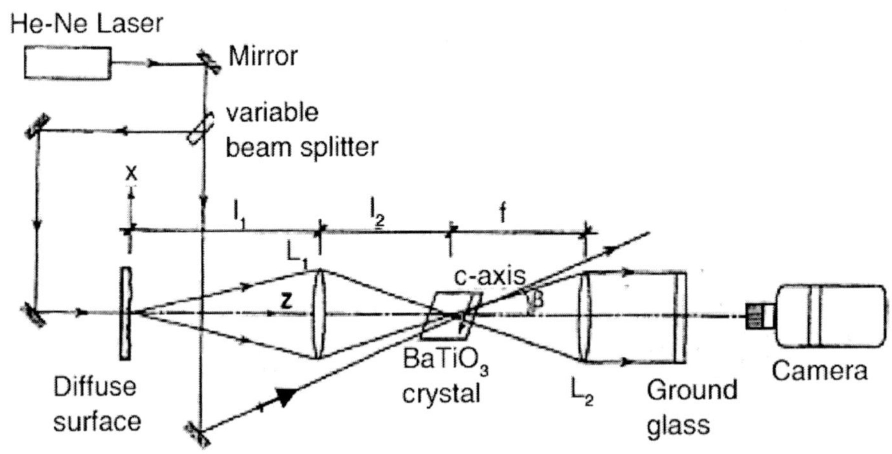


Figure 5. Schematic of a point beam illumination configuration for real-time visualization of speckle correlation fringes.

The basic concept here is that the two beams interfere inside the crystal and produce a dynamic grating. At the observation plane, we have two speckle fields:

i) *the first one constitutes the initial state of the object, that is, due to the diffraction of the pump beam from the grating formed at the crystal plane*

and

ii) *the second scattered beam that represents the directly transmitted object beam which constitutes the deformed state of the object.*

The overlap of these two identical but displaced speckles give rise to the *speckle correlation fringe pattern* in real-time. The resultant Young's fringes recorded directly from the monitor around the notch at two different locations are shown in Fig.6. It should be noted that the optical configuration is re-aligned for each location. However, illuminating the object with number of unexpanded beams allows one to perform deformation studies in real-time without having to re-align the experimental arrangement.

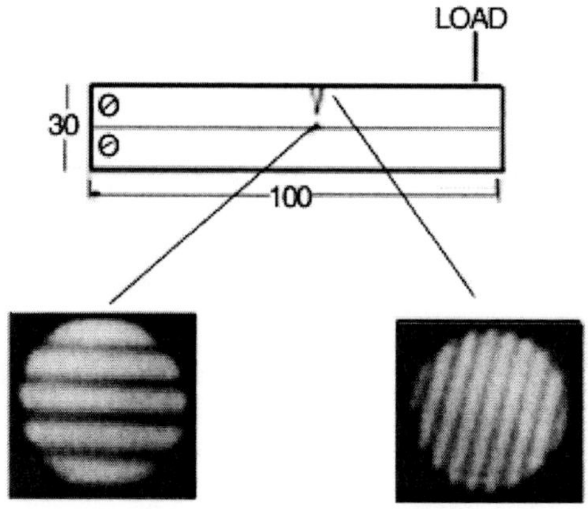


Figure 6. Young's fringes recorded at two different locations from a composite material.

If β is the angle between the zero order and n[th] order fringe with in the halo at the observation plane(GG); the angular separation of the fringe pattern satisfies surface displacement, d, as[8]

$$d_\xi = \frac{n\lambda}{M \sin\beta} \tag{1}$$

where M is the magnification of the imaging system.

## 1.4. Phase Shifting Whole Field Speckle Photography

The phase shifting concept [24] in speckle photography for analysis of the whole field speckle fringe pattern obtained by filtering of dual-plate specklegrams was first put forward by Kaufmann and Jacquot [25]. The method consists of combining the far-field diffraction patterns generated by two single-exposure specklegrams on a photographic plate in a Mach-Zehnder interferometer. A read-out system by combining phase shifting and two-beam coupled photo-refractive tracking novelty filtering techniques is shown in Fig.7. The system integrates a Fourier novelty filtering system[26]. To accomplish the phase shifts in the system, a pre-calibrated air-filled quartz cell connected to a pressure pump and a precision pressure gauge is inserted in the pump beam[27-28]. Phase shift in the system can be implemented by varying the pressure inside the cell, which introduces an additional constant phase variation in the pump beam. The phase shift modifies the position of the bright and dark fringes within the fringe system.

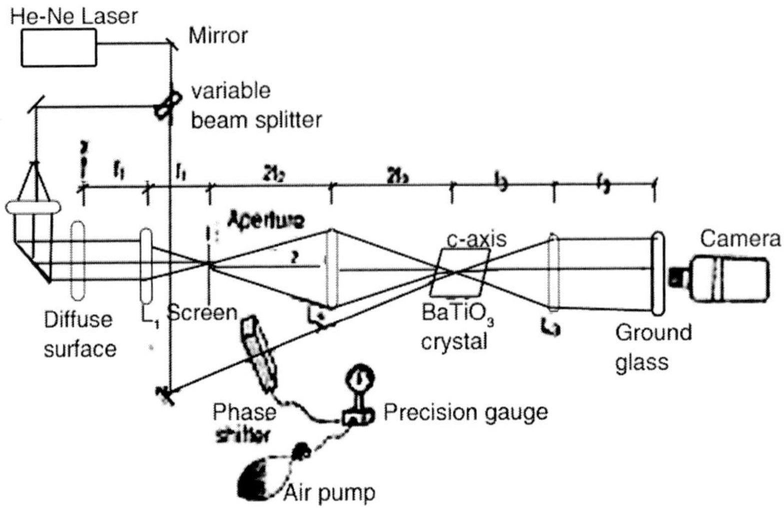


Figure 7. Schematic of a phase shifting whole-field speckle photography setup for measurement of object deformation in real time.

The intensity distribution at a point for each fringe pattern recorded at the observation plane can be expressed as[26]

$$I_N = 2I_0\{1 + \cos(\phi(x, y) + \varphi_N)\} \quad (2)$$

where $I_o$ is the intensity of each speckle field at an instant of time, $\phi(x,y)$ is the phase change due to the object load, $\varphi_N$ is the phase shift value $(0,\pi/2,\pi,3\pi/2)$ and $N = 1,2,3,4$. The phase change $\phi(x,y)$ is responsible for positional shift of the speckles at the observation plane.

The phase at each point can be calculated with the following four-frame algorithm[24]

$$\phi(x, y) = \tan^{-1}\left\{\frac{I_4 - I_2}{I_1 - I_3}\right\} \quad (3)$$

If $\xi$ is the angle between the optical axis and the line joining the center of the transform lens $L_1$ and the off-set aperture location, the fringe pattern satisfies the relation

$$d_\xi = \frac{m\lambda}{\sin\xi} \quad (4)$$

where $d_\xi$ is the component of the in-plane displacement along the aperture direction, m is the fringe order and $\lambda$ is the wavelength of the laser light.

From the evaluated phase $\phi(x,y)$, the surface displacement component $d_\xi$ can be obtained as

$$d_\xi = \frac{\phi(x,y)\lambda}{2\pi \sin\xi} \tag{5}$$

Fig.8 represents the real time speckle fringe pattern, raw phase map and the evaluated three dimensional plot of the in-plane displacement component.

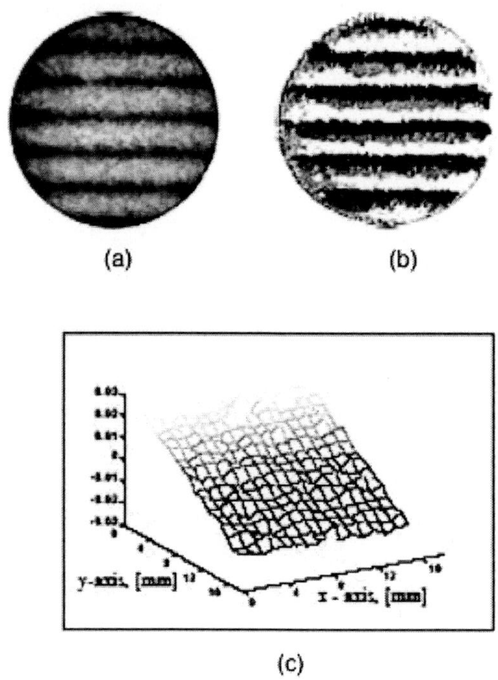


Figure 8. (a) whole-field fringes obtained from a diffuse surface subjected to in-plane rotation in its own plane. The fringe pattern displays the in-plane displacement component along the x-direction. (b) corresponding raw phase map. (c) 3-D plot of the in-plane displacement obtained from the phase map after the removal of the $2\pi$ phase ambiguity.

Rastogi [29] introduced the concept of speckle shear photography for direct measurement of surface strains. It is based on the property that the diffracted field emanating from a coherently lit object point and passing through two different aperture portions on its way to the photographic plate gives rise to two distinct speckle patterns in the image plane when each pattern is viewed independently. Two independent Young's fringes corresponding to displacements at the interrogation points are formed and they are sheared by an amount equal to the distance between the two illuminating beams. The two patterns, therefore, superpose with moderate magnitude of shear and generate a moiré pattern as shown in Fig.9, from which, one can able determine the shear strain components[30].

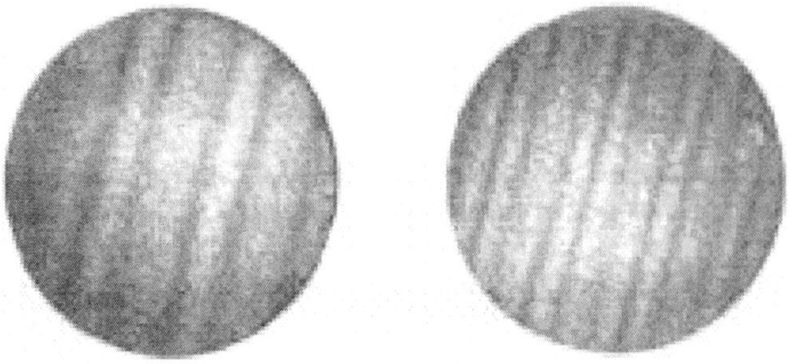

Figure 9. Real-time speckle shear photography with $BaTiO_3$ crystal:Young's moiré patterns representing in-plane strain component at different loadings.

The approach for automating fringe readout whole-field fringes obtained in speckle photography will spur a renaissance of interest in this unique and interesting family of techniques in optical metrology[31].

## 2. Speckle Interferometry

Speckle interferometry is powerful measurement technique involving coherent addition (interference) of the scattered light from the object surface with a reference beam that may be a specular or scattered field not necessarily originating from the object[1,5-8]. Unlike speckle photography, this technique has a sensitivity comparable to that of holographic interferometry. The phase changes within one speckle due to object displacement coded by the reference beam. This displacement is extracted by correlating two speckle patterns one taken before and another after object displacement. The intensity distribution at a point in the observation plane before the object is deformed can be expressed as

$$I_1(x, y) = I_o + I_r + 2\sqrt{I_o I_r} \cos\phi \tag{6}$$

where $\phi$ is the random phase, $I_o$ and $I_r$ are the intensities of object and reference beams at a coordinate point (x,y) respectively.

The intensity distribution after the object is deformed can be written as

$$I_2(x, y) = I_o + I_r + 2\sqrt{I_o I_r} \cos(\phi + \Delta\phi) \tag{7}$$

where $\Delta\phi$ is the phase change introduced due to the deformation of the object. These two intensity distributions are recorded in a photographic plate that is converted to amplitude transmittance upon processing. For $\Delta\phi = 2n\pi$, (n is an integer), the two patterns are fully correlated and for $\Delta\phi = (2n+1)\pi$, they are uncorrelated. Therefore for $\Delta\phi = 2n\pi$, bright fringes are formed. The fringes are termed as correlation fringes; correlation fringes are obtained by the addition of two or more speckle patterns. However, we often use subtraction correlation fringes in Electronic/Digital speckle pattern interferometry.

The phase change because of object deformation can be expressed as

$$\Delta\phi = (\vec{k}_2 - \vec{k}_1).\vec{L} \tag{8}$$

$$\vec{L} = u\hat{i} + v\hat{j} + w\hat{k} \tag{9}$$

where $\vec{k}_1$ and $\vec{k}_2$ are the propagation vectors in the directions of illumination and observation and $(\hat{i}, \hat{j}, \hat{k})$ are the unit vectors, u,v and w are the displacement components along x,y and z directions respectively. Since the fringes are loci of constant phase difference, the deformation vector components can be measured using appropriate optical configurations.

A speckle correlation interferometers for measuring in-plane and out-of-plane displacement components of a deformation vector was first report by Leendertz [1]. For in-plane displacement, two collimated beams symmetric with respect to the surface normal of a diffuse object are used for illumination, and the scattered fields generated are observed along the optical axis of the imaging lens as shown in Fig.10. The inter beam angle between the two illuminating beams determines the sensitivity of the configuration as given in the following relation

$$\Delta\phi_L = (\vec{k}_2 - \vec{k}_1).\vec{L} = \frac{4\pi}{\lambda} u \sin\theta \tag{10}$$

The equation for in-plane displacement, u is

$$u = \frac{n\lambda}{2\sin\theta} \tag{11}$$

Recently a modification of the Leendertz method that leads to twofold increase in the sensitivity has been proposed by Sirohi and Krishna Mohan[32]. This is accomplished by observing the scattered fields independently along the direction of the illumination beams, re-combined at the observation plane. In the interferometer, for each illumination beam there are two scattered fields, that is, the back scattered field along the same direction of illumination and also the scattered field in the specular direction. Therefore, four beams interact at the observation plane. Hence, six phase terms are responsible for fringe formation and the resultant interferogram obtained from this configuration contains the information pertaining to in-plane displacement with sensitivity proposed by Leendertz, as well as a twofold increase in sensitivity. For achieving twofold sensitivity, the object is coated with retro-reflective paint to increase the light scattered back in the direction of the illumination beams, and also to suppress the contributions of the scattered light in the specular directions[33]. Krishna Mohan [34] recently suggested a phase reversal technique to completely eliminate the influence of cross interference terms in a twofold sensitivity arrangement.

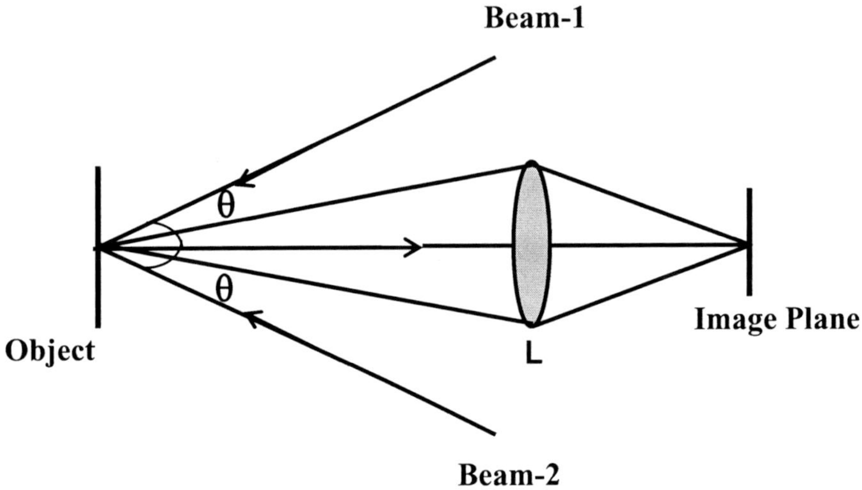


Figure 10. Geometry of Leendertz dual beam illumination arrangement.

**Table 1. Optical configurations in speckle interferometry for displacement and shape measurement**

| Fig. No. | Optical Arrangement | Fringe Formation Equations: In terms of Propagation vectors | Fringe Formation Equations: In terms of Physical parameters |
|---|---|---|---|
| A | 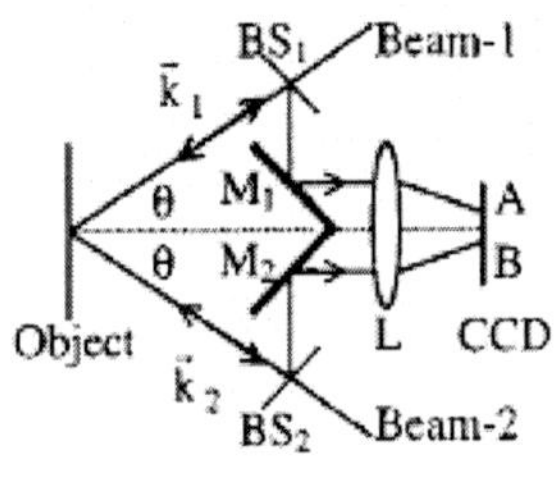  | $\Delta\phi_A = (\bar{k}_2 - \bar{k}_1).\bar{L}$ <br> $\Delta\phi_B = (\bar{k}_2 - \bar{k}_1).\bar{L}$ | In-plane displacement <br> $\Delta\phi_A = \frac{4\pi}{\lambda} u \sin\theta$ <br> $\Delta\phi_B = \frac{4\pi}{\lambda} u \sin\theta$ <br> Shape <br> $\Delta\phi_A = \frac{4\pi}{\lambda} z \sin\Delta\xi \sin\theta$ <br> $\Delta\phi_B = \frac{4\pi}{\lambda} z \sin\Delta\xi \sin\theta$ |
| B | 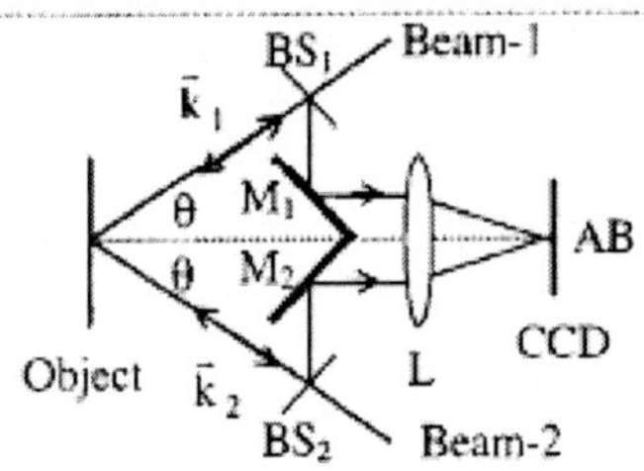  | $\Delta\phi_{AB} = 2(\bar{k}_2 - \bar{k}_1).\bar{L}$ | In-plane displacement <br> $\Delta\phi_{AB} = \frac{8\pi}{\lambda} u \sin\theta$ |
| C | 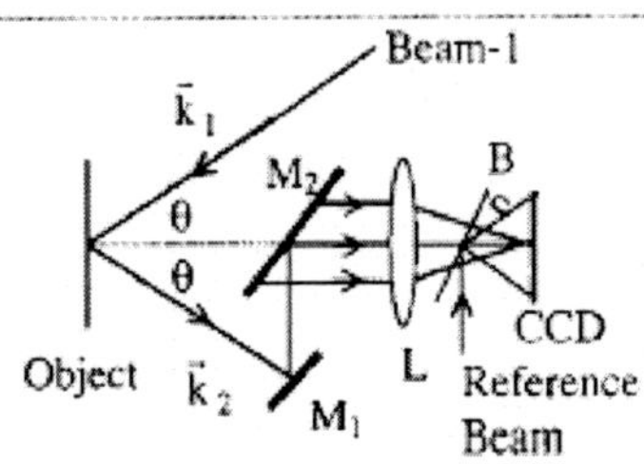  | $\Delta\phi = (\bar{k}_2 - \bar{k}_1).\bar{L}$ | Out-of-plane displacement <br> $\Delta\phi = \frac{4\pi}{\lambda} w \cos\theta$ |

The in-plane sensitive configuration can also be applied for shape measurement. The concept of contouring is based on the fact that for small angles of rotation (tilt), points at the same depth along the viewing direction undergo equal in-plane displacements, and this displacement is directly proportional to the depth of those points[35-36]. The optical configurations for sensing displacement components and shape along with the measuring sensitivities are highlighted in Table.1. The optical configuration in Fig. C senses the out-of-plane displacement component (w) with reduced measuring sensitivity. A comparative speckle interferometric configuration in which two objects (master and test objects) are compared with each other for non destructive testing is also reported in literature [37-40]. All these interferometers have been extensively used with ESPI for various applications of optical metrology.

## 2.1. Duffy's Two-Aperture Technique

A method for measuring in-plane displacement has been described by Duffy[41-42]. The technique consists of imaging the object surface through a lens fitted with a double aperture located symmetrically with respect to the optical axis. The imaging system is illustrated in Fig.11. The lens focused on the surface sees an arbitrary point on the object surface along two different directions. As a result of the coherent superposition of the beams emerging from the two apertures, each speckle is modulated by a grid structure running perpendicular to the line joining the apertures. The pitch of the fringe system embedded in each speckle depends on the separation of the two aperture. During filtering, the grid structure diffract light in well-defined directions. Because there is considerably more light available in these halos, there is an enormous gain in the signal-to-noise ratio, resulting in almost unit contrast fringes [43].

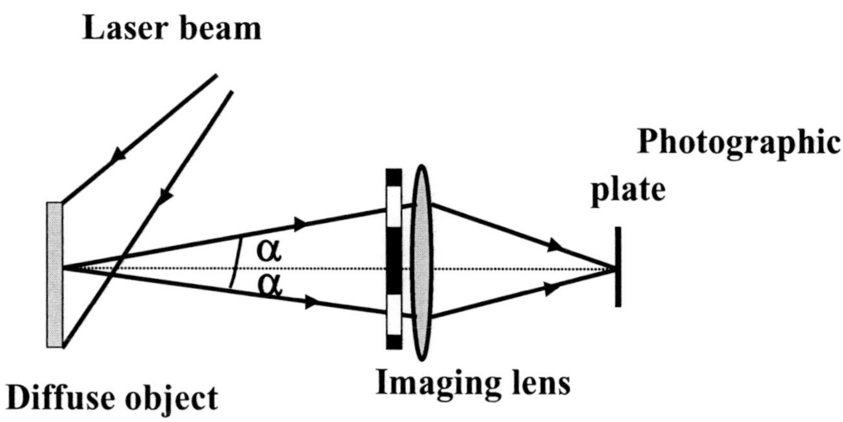


Figure 11. Geometry of Duffy's double aperture arrangement.

The in-plane displacement (u) can be expressed in terms of optical geometry as

$$u = \frac{n\lambda}{2\sin\alpha} \qquad [12]$$

The Eq.(12) is similar to Eq.(11) except that $\theta$ is replaced by $\alpha$. The angle $\alpha$ is governed by the aperture of the lens, and hence the sensitivity of Duffy's arrangement is inherently poor. Chiang et.al.[43] extended this technique by using four apertures, two along x-axis and two along the y-axis. Tai[44] extended and shown that the sensitivity to in-plane displacement can be increased. An optical arrangement with two apertures in front of the imaging lens to achieve measuring sensitivity equal to Leendertz's dual beam illumination configuration is illustrated in Fig. A in Table.2. The object is illuminated normally and the scattered fields are collected symmetrically at an angle $\theta$ with respect to the optical axis, and recombined at the image plane using a two-aperture imaging system[45]. Fig.12(a) and Fig.12(b) show the high contrast u and v family displacement fringes of a cantilever beam and contour fringes of a bulb, respectively using this arrangement. Two-aperture arrangement with other configurations are also examined for in-plane displacement, and contouring and they are highlighted in Table.2 [46]. Novelty of the multi-aperture techniques is in multiplexing using theta modulation or frequency modulation[47].

**Table2. Two-aperture optical configurations in speckle interferometry for displacement and shape measurement**

| Fig. No. | Optical configurations | Fringe formation equations: In terms of propagation vectors | Fringe formation equations: In terms of physical parameters |
|---|---|---|---|
| A. | 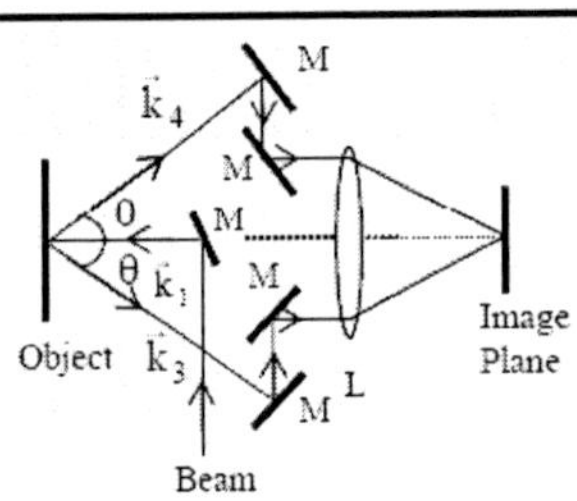  | $\Delta\phi_{KM} = (\vec{k}_4 - \vec{k}_3).\vec{L}$ | In-plane displacement<br>$\Delta\phi_{KM} = \frac{4\pi}{\lambda} u \sin\theta$<br>Shape<br>$\Delta\phi_{KM} = \frac{4\pi}{\lambda} z \sin\Delta\xi \sin\theta$ |
| B. | 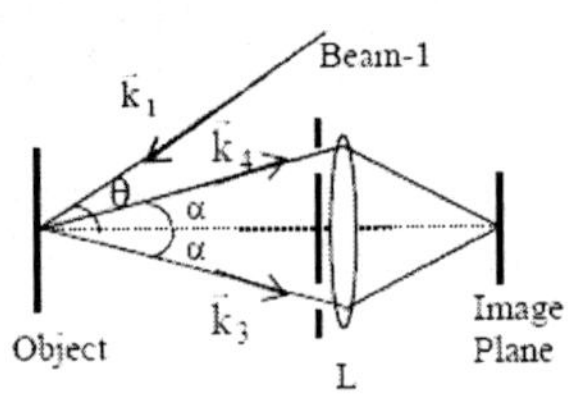  | $\Delta\phi_D = (\vec{k}_4 - \vec{k}_3).\vec{L}$ | In-plane displacement<br>$\Delta\phi_D = \frac{4\pi}{\lambda} u \sin\alpha$<br>Shape<br>$\Delta\phi_D = \frac{4\pi}{\lambda} z \sin\Delta\xi \sin\alpha$ |
| C | 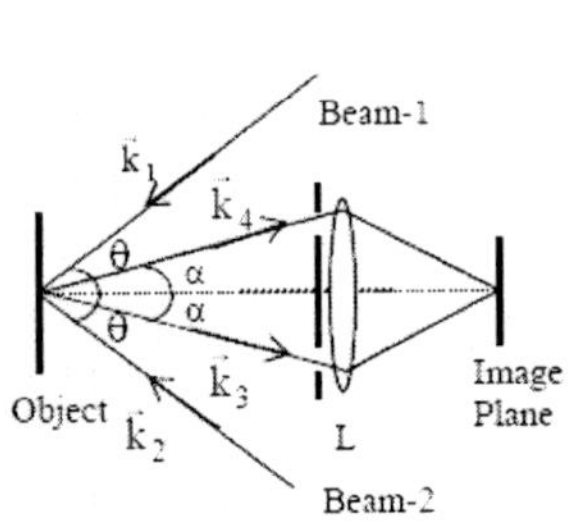  | $\Delta\phi_{LD} = \begin{matrix} (\vec{k}_2 - \vec{k}_1).\vec{L} \\ (\vec{k}_4 - \vec{k}_3).\vec{L} \end{matrix}$ | In-plane displacement<br>$\Delta\phi_{LD} = \begin{matrix} \frac{4\pi}{\lambda} u \sin\theta \\ \frac{4\pi}{\lambda} u \sin\alpha \end{matrix}$<br>Shape<br>$\Delta\phi_{LD} = \begin{matrix} \frac{4\pi}{\lambda} z \sin\Delta\xi \sin\theta \\ \frac{4\pi}{\lambda} z \sin\Delta\xi \sin\alpha \end{matrix}$ |

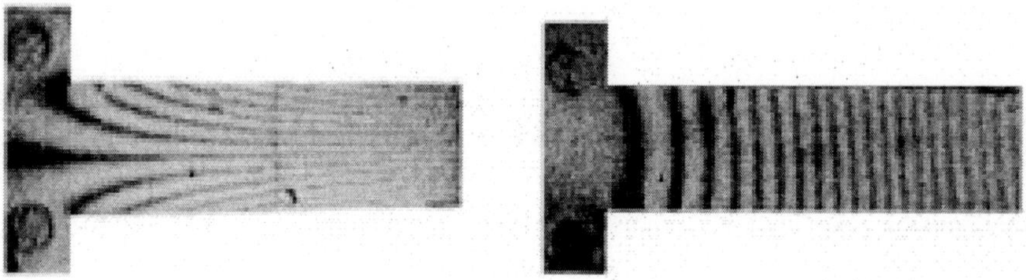

(a)

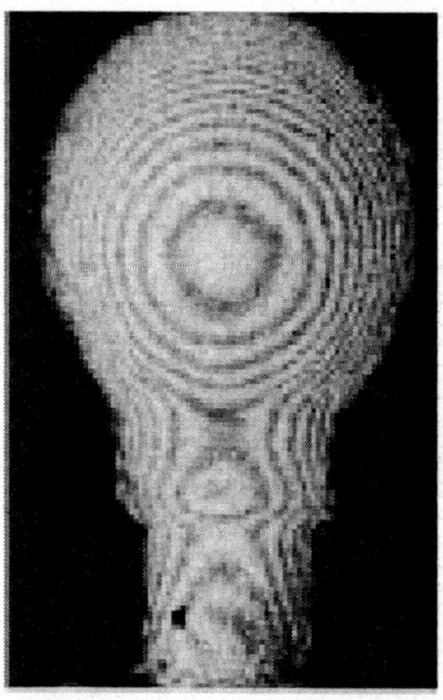

(b)

Figure 12. (a) u and v in-plane displacement fringes,(b) Contour fringes of a light bulb and the contour interval is 0.564 mm. The results obtained using the configuration illustrated in Fig. A of Table.2

## 3. SPECKLE SHEAR INTERFEROMETRY

The potential speckle techniques for the measurement of the derivatives of surface displacement was first put forward independently by Leendertz and Butters [48] and by Hung and Taylor [49]. Speckle shear interferometry also referred as *shearography*, a point in the image plane receives contributions from two or more points on the object. Alternatively, a point on the object may be imaged as two or more points in the image plane. A speckle shear interferometer consist of an imaging element and one or two shearing elements for obtaining the displacement derivatives (strain, slope and curvature components). In speckle interferometry, the object surface is compared with a reference beam from a mirror or a scattered beam, while in shearography, an object point is compared with a nearly point, but sheared by known amount. Leendertz [48] introduced a Michelson shear interferometer and the shear is introduced by tilting one of the mirrors in the interferometer. This configuration is widely exploited with ESPI for optical metrology. Hung set-up relies on Duffy's [41-42] two-aperture arrangement, where a shear element is introduced in front of any one of the apertures (image plane shear) [49].

The phase changes of the wave front from two laterally sheared points can be written as

$$\Delta\phi_1 = (\vec{k}_2 - \vec{k}_1).\vec{L}(x, y)$$

and

$$\Delta\phi_2 = (\vec{k}_2 - \vec{k}_1).\vec{L}'(x + \Delta x, y) \qquad (13)$$

where $\vec{L}(x, y)$ and $\vec{L}(x + \Delta x, y)$ are the deformation vectors at the two object points $(x, y)$ and $(x + \Delta x, y)$ respectively. The phase difference $\Delta\phi$ in terms of physical parameters can be written as

$$\Delta x = \frac{2\pi}{\lambda}\left[\frac{\partial u}{\partial x}\sin\theta + \frac{\partial w}{\partial x}(1 + \cos\theta)\right]\Delta x \qquad (14)$$

where the illumination beam lies in the xz plane and makes an angle $\theta$ with the z axis.

The bright fringes are formed wherever

$$\frac{\partial u}{\partial x}\sin\theta + \frac{\partial w}{\partial x}(1 + \cos\theta) = \frac{n\lambda}{\Delta x} \qquad [15]$$

For normal illumination $\theta$ =0, the Eq.(15) reduces to

$$\frac{\partial w}{\partial x} = \frac{n\lambda}{2\Delta x} \qquad [16]$$

The fringe pattern represents the partial x-slope. The fringes representing partial y slope are obtained by introducing shear along the y-direction[50]. A three aperture shear interferometer placed in front of an imaging system illustrated in Fig.13 yields the first and second-order derivatives of out-of-plane displacements simultaneously( Fig.14) [51-53]. The multi-aperture shear interferometers are intrinsically sensitive to in-plane displacement, the slope fringes will have (i) in-plane contributions, (ii) its derivatives and (iii) the derivative of the out-of-plane displacement component (slope). Therefore, one should implement this technique with extreme caution [54-58]. A modified two aperture speckle shear interferometer to eliminate the contributions of the in-plane displacement component and its derivatives to the phase change and yields only a fringe pattern depict to the first order partial derivatives of the out-of-plane displacement is proposed by Krishna Mohan [59] and shown in Fig.15. In the optical arrangement, two laterally sheared object points are viewed axially and the scattered fields from these points are independently combined at the image plane of the two-aperture imaging system. In addition to the lateral shear configurations, the classical shear interferometers such as radial, rotational, inversion and folding shear are also implemented for wide range of applications[50].

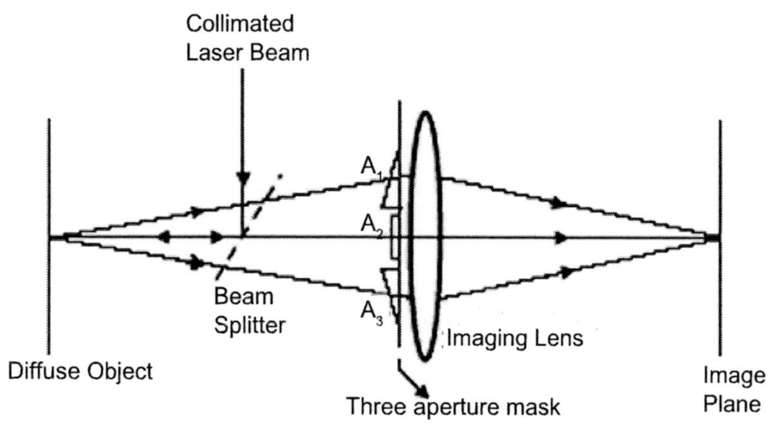


Figure 13. Schematic of a three-aperture speckle shear interferometer for measuring slope and curvature.

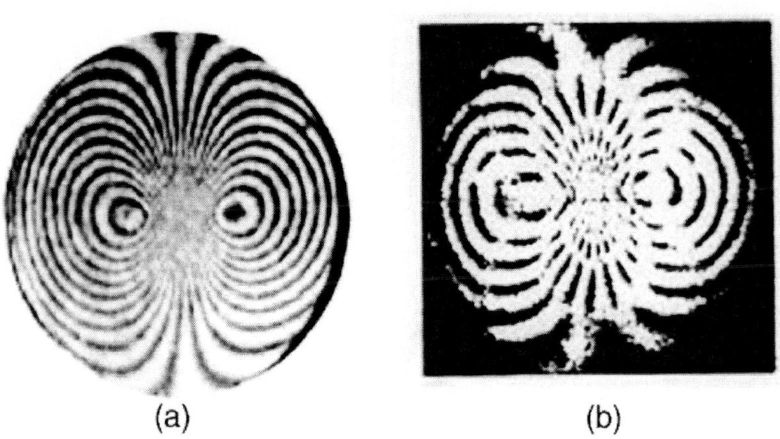


Figure 14. (a) Slope fringes ($\partial w / \partial x$) and (b) Curvature fringes ($\partial^2 x / \partial x^2$) for a centrally loaded diaphragm. The curvature fringes appearing as a moiré between the two slope patterns.

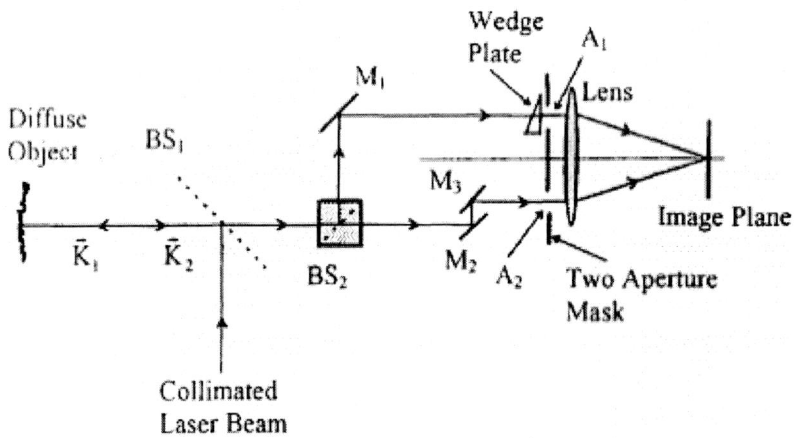


Figure 15. Schematic of an in-plane insensitive multi-aperture speckle shear interferometer.

Simultaneous measurement of displacement components, slope and curvature from a single optical arrangement based on the speckle photography, speckle and speckle shear interferometric techniques discussed earlier are also reported[60-65]. Holographic optical elements(HOEs) such as gratings, lenses and shear and double shear lenses also played a key role in speckle techniques for optical metrology. Chandra Shakher reported various novel techniques for extracting in-plane displacement components using a hololens as an imaging system in speckle photography[66-67]. The HOEs are also extensively implemented in speckle interferometry and speckle shear interferometry configurations for the measurement of displacement and its derivatives [68-69].

## 4. Electronic Speckle Pattern Interferometry

The origin of performing speckle interferometry by replacing the photographic plate with a detector and a image processing unit for real-time visualization of correlation fringes was first appeared in a famous article of Butters and Leendertz [70] in 1971. This pioneering contribution today referred in literature as electronic speckle pattern interferometry (ESPI), digital speckle pattern interferometry (DSPI), phase shifting speckle interferometry (PSSI) or TV holography. Staring from these basic ideas, contributions on various refinements, extensions and novel developments have been proposed. They have continued to pour in over the years from researchers across the globe, and have contributed to the gradual evolution of the concept into a routinely used reliable and versatile technique for quantitative measuring displacement components of a deformation vector of diffusely scattering object under static or dynamic load. The ability to measure three dimensional surface shapes by generating contours of constant depth of an object is yet another major achievement of ESPI. The ESPI also measures the derivatives of surface displacements which has a potential application in industry for non destructive testing. This technique is classified as electronic speckle shear pattern interferometry (ESSPI) or shearography. The past 10 years have seen a significant increase in the breadth and scope of applications of these two methods in areas such as experimental mechanics, vibration analysis, non destructive testing and evaluation and three-dimensional analysis. In recent years, potential areas of optical metrology are to study the mechanical behaviour of miniaturized structures like microelectromechanical systems (MEMS) and microoptoelectromechanical systems (MOEMS) [71-72] and biological specimens such as *In Vitro* and *In Vivo*[72-74]. These investigations require a microscopic/endoscopic ESPI systems.

The contributions on ESPI are extremely large and the rapid growth observed in the number of publications in the field, in the last few years, reflects the vigour of this rich and growing field. Three special issues on developments in speckle, speckle shear interferometry and digital speckle pattern interferometry [75-77], SPIE Milestone series on electronic speckle pattern interferometry: principles and practice [78] and two recent books [79-80] show the prominent importance of the ESPI systems for optical metrology. The focus in this section, is on application oriented procedures rather than on the well established fundamentals.

## 4.1. ESPI Measuring System and Data Processing

The basic ESPI system consists of an optical head, CCD camera, host computer, imaging processing system and a TV monitor. Here the CCD camera stores the imaged speckle pattern with a reference pattern and employs an electronic processing unit for rectification and high pass filtering of the video signal to convert it into a correlation speckle fringe pattern for display on a TV monitor in real time. These techniques are normally used in two modes of operation: the addition or subtraction mode, which is used in (i) static deformation analysis, and the time-average method, that is applied for (ii) vibration analysis[81-87]. In either case, fringe overlaid images of the object are obtained, where the fringes represent contours of constant surface displacement or constant vibration amplitude which is related to qualitative data. A further fringe analysis process with the support of phase shifting facility[24] in the optical head is required for intensity data acquisition and characteristic omnipresence of speckle noise reduction in the fringe pattern to yield quantitative data.

For static deformation analysis, the diffuse surface in its initial state, one frame of the interference of the speckle fields (object and reference beams) is recorded on a video store. The object is deformed and the current frame of the speckle fields is subtracted electronically pixel by pixel from the stored frame. The correlation speckle fringes are displayed in the monitor in real time. The equation for the fringe formation is related to the brightness of the monitor as

$$B \propto I_2 - I_1 = C\left|I_o I_r \sin^2(\phi + \Delta\phi/2)\sin^2(\Delta\phi/2)\right| \tag{17}$$

where C is the proportionality constant and $\sin(\phi + \Delta\phi/2)$ represents the speckle noise which varies randomly between 0 and 1 across the object image. Eq.(17) describes the modulation of the high frequency noise by a low frequency speckle interference pattern related to the phase change $\Delta\phi$.

The brightness will be minimum, whenever

$$\Delta\phi = 2n\pi \text{ , } n = 0,1,2..... \quad (18)$$

The condition corresponds to a dark fringe and denotes all of those regions where the speckle patterns before and after deformation are correlated. The brightness will be maximum whenever

$$\Delta\phi = (2n+1)\pi \; n = 0,1,2..... \quad (19)$$

The condition corresponds to a bright fringe when the speckle patterns are uncorrelated. As a result, correlation fringes representing contour of constant phase change $\Delta\phi$ are seen on the monitor in the form of bright and dark fringes modulating the object.

For vibration analysis using a continuous illuminating source, the fringe patterns observed in time-average mode, the equation for the intensity of speckle pattern in each frame can be expressed as[83]

$$I_1 = I_o + I_r + 2\sqrt{I_o I_r}\cos(\phi) J_0(\Delta\phi) \quad (20)$$

where $J_0(\Delta\phi)$, the characteristic fringe function is the zero order Bessel function of the first kind which modulates the interference of two speckle fields due to object vibration.

The phase change $\Delta\phi$ is directly related to propagating vectors in the direction of illumination and observation, and deformation vector $\vec{L}(x,y)$; appropriate optical configuration in the optical head plays an important role in extraction of the desired information. The optical head of the system can be arranged in several different ways for measuring displacement components and its derivates (*shearography*). This concept is already described in section.2 and section.3.

## 4.2. Phase Shifting Facility in ESPI

Phase shifting interferometry [24] has been applied to ESPI for obtaining quantitative data from the system. The method involves the capture of two or multiple images of the object. As we have seen, the phase varies randomly in ESPI because of the random phase of the individual speckles. Hence, attempts have also been made to suppress the random speckle phase by the use of suitable filters and algorithms. Several approaches have been described for calculating the phase maps in ESPI and also the software package for speckle fringe analysis is now available commercially. In ESPI, the path length of the reference beam is varied in controlled manner. This is usually achieved by piezo electric driven mirror inserted in either object or reference beam of the setup. The other schemes available for introducing phase shifts are, polarization based devices, liquid crystal phase shifters, source wavelength modulation [79,86], air filled quartz cell [26] etc.,. The phase shifting allows for the measurement of phase values at individual pixels.

A four frame phase shifting technique widely implemented in literature[6-8] for calculating the phase change $\Delta\phi$ is given here:

$$I_1(x,y) = I_{dc}\left[1+\gamma(x,y)\cos\ \Delta\phi\right]$$

$$I_2(x,y) = I_{dc}\left[1+\gamma(x,y)\cos\left(\Delta\phi+\frac{\pi}{2}\right)\right]$$

$$I_3(x,y) = I_{dc}\left[1+\gamma(x,y)\cos\left(\Delta\phi+\pi\right)\right]$$

$$I_4(x,y) = I_{dc}\left[1+\gamma(x,y)\cos\left(\Delta\phi+\frac{3\pi}{2}\right)\right] \tag{21}$$

The phase distribution $\Delta\phi$ can be obtained from the following arctan function[80]

$$\Delta\phi = \arctan\left[\frac{I_4(x,y)-I_2(x,y)}{I_1(x,y)-I_3(x,y)}\right] \tag{22}$$

Likewise a self calibrating five frame algorithm [88-90] which is less prone to errors is given as

$$\Delta\phi = \arctan\left[2\frac{I_2(x,y)-I_4(x,y)}{2I_3(x,y)-I_5(x,y)-I_1(x,y)}\right] \tag{23}$$

The "phase-of-differences" method[78], represents a useful way of generating visible phase shifted fringe patterns. The Eq.(22) and Eq.(23) are derived assuming that the phase $\Delta\phi$ varies much slower in comparison to $\phi$. There is an alternative method known as "difference-of-phases" where the phase $\phi$ evaluated from the phase shifted "before" load images is subtracted from the phase $\phi+\Delta\phi$ from the phase shifted "after" load images to yield the phase $\Delta\phi$ [78]. In each case, the input datasets were identical, and after smoothing, one can adopt a standard phase-shifting algorithm [24] to obtain the phase map.

Fast phase "*unwrapping*" algorithms have been reported in literature for obtaining quantitative data[80,91-98]. Quantitative data can also be extracted from time-average vibration fringes using phase shifting techniques. Stetson introduced an additional phase modulation mirror in the setup for optical shifting $J_0$ fringes according to ($J_0$+B), where B is the amplitude of the in phase vibration. The phase modulation mirror vibrates sinusoidally at a frequency and phase identical to the object vibration. The shift in $J_0$ fringes follows similar to that in which phase modulation shifts cosinusoidal fringes in the double exposure methods[81-84]. Minimum three images are required, without bias, with positive bias and negative bias for quantitative evaluation. The other prominent vibration analysis methods are based on stroboscopic illumination, synchronized acoustic optical modulator and source modulation[80,86,99-101].

## 4.3. ESPI System for Static and Vibration Analysis

Fig.16 represents the schematic of out-of-plane sensitive measuring system for static and vibration analysis. The laser beam is split into an object and a reference beam. The reference beam with the help of a fibre illuminates the CCD detector, via a piezo mounted mirror (PZT). The object beam illuminates the object and a zoom video lens system images the object on the CCD detector. The PS in the reference beam permits phase shifting, while the PM in the object beam is used for modal analysis of vibrating object. It is possible to implement simultaneously both phase shifting and phase modulation on a single phase shifting device. Krishna Mohan et.al. described a simple circuit which adds electronically the two required signals (phase-shifting at the camera framing rate and sinusoidal modulation at the vibration frequency) and drives a single piezoelectric translator with the combined signal[102]. Operating the system in static mode, the out-of-plane displacement results for an aluminum plate supported along two sides are shown in Fig.17. Fig.18 non destructive evaluation of debonds detection in a honeycomb panel. Fig.19 illustrates an Indian classical musical instrument *Veena* using the system in time-average vibration mode[103].

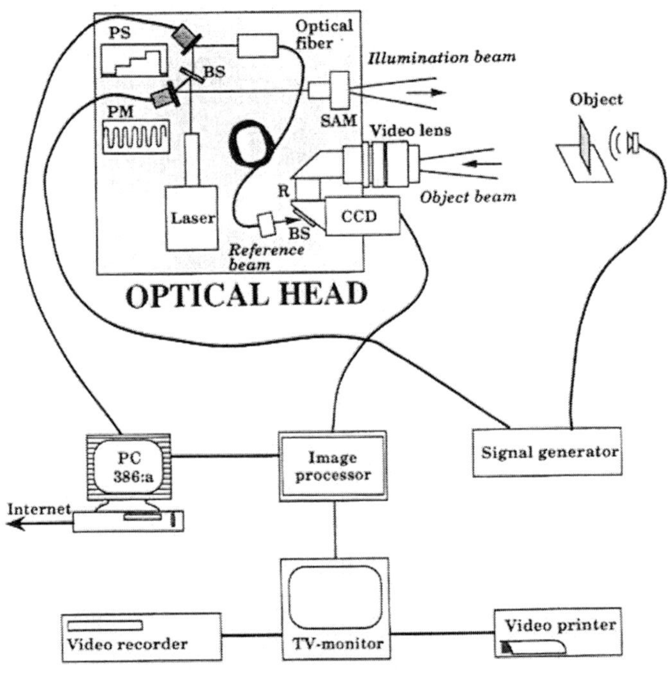


Figure 16. Schematic of a ESPI measuring system

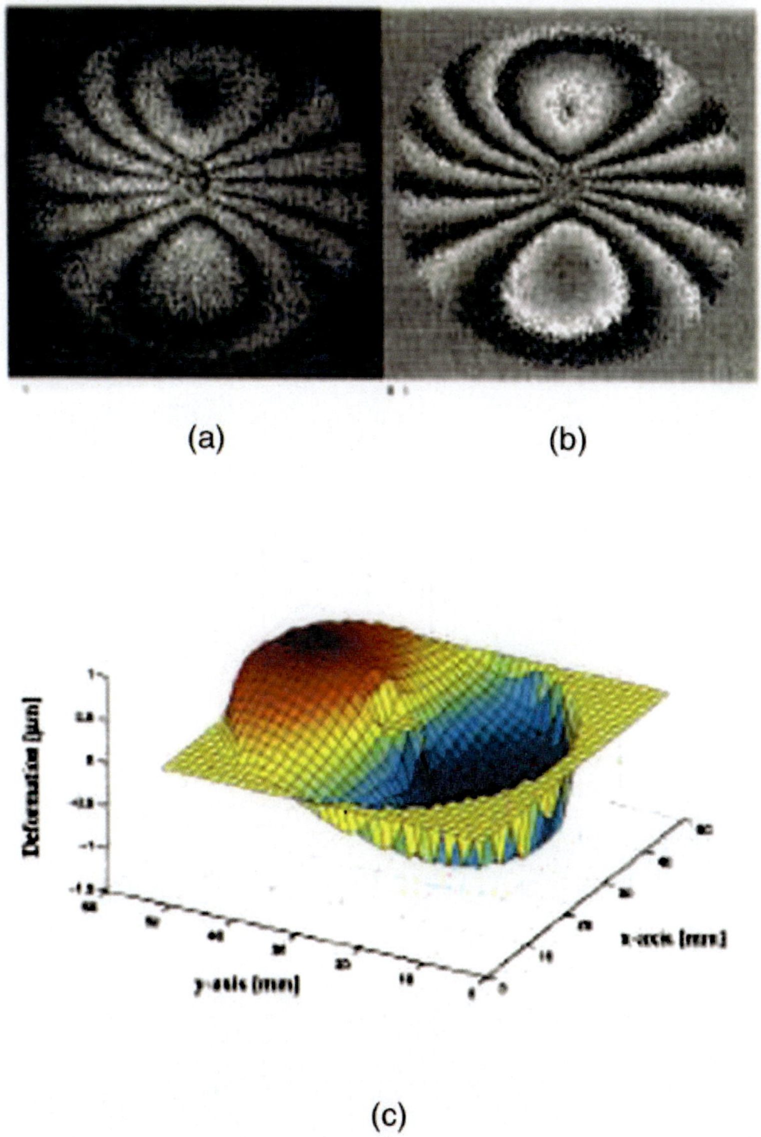


Figure 17. Static speckle fringe analysis: (a) out-of-plane displacement fringes (b) phase map for out-of-plane displacement. (c) quantitative measured out-of-plane displacement using four step algorithm.

Several optical configurations have also been developed for out-of-plane displacement measurements to cater to different measurement needs. These methods, which also feature the possibility of modifying the sensitivity of measurements, are based on (i) in-line reference wave configuration [104-105], and (ii) off-axis reference wave configuration [106-107]. A desensitized ESPI for large out-of-plane displacement is also proposed [102,108]. Further development of portable ESPI systems have been reported with the possibility for the system to be implemented in remote conditions [109-113].

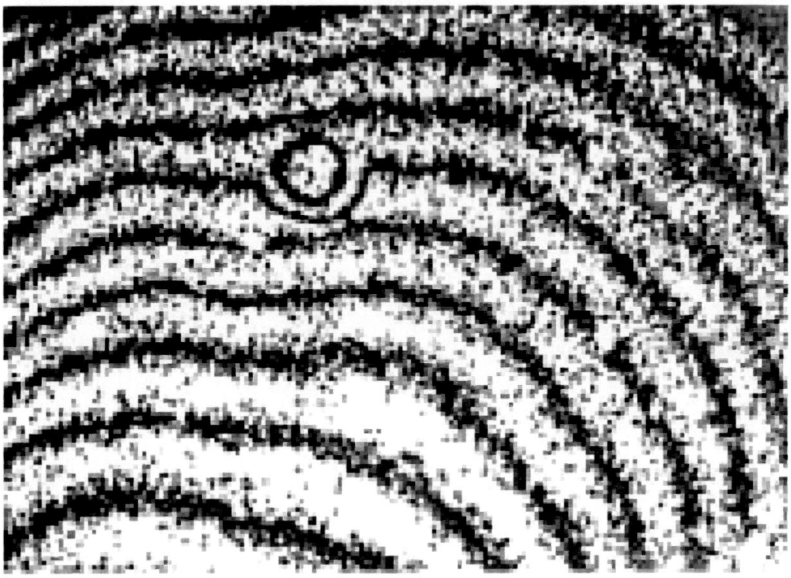

Figure 18. Non destructive evaluation of debonds detection in a honeycomb panel.

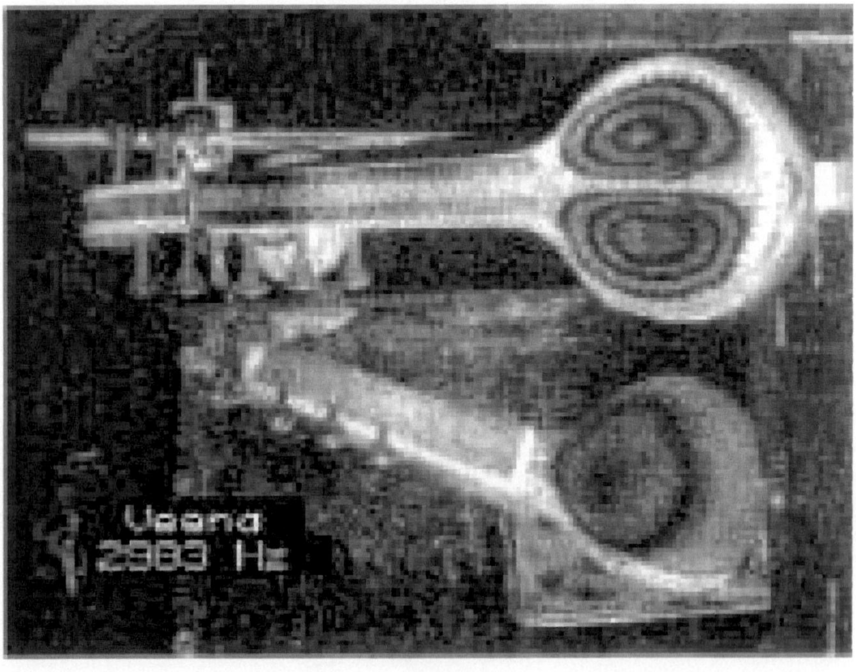


Figure 19. Time-average pattern on Veena, an Indian classical music instrument.t

Leendertz dual beam illumination and normal observation which is used for the measurement of in-plane displacement component along x-direction is shown in Fig.10 is widely used configuration in ESPI. The other configurations emerged in recent years are (i) oblique illumination-observation [114], (ii) normal illumination-dual direction observation [115] and (iii) dual beam symmetric illumination-observation [116-118]. The optical arrangement based on dual beam symmetric illumination-observation (*stereovision*) which is essentially two separate interferometers into one system is shown in Fig.20. The object is symmetrically illuminated with two collimated beams, making an angle θ with respect to the symmetric axis. The scattered fields are collected by means of a bi-mirror ($M_1$ and $M_2$) and a pair of beam splitters ($BS_1$ and $BS_2$). In the arrangement, for each illumination beam there are two scattered fields: namely, the scattered field along the same direction of illumination (back scatterd), and the scattered field in the symmetrically opposite direction (specular reflection direction). The scattered fields enter each half of a zoom video lens and relay lenses, and are imaged as two separate images (channel A and channel B) onto the CCD camera. One can observe two sets of correlation fringe patterns on the monitor with identical measuring sensitivities similar to Leendertz method[70], and the results on a plate subjected to in-plane rotation is shown in Fig.21. This arrangement can also be implemented for shape evaluation. Some novel approaches for evaluating displacement components from a single optical head are also described [118-122]: A photograph that represents the general view of the optical head developed at Luleå University of Technology, Luleå, Sweden by Krishna Mohan et.al. The schematic of the system, theory of fringe formation and the results for shape measurement obtained are given in Fig.22 to Fig.24 respectively.

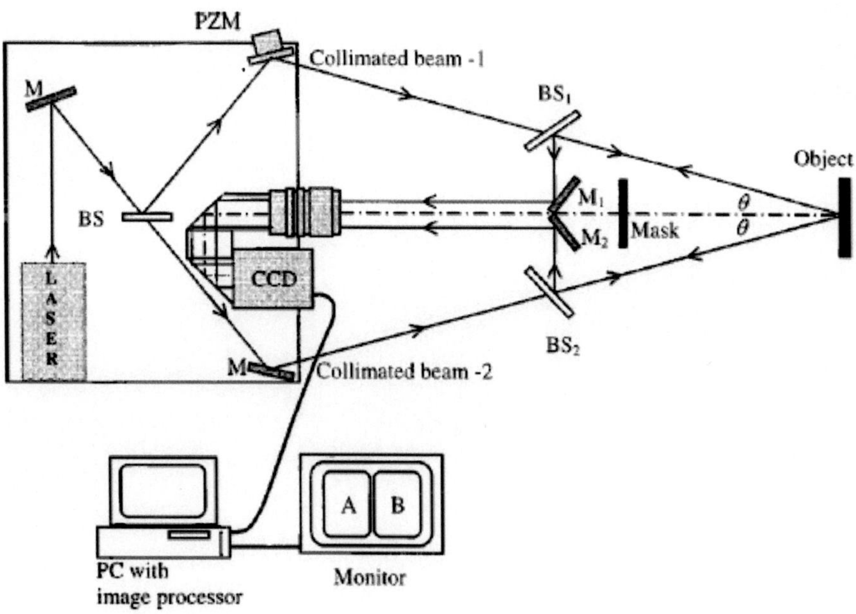


Figure 20. Schematic arrangement of a dual beam symmetric illumination-observation ESPI system.

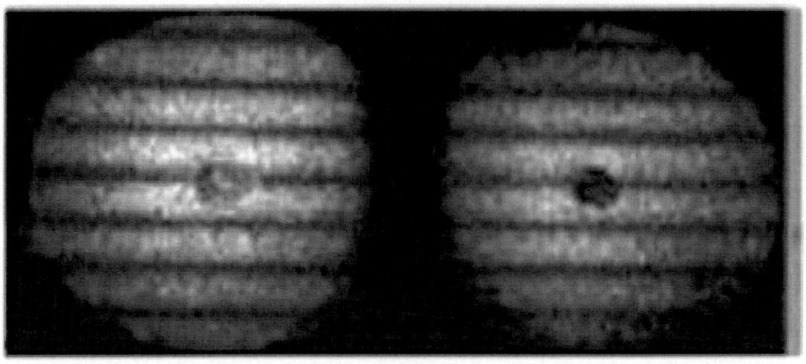

(a)

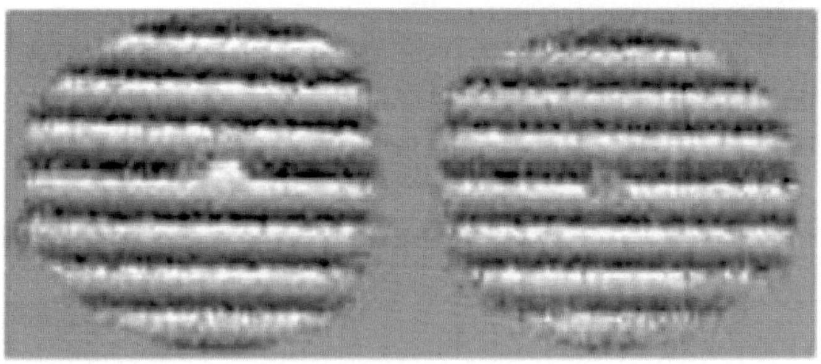

(b)

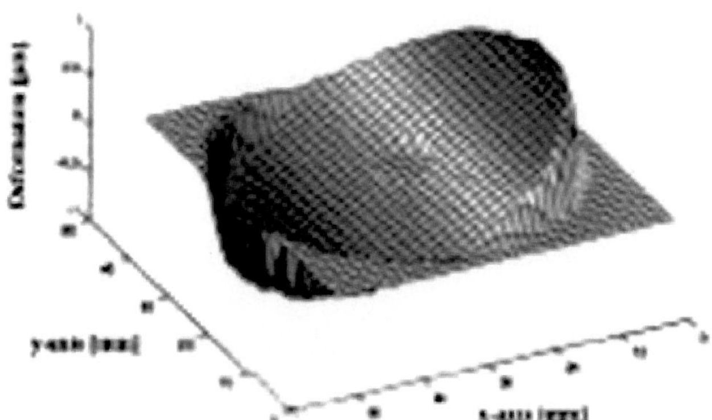

(c)

Figure 21. (a) Speckle correlation fringes and (b) phase maps obtained simultaneously from channels A and B for a plate subjected to rotation (c) 3-D plot corresponding to one of the channels.

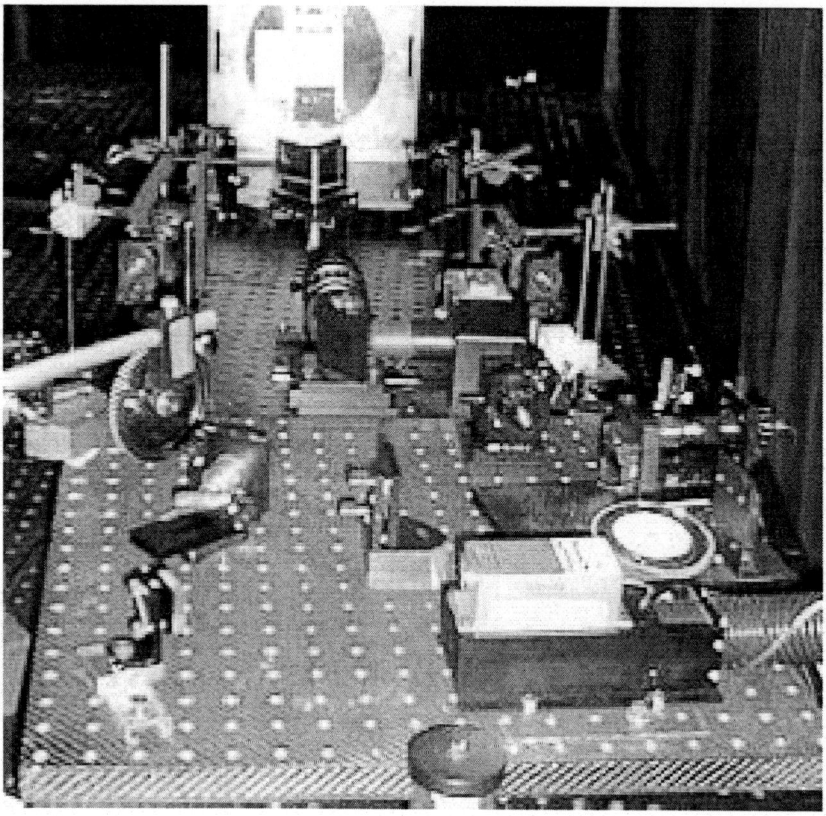

Figure 22. A photograph represents the general view of the optical head of ESPI system for sensing (a) the in-plane and the out-of-plane displacement or (b) shape and tilt component from a single optical setup. [Ref.118].

## 4.4. ESSPI System (Shearography)

Electronic speckle shear pattern interferometry (ESSI) has been widely used for measuring the derivatives of surface displacements of rough objects undergoing deformation. The technique has found a prominent use in aerospace industry for non destructive testing of space craft structures[123-125]. A Michelson type of shear interferometer[48,50] is extensively employed to introduce lateral shear in an optical configuration [126-129]. A universal shear interferometer has been proposed to implement lateral, radial, rotational and inversion shear in ESPI[130]. Dual-beam symmetric illumination in a Michelson shear arrangement for obtaining the partial derivatives of in-plane displacements (strain) is also demonstrated[131-133]. The other prominent shearing devices reported are 'split lens shear element'[134], 'bire-fringent crystals'[135-136] and holographic optical elements such as 'Holo-Shear Lens' (HSL)[137] and 'Holo-Gratings' (HG)[138]. Fig. 25 shows a shearography system for the measurement of first order derivative of out-of-plane displacement (slope) for static and vibrating objects. As an example of the resulting Bessel fringe patterns, with zero bias modulation, positive shift bias modulation and negative shift bias modulation [81-

83,126] for a edge-clamped aluminium plate vibrating at fundamental mode are shown in Fig.26. These patterns depict horizontal slope of vibration amplitudes. The three-dimensional plot is also shown in the same figure, while the pattern in Fig.26(e) represents the vibration amplitudes obtained by numerical integration. The maximum amplitude in Fig.26(e) is as high as 8λ that corresponds to 32 dark fringes in a conventional out-of-plane time-average fringes which is in practice limited to few fringes. The application of this method for evaluating slope change of vibration amplitudes at a higher frequency is demonstrated in Fig.27.

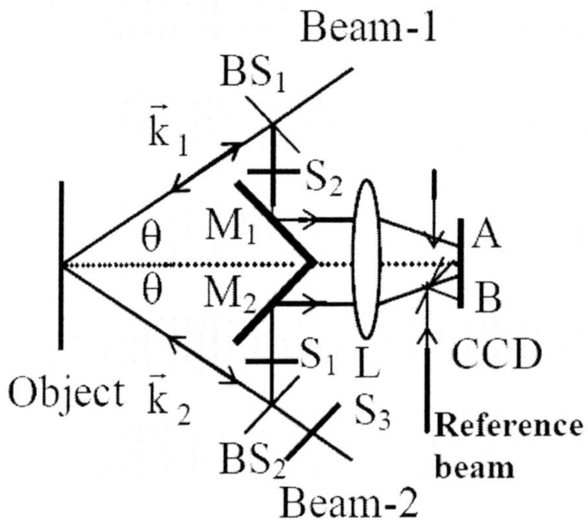


***Propagation Vectors*:**

$$\Delta\phi_A = (\bar{k}_2 - \bar{k}_1).\bar{L}$$

$$\Delta\phi_B = -(\bar{k}_2 + \bar{k}_1).\bar{L}$$

***Physical parameters*:**

**Displacement**

$$\Delta\phi_A = \frac{4\pi}{\lambda} u \sin\theta$$

$$\Delta\phi_B = \frac{4\pi}{\lambda} w \cos\theta$$

**Shape**

$$\Delta\phi_A = \frac{4\pi}{\lambda} z \sin\Delta\xi \sin\theta$$

$$\Delta\phi_B = \frac{4\pi}{\lambda} x \sin\Delta\xi \cos\theta$$

Figure 23. Schematic of a dual beam illumination –observation configuration with a reference added in channel B for the evaluation of in-plane and out-of-plane displacement components. Fringe formation equations are also presented in the figure.

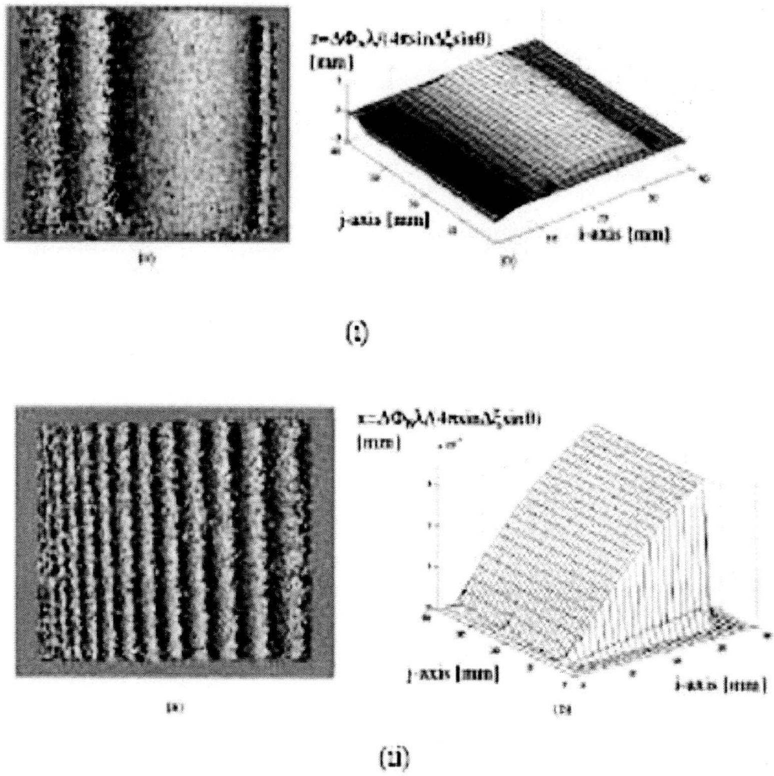


Figure 24. (i) Channel A: (a) phase map (b) 3-D surface profile. The results are obtained for a cylindrical surface subjected to rotation around the y axis from an in-plane sensitive configuration. Fig.24 (ii) Channel B: (a) phase map and (b) 3-D surface profile. The results are obtained for cylindrical surface subjected to rotation around the y axis. The results are extracted by using a out-of-plane sensitive configuration.

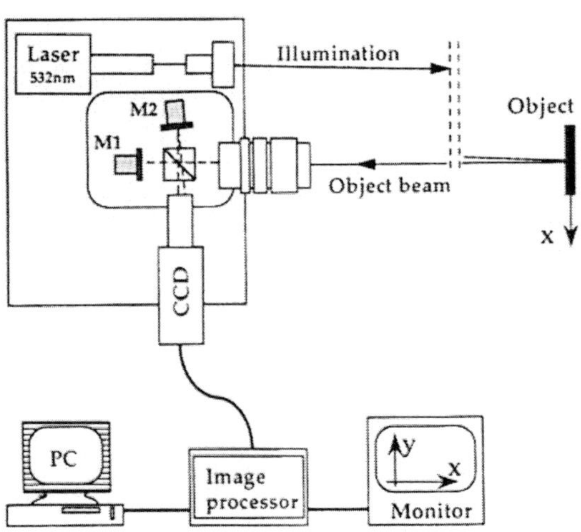


Figure 25. Schematic of electronic speckle shear pattern interferometer for static and vibration analysis.

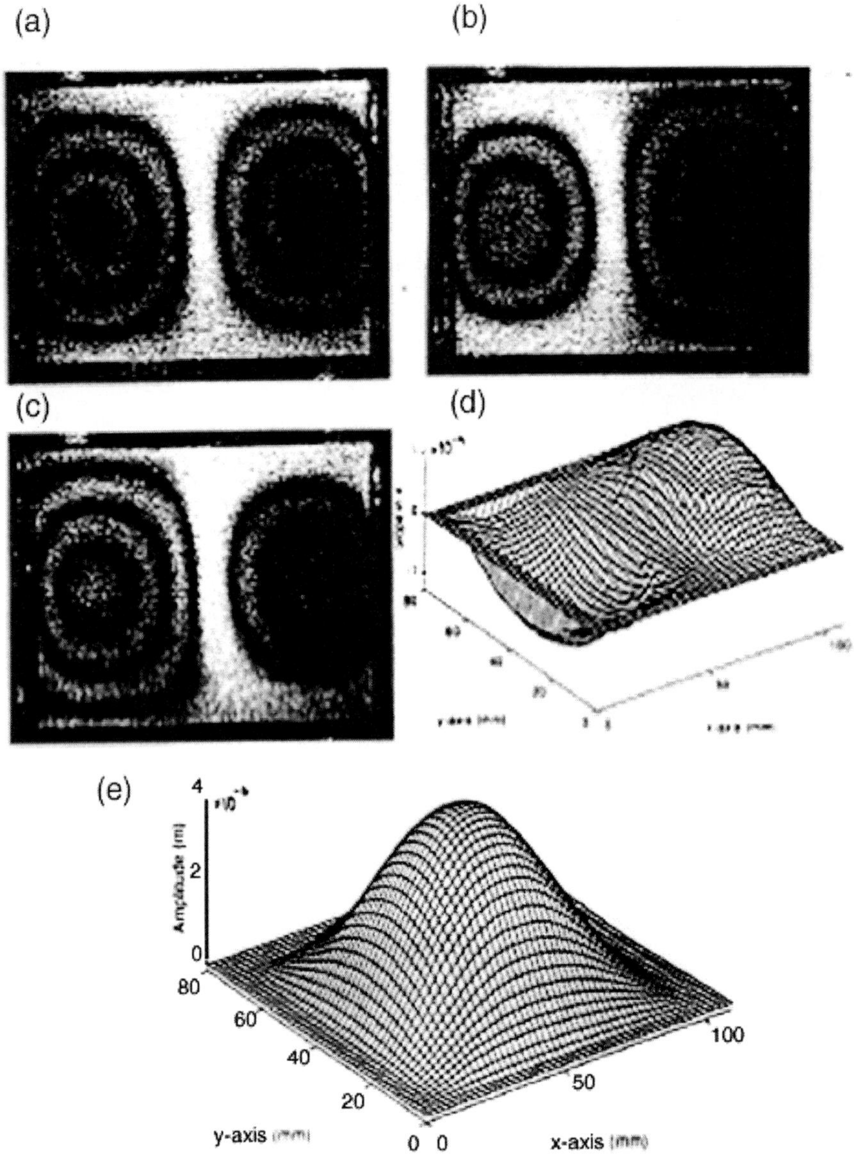


Figure 26. Modal analysis of an edge-clamped aluminium plate using electronic shearography, where the plate is vibrating at its first eigen frequency at 600 Hz with one anti-node at the center of the plate: (a) fringe pattern produced during the time-average recording of the plate vibrating without bias modulation, (b) with a positive bias modulation (c) with a negative bias modulation, (d) 3-D plot of the slope change of the vibration amplitude along the x-direction, computed from the three images in (a),(b) and (c). (e) The vibration amplitude field evaluated by numerical integration of (d).

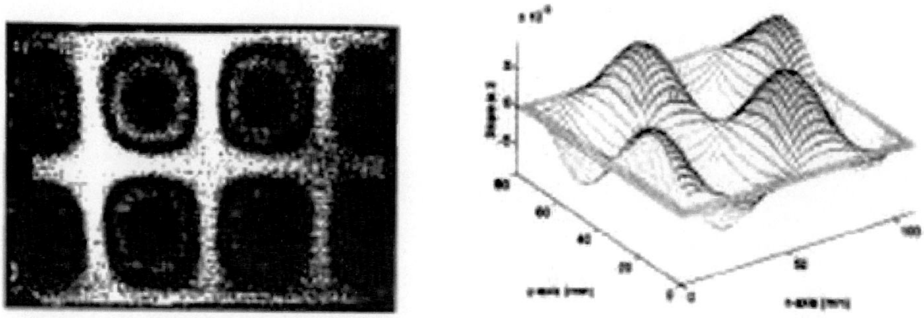

Figure 27. The slope of a vibration amplitude at a higher frequency and corresponding 3-D plot of the slope change of the vibration amplitude.

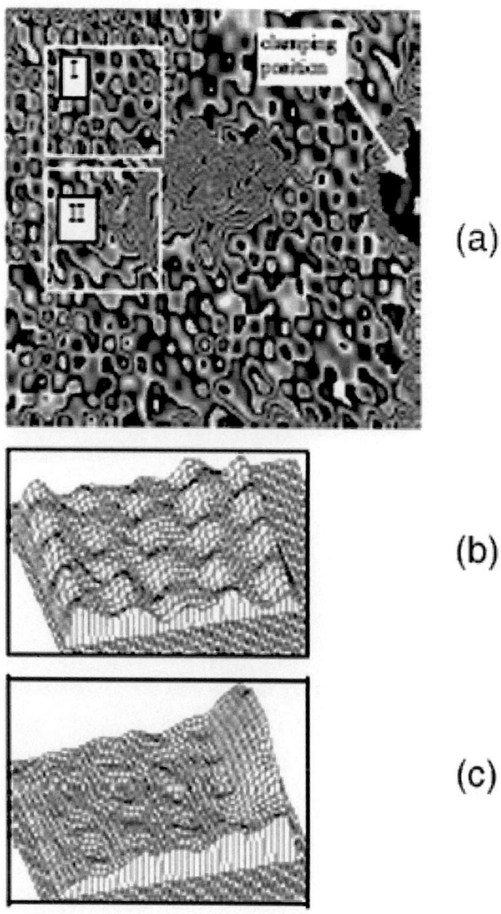


Figure 28. (a) NDT of a honeycomb aluminium panel under thermal load with both the displond and honeycomb structures displayed clearly,(b) 3D plot for the area I, (c) 3D plot for the area II. (Lianxlang Yang)

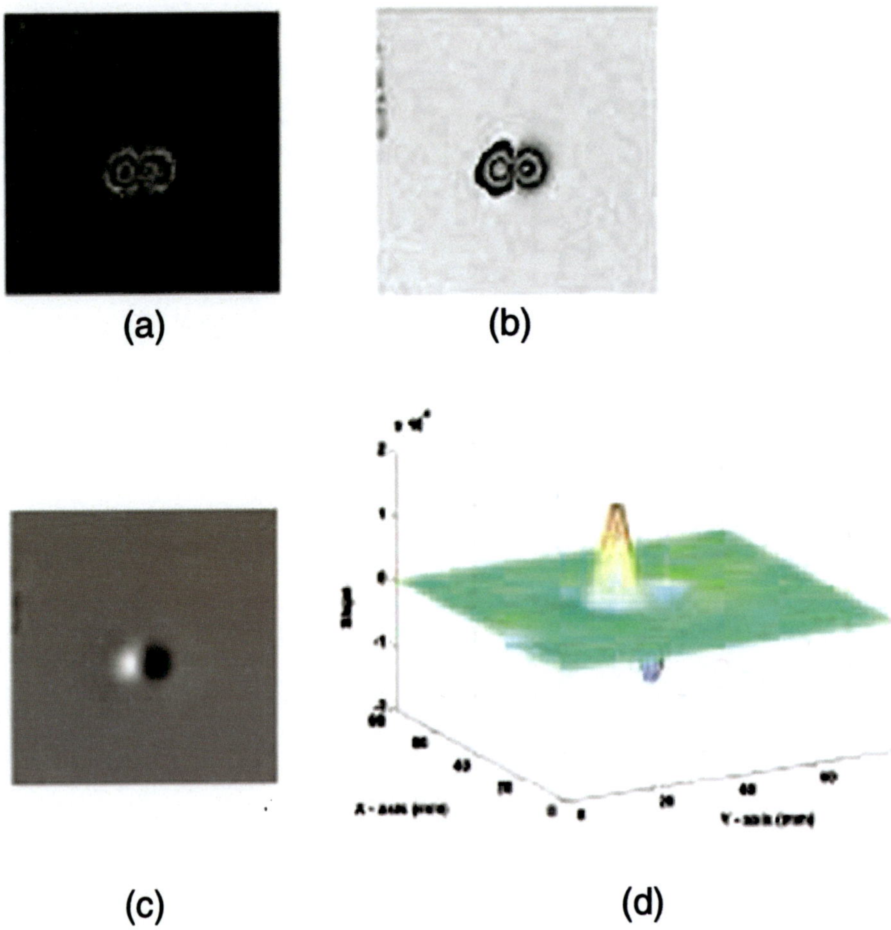


Figure 29. Application of shearography for NDT(a) fringe pattern near the defect site of a honeycomb structure (b) wrapped phase map (c) unwrapped phase map (d) 3D plot.

The shearography as a nondestructive testing of aerospace structures and disband in a honeycomb panels under thermal stressing is reported by Yang [124] and few novel applications with shearography for NDT is shown in Fig.28. A software based on LabVIEW [139] has also been developed in ESPI[139-140], which is more convenient with National Instruments frame grabber cards, for example NI1407, NI1409, etc. for phase shifting, grabbing and visualization of the speckle and speckle shear correlation fringes in real time. Non destructive testing using five steps phase stepping algorithm on a honeycomb structure is demonstrated in Fig.29.

Waldner [141-143] (Fig.30) developed a portable compact shearography system, and the illumination module consists of a laser, an electronic shutter, a microscopic objective and pin-hole is mounted on the top of the interferometer. The box in the lower half contains a Michelson interferometer for introducing shear in any direction, imaging lens and CCD. ESPI systems for measurement of 3D displacement and its derivates are also now commercial available in the market [144-146]. Few of them are (i) TV Holography/Shearography with zoom and relay lenses from Karl Stetson Associates,USA,(ii)3D-ESPI/Shearography system(Microstar) from Dr.Ettemeyer Gm BH&Co, Germany and (iii) Optonor, Norway. A multi wavelength shearography where three different wavelengths are also reported for extracting the in-plane and out-of-plane strains from a single arrangement[147-148]. A large area composite helicopter structures for NDT inspection is also examined by Pezzoni and Krupka [149] to show the potential application of the shearography for large size structures.

Figure 30. An example of a compact electronic speckle shearography system.[S. Waldner]

Of the many salient developments, the two that one could probably cite here concern the simultaneous measurement [150] of out-of-plane displacement and their derivatives, and the measurement of second order partial derivatives [151] of out-of-plane displacement (curvature and twist). The optical arrangement and the real-time out-of-plane displacement and its first order partial derivative obtained from the setup is illustrated in Fig.31. The other methods or measuring the out-of-plane and its slope change from a single set-up are also reported [152-153]. The optical arrangement and the applications of a dual function ESPI system are given in Fig.32 and Fig.33 respectively. The compact system is convenient and also efficient to switch over from an out-of-plane sensitive configuration to shearography. The shearography is also exploited for slope contouring of 3D objects either by varying refractive index, wavelength or rotation of the 3D surface between the frames [154-156].

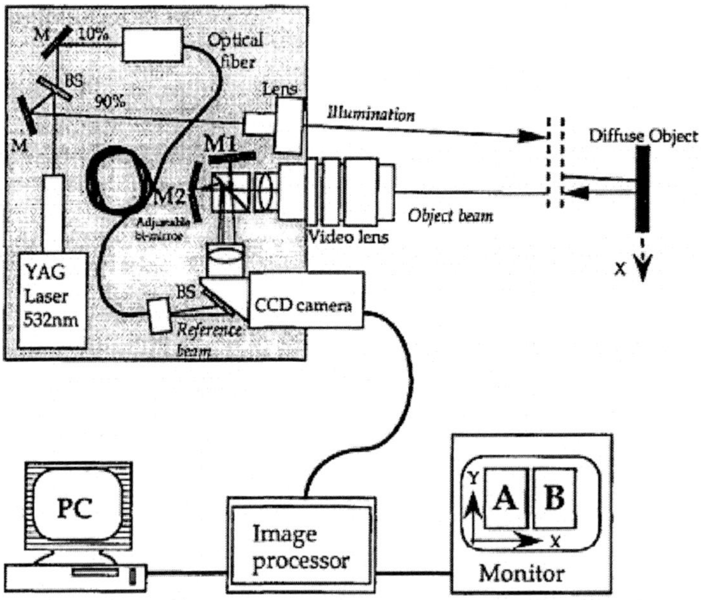


(a)

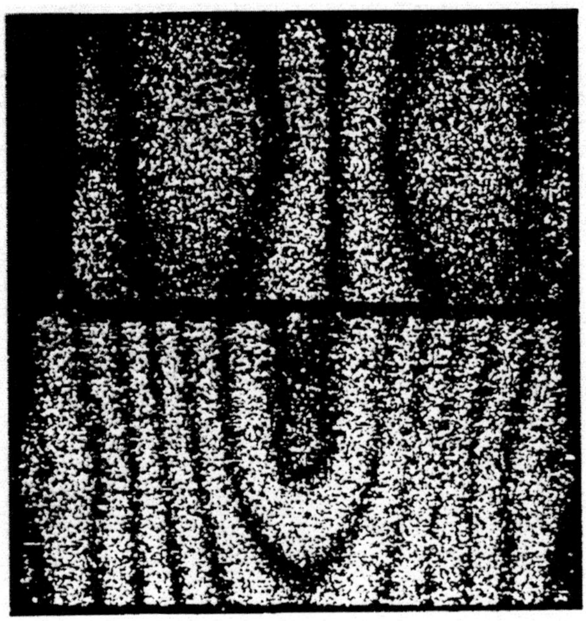

(b)

Figure 31. (a) Schematic for simultaneous measurement of out-of-plane displacement and slope (b) Fringe patterns depict to slope (upper half) and out-of-plane displacement (lower half.)

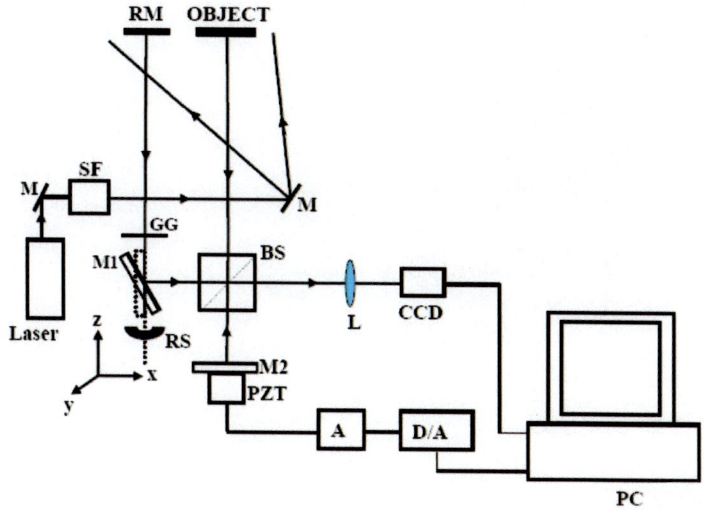


Figure 32. Schematic of a dual function ESPI system: BS:Beam splitter, RM:Reference mirror, GG:Ground glass, RS: Rotation stage, L: Imaging lens, A: Amplifier, D/A: Digital to Analog converter card

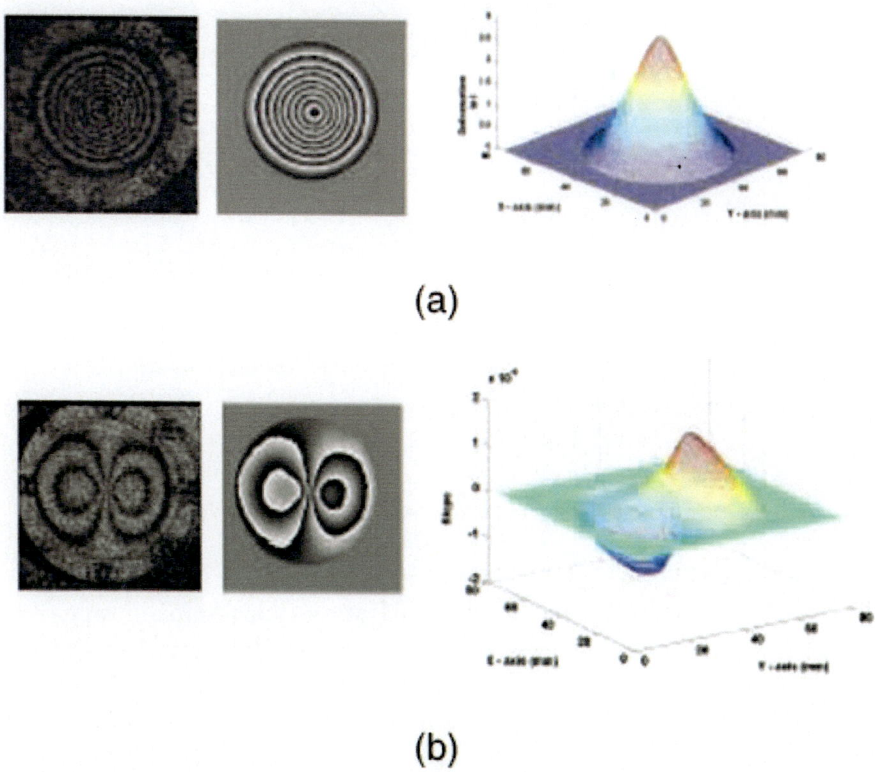


Figure 33. (a)correlation fringes, phase map and 3D plot of a circular diaphragm rigidly clamped along the edges and loaded at the center. obtained from out-of-plane sensitive configuration, (b)correlation fringes, phase map and 3D plot of a circular diaphragm rigidly clamped along the edges and loaded at the center obtained from shearography configuration. The data is extract from a dual function ESPI system shown in Fig.32.

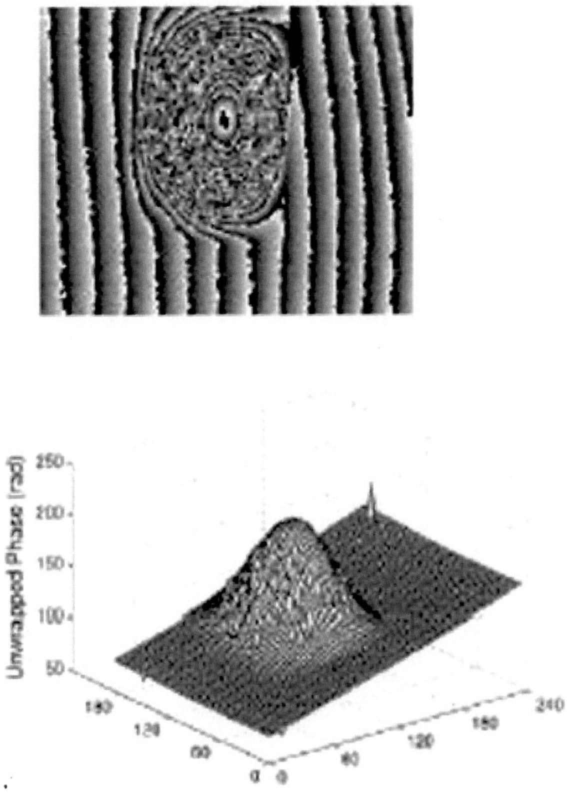


Figure 34. The evaluated phase map and 3D plot for the carbon fibre compsite using high-speed phase shifted speckle interferometery. [Davila]

Notable application in ESPI is in recording dynamic events using high-speed phase shifting speckle interferometry [97,157-159]. The high speed camera used in these investigations can capture and store up to 1600 images at a resolution of 239x192 pixels with a frame rate up to 1K frames/sec. Temporal phase unwrapping through sequences of more than 1K frames allows the determination of time-varying dynamic events. An evaluated phase map and 3D plot in carbon fibre composites are shown in Fig. 34.

#### *4.4.1. Microscopic ESPI Systems for MEMS and (MOEMS)*

Micro electro mechanical systems (MEMS) and Micro-optoelectromechanical systems (MOEMS) combining micro size electrical and mechanical mechanism are finding increasing applications in various fields. In recent years the technology of fabricating MEMS has been adopted by a number of industries to produce sensors, actuators, visual display components, AFM probe tips, semiconductor devices etc., Aided by photolithography and other novel methods, large number of MEMS devices are fabricated simultaneously. So, high quality standards are a must for the manufacturers in MEMS Technology. The materials behavior in combination with new structural design can not be predicted by theoretical simulations. A

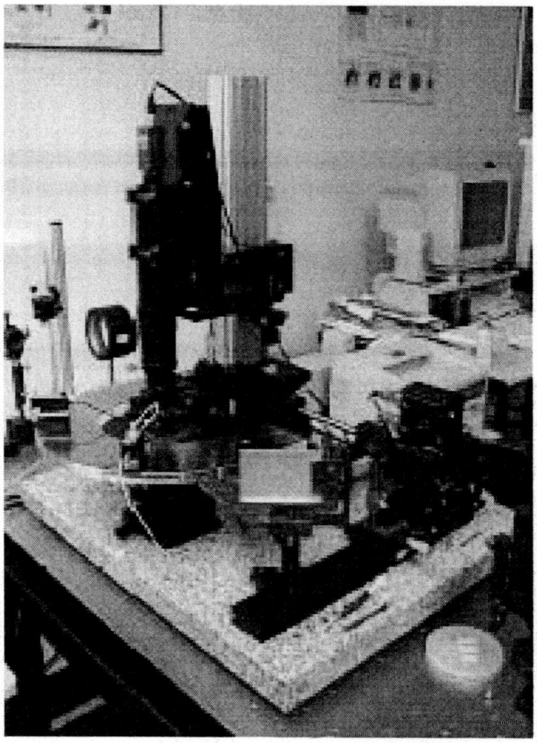

Figure 35. Photograph of a long working distance microscope ESPI system. [Jacek Kacperski and Malgorzata Kujawinska]

possible reason for wrong predictions made by FEM and other model analysis with respect to the loading conditions or in practical applications of micro devices is for instance the lack of reliable materials data and boundary conditions in microscale. The MEMS devices require a measurement and inspection techniques that are fast non-contact and robust. The reasons for such a technology is obvious, properties determined on much larger specimens cannot be scaled down from the bulk material without any experimental evaluation. Further on in micro structures, the materials behavior is noticeably affected by the production technology. Therefore simple and robust optical measuring methods to analyze the shape and deformation of the micro-components under static and dynamic conditions are highly desired. The optical measuring system for the inspection and characterization of M(O)EMS should have the following conditions. First it should not alter the integrity and the mechanical behaviour of the device. Since the microelements have an overall size up to a millimeters, a high spatial resolution measuring system is required. The measuring system should be able to perform both static and dynamic measurements. As the deflections, motions or vibration amplitudes of MEMS are typically in the nanometer to a few microns range, a high sensitivity optical techniques are highly desired. Finally, the measurement must be quantitative and reliable with a low sensitivity to environmental conditions. Microscopic optical techniques for MEMS are mostly dependent on the optical system such as long working distance microscope with the combination of different magnification objective lens. The interferometry of various types, such as monochromatic, white light, digital holography and microscopic ESPI, is capable of providing full-field non contact quantitative information about deformations, strains and surface profile of microscale specimens[71,160-161]. Lokberg[162], first proposed feasibility of microscopic ESPI system in 1997 on microstructures. Kujawinska and Goreck proposed an interferometric system using Model K2 – long working distance microscope (Fig.35) from Infinity – Photo Optical company for MEMS testing[163]. Aswendt reported a new deep ultraviolet micro speckle interferometer operating at 266 nm of wavelength for the measurement of both, in-plane and out-of-plane movements MEMS with nanometer accuracy[164]. Patorski et. al have carried out experimental studies of vibration resonant mode patterns of silicone micro-elements and micro-beams[165-166]. Yang and Colbourne[72] designed a digital laser micro-interferometer using a long workdistance microscope (LDM) from OPTEM, Model zoom 100 series. The system is capable of evaluating specimens from 5x7 mm to 0.5x.0.7mm for specular or scattered microelements. It is also reported that the system can be used to investigate the microelements as small as 0.05x0.05 mm which is ideal for MEMS technology. The system also equipped with a phase shifting mirror and a shearwin software for quantitative analysis. Fig.36 represents a microscopic ESPI for out-of-plane and in-plane displacement measurement of microstructures. Fig.36(a) shows the out-of-plane sensitive configuration for sensing the deformation vector component in z-direction. In this arrangement, the speckle field generated from the specimen interferes with the smooth reference beam after they emerge from the beam splitter(BS). The geometry shown in Fig.36(b) is to obtain the in-plane component (deformation vector component along x, or y-direction). In the optical arrangement, the PZT mirror redirects half of the collimated beam onto the specimen to produce dual-beam symmetric illumination at an angle 45° with respect to the normal. The fringe sensitivity for out-of-plane is 0.5λ, while, for in-plane displacement it is 0.707λ. As an example, based on difference-of-phase method, Fig.37 shows the application of the ESPI microscopic system for out-of-plane displacement for a sample of a circularly metal plate clamped at the edges and centrally loaded[72]. The current research is focused at various research laboratories for designing a full field reliable system for MEMS which itself is a front runner in modern technology. A microscopic ESPI system (Micro Map 5010) is also commercially available from Optonor, Norway for MEMS static and vibration analysis.

## 4.5. Endoscopic ESPI system

ESPI system is also exploited for biology and medical applications[167-168,73-74]. A compact endoscopic ESPI camera system that can be applied for medical intra-cavity inspection is reported by Von Bally group and shown in Fig.38. For illumination, the system uses optical fibers and for endoscopic imaging, a Hopkins rod-lens fitted with a progressive scan CCD is developed. Applications for technical membrane and endoscope intra-cavity investigations of a porcine stomach *In Vitro* at two different areas of the mucous membrane are shown in Fig.39 and Fig.40 respectively.

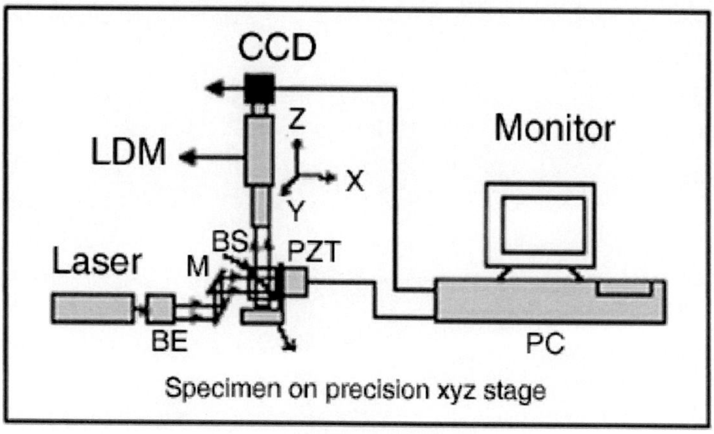


(a)

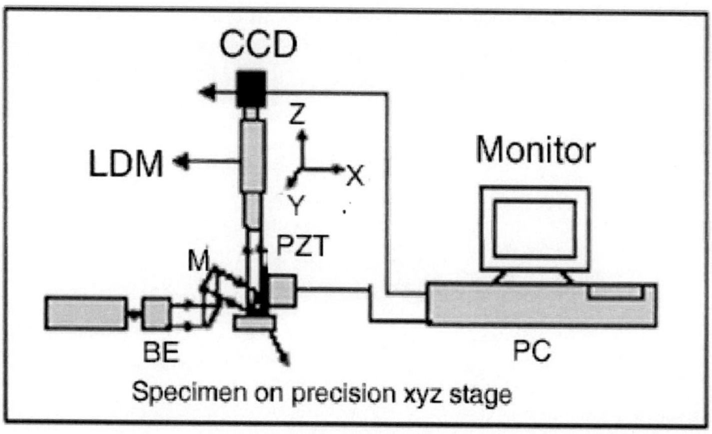


(b)

Figure 36. Schematic of a microscopic ESPI system for measuring (a) out-of-plane displacement (b) in-plane displacement of MEMS.

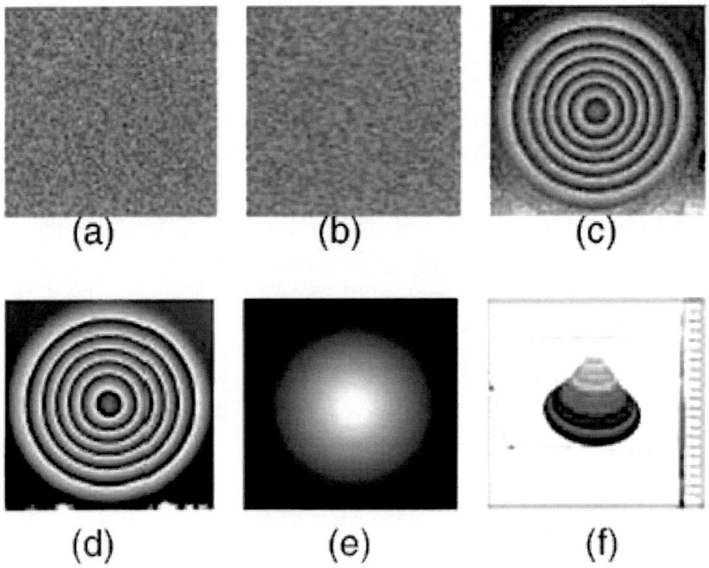

Figure 37. Measuring procedure for a sample circularly clamped and centrally loaded:(a) phase distribution before deformation, (b) phase distribution after loading, (c) relative-phase map obtained by subtracting phase distribution in Fig.(b) from phase distribution in Fig.(a),(d) smooth image of phase map (e) unwrapped image of the phase difference (e) 3D display of the evaluation. [Lianxiang Yang, Ref.72].

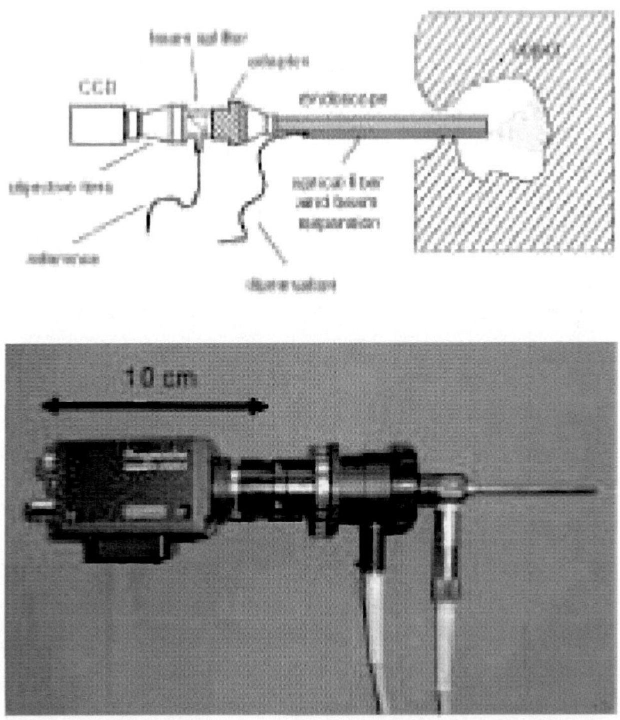


Figure 38. Schematic of the endoscope ESPI camera system as developed at the Laboratory of Biophysics University of Münster, Germany. [Björn Kemper and Gert Von Bally]

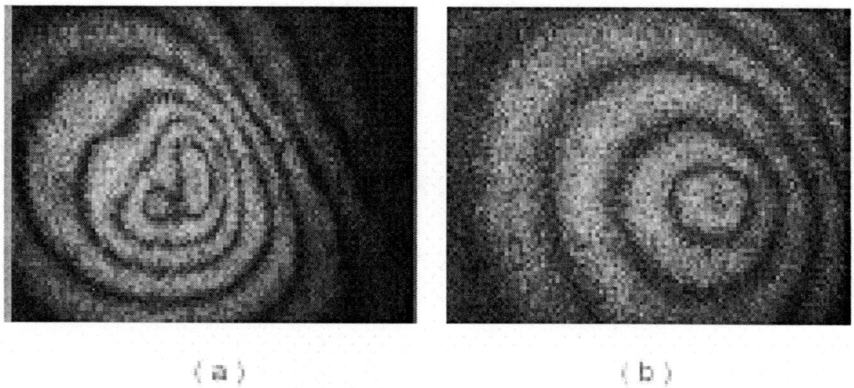


Figure 39. Investigations of a technical membrane to find the optimum contrast of the correlation fringes. [Björn Kemper and Gert Von Bally]

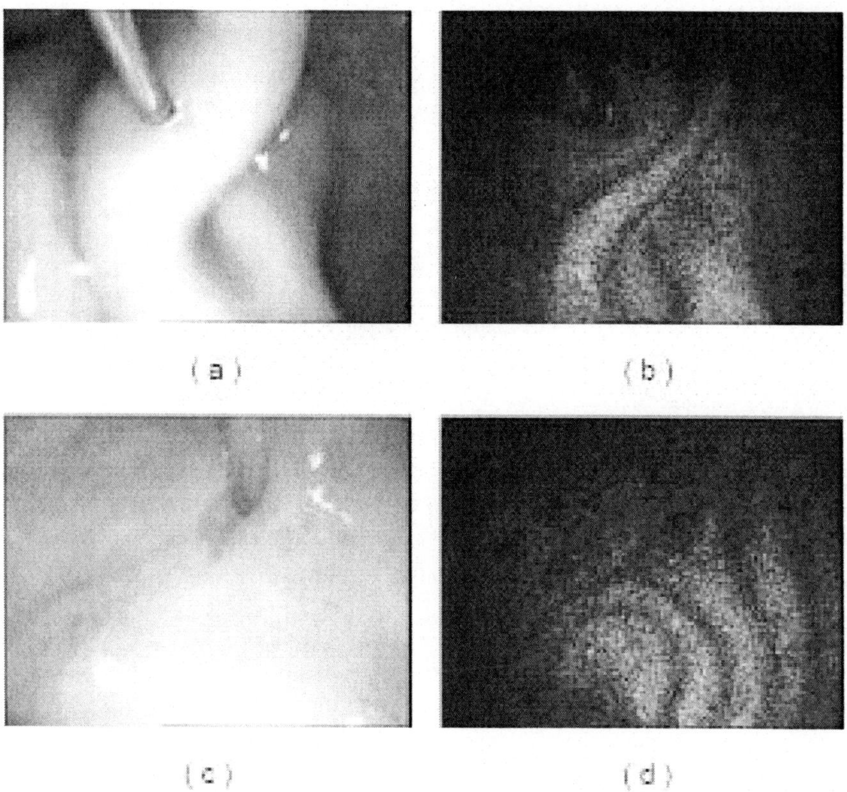


Figure 40. Intracavity inspection of biological specimen in vitro (porcine stomach, mucous membrane): (a)and(C) white-light endoscoic images of the mucous membrane, (b) and (d) correlation ESPI fringe patterns of the stimulated areas in (a) and (c). [Björn Kemper and Gert Von Bally]

## 5. Conclusion

The chapter has presented a concise developments of speckle techniques for optical metrology. These techniques can be classified into three categories: speckle photography, speckle interferometry and speckle shear interferometry (shearography). The applications for measurement covered include (i) static and dynamic deformation and its gradient measurements, (ii) the shape and its slope change of the three-dimensional objects, and non-destructive testing (NDT) of engineering structures with results for some typical applications. It is also described the potential applications of ESPI with latest developments and refinements in terms of optical systems, fast and reliable image processing systems, modern fringe analysis methods. Particular emphasis is also provided for the measurement of mechanical behavior of MEMS structures. Novel applications of speckle techniques, especially, DSP, ESPI and shearography, with the support of high-resolution CCD cameras, and faster, efficient and robust algorithms continue to be published, thus testifying to the vitality of the technique. It is therefore, reasonable to expect that in the coming years, the speckle techniques will find a widespread use in the field of aerospace engineering, biological and medical science, bio-photonics, electronic and automotive industries.

## References

[1] Leendertz, J. A.; Interferometric displacement measurement on scattering surfaces utilizing speckle effect. *J Phys.E: Sci. Instrum.* 1970,3,214-218.

[2] Archbold, E.; Burch, M. Ennos, A.E. Recording of in-plane surface displacement by double-exposure speckle photography. *Opt. Acta.* 1992,17,883-889.

[3] Archbold, E.; Ennos, A.E. Displacement measurement from double exposure laser photographs. *Opt. Acta.* 1992,19,253-271.

[4] Khetan, R.P.; Chang, F.P. Strain analysis by one-beam laser speckle interferometry. 1. Single aperture method. *Appl. Opt.* 1976,15,2205-2215.

[5] Erf, R.K.; (Ed.), *Speckle Metrology*, Academic press, New York, 1978.

[6] Sirohi, R.S.; (Ed.), *Speckle Metrology*, Marcel Dekker, New York, 1993.

[7] Gåsvik, K.J.; *Optical Metrology*, 2nd Edition, John Wiley & Sons, 1995.

[8] Rastogi, P.K.; (Ed.), *Optical Measurement Techniques and Applications* , Artech House, Boston, 1997.

[9] Sutton, M.A.; Wolters, W.J. Peters, W.H. Ranson, W.F. McNeill, S.R. Determination of displacements using an improved digital correlation method. *Image Vision Comput.* 1983,1, 133-139.

[10] Chen, D.J.; Chiang, F.P. Optimal sampling and range of measurement in displacement-only laser-speckle correlation. *Exp. Mech.* 1992,32,145-153.

[11] Noh, S.; Yamaguchi, I. Two-dimensional measurement of strain distribution by speckle correlation. *Jpn. J. Appl. Phys.* 1992, 31,L1299-L1301.

[12] Sjödahl, M.; Benckert L.R. Electronic speckle photography analysis of an algorithm giving the displacement with subpixel accuracy. *Appl Opt.*1993,32, 2278-2284.

[13] Sjödahl, M.; Benckert L.R. Systematic and random errors in electronic speckle photography.. *Appl Opt.* 1994,33,7461-7471.

[14] Sjödahl, M. Electronic speckle photography: increased accuracy by nonintegral pixel shifting. *Appl Opt.*,1994,33,6667-6673.

[15] Andersson, A.; Krishna Mohan, N. Sjödahl, M. Molin, N.-E. TV shearography: Quantitative Measurement of shear magnitudes using digital speckle photography. *Appl. Opt.* 2000, 39,2565-2568.

[16] Molin, N.-E.; Sjödahl, M. Gren, P. Svanbro A. Speckle photography combined with speckle interferometry *Opt. Laser Eng.* 2004,41,673-686.

[17] Sjödahl, M.; *Digital speckle photography* in *Digital speckle pattern interferometry and related techniques.* Rastogi, P.K (Ed.). John Wiley, England, 2001, chapter 5. 289-336.

[18] Yeh, P.; Gu, C (Eds.), *Landmark Papers on Photorefractive Nonlinear Optics*, World Scientific, New Jersey, USA,1995.

[19] Tiziani, H.J.; Leonhardt, K. Klenk, J. Real-time displacement and tilt analysis by a speckle technique using $Bi_{12}SiO_{20}$ crystal. *Opt. Commun.* 1980,34,327-331.

[20] Liu, L.; Helmers, H.; Hinsch, K.D. Speckle metrology with novelty filtering using photorefractive two-beam coupling in $BaTiO_3$. *Opt. Commun.* 1993, 100,301-305.

[21] Sreedhar, P.R.; Krishna Mohan, N. Sirohi, R.S. Real-time speckle photography with two-wave mixing in photorefractive $BaTiO_3$ crystal. *Opt. Eng.* 1994, 33,1989-1995.

[22] Sreedhar, P.R.; Krishna Mohan, N. Sirohi, R.S. Time evaluation of fringes obtained with a two-beam coupled photorefractive novelty filter. *J. Optics* (India). 1995,24,15-22.

[23] Krishna Mohan, N.; Darlin, J.S. Ara, M.H. Kothiyal, M.P. Sirohi, R.S. Speckle photography with $BaTiO_3$ crystal for the measurement of in-plane displacement field distribution of distant object. *Opt. Lasers Eng.* 1998,29,211-216.

[24] Creath, K.; *Phase-measurement interferometry techniques* in *Progress in Optics*, Wolf, E. (Ed.), North-Holland, Amsterdam, Vol.XXV,pp.349-393,1988

[25] Kaufmann, G.H.; Jacquot, P. Phase shifting of whole field speckle photography fringes. *Appl. Opt.*, 1992,29,3570-3575.

[26] Krishna Mohan, N.; Rastogi, P.K. Phase shifting whole field speckle photography technique for the measurement of object deformations in real-time. *Opt. Lett.* 2002,27,565-567.

[27] Krishna Mohan, N.; Rastogi, P.K. Phase shift Calibration in Speckle Photography – A First Step to Practical Real Time Analysis of Speckle Correlation Fringes. *J. Mod. Opt.* 2003,50,1183-1188

[28] Krishna Mohan, N.; Rekha Mathews. Photorefractive phase reversal speckle photography for evaluation of in-plane displacement. *Lasers Eng.* 2005,15,91-105.

[29] Rastogi, P.K.; Speckle shearing photography: a tool for direct measurement of surface strains. *Appl. Opt.* 1998,37,1292-1298.

[30] Ara, M.H.; Vijayan, C, Krishna Mohan, N. Sirohi, R.S. Real-time speckle shear photography using $BaTiO_3$ for in-plane strain and displacement measurements at multiple wavelengths. *Proc. SPIE* 1999,3801,180-189.

[31] Rastogi, P.K.;(Ed.) *Special issue on "Speckle Photography".* Opt. Lasers Eng. 1998, Vol.29 (2 and 3).

[32] Sirohi, R.S.; Krishna Mohan, N. In-plane displacement measurement configuration with twofold sensitivity. *Appl. Opt.* 1993.32,6387-6390.

[33] Solmer, A.; Joenathan, C. Twofold increase in sensivity with a dual beam illumination arrangement for electronic speckle pattern interferometry. *Opt. Eng.* 1996,35,1943-1948.

[34] Krishna Mohan, N.; Measurement of in-plane displacement with twofold sensitivity using phase reversal technique. *Opt. Eng.* 1999,38,1964-1966.

[35] Ganesan, A.R.; Sirohi, R.S. New method of contouring using digital speckle pattern interferometry (DSPI). *Proc. SPIE* 1988, 954,327-332.

[36] Joenathan, C.; Pfister, B. Tiziani, H.J. Contouring by electronic speckle pattern interferometry employing dual beam illumination. *Appl. Opt.* 1990,29,1905-1911.

[37] Joenathan, C.; Ganesan A.R. Sirohi, R.S. Fringe compensation in speckle interferometry – application to non destructive testing. *Appl. Opt.* 1986,25,3781-3784.

[38] Rastogi, P.K.; Jacquot, P. Measurement of difference deformation using speckle interferometry. *Opt. Lett.* 1987,12,596-598.

[39] Ganesan A.R.; Joenathan, C. Sirohi, R.S. Real-time comparative digital speckle pattern interferometry. *Opt. Commun.* 1987,64,501-506.

[40] Saldner, H.O.; Krishna Mohan, N. Molin, N.-E. Comparative TV holography for vibration analysis. *Opt. Eng.* 1995,34,486-492.

[41] Duffy, D.E. Moiré gauging of in-plane displacement using double aperture imaging. *Appl. Opt.* 1972,11,1778-1781.

[42] Duffy, D.E.; Measurement of surface displacement normal to the line of sight. *Exp. Mech.* 1974,14,378-384.

[43] Khetan, R.P.; Chang, F.P. Strain analysis by one-beam laser speckle interferometry. 2 Multiaperture aperture method. *Appl. Opt.* 1979,18,2175-2186.

[44] Tai, A.; Speckle interferometry with an imaging grating interference. *Opt. Lett.* 1980,5,552-554.

[45] Sirohi, R.S.; Krishna Mohan, N. Santhanakrishnan, T. Optical configurations for measurement in speckle interferometry. *Opt.Lett.* 1996,21,1958-1959.

[46] Krishna Mohan, N.; Santhanakrishnan, T. Senthilkumaran, P. Sirohi, R.S. Simultaneous implementation of Leendertz and Duffy's methods for in-plane displacement measurement. *Opt. Commun.* 1996,124,235-239.

[47] Joenathan,C.; Mohanty, R.K. Sirohi, R.S. Multiplexing in speckle shear interferometry. *Opt. Acta.* 1984,31,681-692.

[48] Leendertz, J. A.; Butters, J.N. An image-shearing speckle pattern interferometer for measuring bending moments. *J. Phys.E: Sci. Instrum.* 1973,6,1107-1110.

[49] Hung, Y.Y.; Taylor, C.E. Speckle-shearing interferometric camera – a tool for measurement of derivatives of surface displacement. *Proc. SPIE* 1973,41, 169-175.

[50] Sirohi, R.S.; Speckle shearing interferometry – A review. *J. Optics.* 1984,33,95-113.

[51] Sharma, D.K.; Sirohi, R.S. Kothiyal, M.P. Simultaneous measurement of slope and curvature with a three-aperture speckle shearing interferometer. *Appl. Opt.* 1984,23,1542-1546.

[52] Sharma, D.K.; Krishna Mohan, N. Sirohi, R.S. A holographic speckle shearing technique for the measurement of out-of-plane displacement, slope and curvature. *Opt. Commun.* 1986,57,230-235.

[53] Krishna Mohan, N.; The influence of multiple-exposure recording on curvature pattern using multi-aperture speckle shear interferometry. *Opt. Commun.* 2000,186,259-263.

[54] Brdicko, J.; Olson, M.D. Hazell, C.R. Theory for surface displacement and strain measurements by laser speckle interferometry. *Opt. Acta.* 1978,25,963-989.

[55] Iwahashi, Y.; Iwata, K. Nagata, R. Influence of in-plane displacement in single aperture and double aperture speckle shear interferometry. *Appl. Opt.* 1985,24, 2189-2192.

[56] Krishna Mohan, N.;Masalkar, P.J. Murukeshan, V.M. Sirohi, R.S. Separation of the influence of in-plane displacement in multiaperture speckle shear interferometry. *Opt. Eng.* 1994,33,1973-1982.

[57] Krishna Mohan, N.; Sirohi, R.S. Fringe formation in multiaperture speckle shear interferometry. *Appl. Opt.* 1996,35,1617-1622.

[58] Krishna Mohan, N.; Sirohi, R.S. Fringe formation in symmetric three-aperture speckle shear interferometry: An analysis. *Opt. Lasers in Eng.* 1997, 26,437-447.

[59] Krishna Mohan, N.; Double and multiple-exposure methods in multi-aperture speckle shear interferometry for slot and curvature measurement. *Proceedings of Laser Metrology for Precision Measurement and Inspection in Industry*, Brazil.*Module 4,* Holography and Speckle Techniques, 1999,4.87-4.98.

[60] Mohanty, R.K.; Joenathan, C. Sirohi, R.S. Speckle and speckle shearing interferometers combined for the simultaneous determination of out-of-plane displacement and slope. *Appl. Opt.* 1985,24,3106-3109.

[61] Mohanty, R.K.; Joenathan, C. Sirohi, R.S. Speckle interferometric method of measuring small out-of-plane displacements. *Opt. Lett.* 1985,9,475-477.

[62] Sharma, D.K.; Krishna Mohan, N. Sirohi, R.S. A holographic speckle shearing technique for the measurement of out-of-plane displacement, slope and curvature. *Opt. Commun.* 1986,57,230-235.

[63] Sirohi, R.S.; Krishna Mohan, N. Speckle interferometry for deformation measurement. *J. Mod. Opt.* 1992,39,1293-1300.

[64] Sirohi, R.S.; Krishna Mohan, N. Role of lens aperturing in speckle metrology. *J. Sci. Indust. Res.* 1995,54,67-74.

[65] Krishna Mohan, N.; An optical configuration in multi-aperture speckle interferometry for simultaneous measurement of in-plane and out-of-plane deformation. *J. Instru.* 1999.29,147-151.

[66] Shaker, C.; Rao, G.V. Speckle metrology using hololenses, Part 2. *Appl. Opt.* 1987,26,654-657.

[67] Shaker, C.; Rao, G.V. Speckle metrology using hololenses. *Appl. Opt.* 1984, 23,4592-4595.

[68] Joenathan, C.; Mohanty, R.K Sirohi, R.S. Hololenses in speckle and speckle shear interferometry. *Appl. Opt.* 1985,24,1294-1298.

[69] Joenathan, C.; Sirohi, R.S. Holographic gratings in speckle shear interferometry. *Appl. Opt* 1985,24,2750-2751.

[70] Leendertz, J.N.; Butters, J.N. Holographic and video techniques applied to engineering measurements. *J. Meas. Control.* 1971,4,349-354.

[71] Bosseboeuf, A.; Petitgrand, S. Characterization of the static and dynamic behaviour of M(O)EMS by optical techniques: status and trends. *J. Micromech. Microeng.* 2003,13,823-833.

[72] Yang, L.; Colbourne, P. Digital laser microinterferometer and its applications. *Opt. Eng.* 2003,42,1417-1426.

[73] Kemper,B.; Dirksen, D. Avenhaus,W. Merker, A. VonBally, G. Endoscopic double-pulse electronic-speckle-pattern interferometer for technical and medical intracavity inspection. *Appl. Opt.* 2000,39,3899-3905.
[74] Kandulla, J.; Kemper,B.; Knoche, S. VonBally, G. Two-wavelength method for endoscopic shape measurement by spatial phase-shifting speckle-interferometry. *Appl. Opt.* 2004, 43,5429-5437.
[75] Rastogi, P.K.(Ed.). *Special issue on Speckle and speckle shearing interferometry -1. Opt. Lasers Eng.* 1997,26,83-278.
[76] Rastogi, P.K.(Ed.). *Special issue on Speckle and speckle shearing interferometry -2. Opt. Lasers Eng.* 1997,26, 279-460.
[77] Krishna Mohan, N.; Rastogi, P.K.(Eds.). *Special issue in Recent developments in digital speckle pattern interferometry*. *Opt.Lasers.Eng.* 2003,40, 439-588.
[78] Meinlschmidt,P. Hinsch, K.D. Sirohi, R.S. (Eds.). *Selected papers on Electronic speckle pattern interferometry – Principles and Practice.* SPIE: Milestone series, MS132. SPIE, Optical Engineering Press. Washington, DC.1996.
[79] Rastogi, P.K.; (Ed.) *Digital speckle pattern interferometry and related techniques*. John Wiley, England, 2001.
[80] Steinchen, W.; Yang, L. *Digital shearography: Theory and Application of digital speckle pattern shearing interferometry*. PM100, SPIE, Optical Engineering Press. Washington, DC.2003.
[81] Stetson, K.A.; Electro-optic holography system for vibration analysis and non-destructive testing. *Opt. Eng.* 1987,18,1861-1883.
[82] Stetson, K.A.; Brohinsky, W.R. Wahid, J. Bushman,T. An electro-optic holography system with real-time arithmetic processing. *J. Non. Destruct. Eval.* 1989,8,69-77.
[83] Stetson, K.A.; Pryputniewicz, R.J. Measurement of vibration patterns using electro-optic holography. Measurement of vibration patterns using electro-optic holography. *Proc. SPIE* 1989,1162,456-467.
[84] Bushman, T.; Development of a holographic computing system. *Proc. SPIE* 1989.1162,66-77.
[85] Lokberg, O.J.; Høgmoen, K. Use of modulated reference wave in electronic speckle pattern interferometry. *J. Phys.E.* 1976,9,847-851.
[86] Tatam, RP.; Optoelectronic developments in speckle interferometry. *Proc. SPIE* 1996,2860,194-212.
[87] Doval, A.F.; A systematic approach to TV holography. *Meas. Sci. Technol.* 2000,11,R1-R36.
[88] Hariharan, P.; Oreb, B.F. Eiju,T. Digital phase shifting interferometry – a simple error-compensating phase calculation algorithm, *Appl. Opt.* 1987,26,2504-2506.
[89] Schwider, J.; Burow, R. Elssner, K.E. Grzanna, J, Spolaczyk, R. Merkel, K. Digital wave-front measuring interferometry – some systematic error sources. *Appl. Opt.* 1983,22,3421-3432.
[90] Kao, C.C.; Yeh, G.B. Lee, S.S. Lee, C.K. Yang, C.S. Wu, K.C. Phase shifting algorithms for electronic speckle pattern interferometry. *Appl. Opt.* 2002. 41,46-54.
[91] Haible, P.; Kothiyal, M.P. Tiziani, H.J. Heterodyne temporal speckle pattern interferometry. *Appl. Opt.* 2000,39,114-117.
[92] Kujawinska, M.; *Spatial phase shifting methods*. in Robinson, D.W..; Reid.G.T. (Eds.). *Interferogram Analysis.* Bristol and Philadelphia. IOP Publishing Ltd. 1993,141-193.

[93] Surrel, Y.; Design of algorithms for phase measurements by the use of phase-stepping. *Appl. Opt.* 1996,35,51-60.

[94] Joenathan, C.; Franze, B, Haible, P. Tiziani, H.J. Speckle interferometry with temporal phase evaluation for measuring large-object deformation. *Appl. Opt.* 1998,37,2608-14.

[95] Huntley, J.M.; Saldner, H. Temporal phase-unwrapping algorithm for automated interferogram analysis. *Appl. Opt.* 1993.32,3047-3052.

[96] Pouet, B.F.; Krishnaswamy, S. Techniques for the removal of speckle phase in electronic speckle interferometry. *Opt. Lett.* 1995.20,318-320.

[97] Davila, A.; Ruiz, G.H.; Kaufmann,G.H. Huntley, J.M. Measurement of sub-surface delaminations in carbon fibre composites using high-speed phase shifted speckle interferometry and temporal phase unwrapping. *Opt. Lasers Eng.* 40 2003, 447-458.

[98] Arevallilo Herràez, M.; Burton, D.R. Lalor, M.J. Gdeisat, M.A. Fast two-dimensional phase-wrapping algorithm based on sorting by reliability following a non continuous path. *Appl. Opt.* 2002,41,7437-7444.

[99] Chatters,T.C.; Krishnaswamy, S. Additive-subtractive phase-modulated shearography with synchronized acoustic stressing. *Exp. Mech.* 1995,30,159-165.

[100] Pouet, B.F.; Krishnaswamy, S. Additive-subtractive phase-modulated with electronic speckle pattern interferometry: Analysis of fringe visibility. *Appl. Opt.* 1994,33,6609-6616.

[101] Yang, L.; Steinchen, W. Kupfer, G. Mäckel, P. Vössing, F. Vibration analysis by means of digital shearography. *Opt. Lasers Eng.* 1998,30,199-212.

[102] Krishna Mohan, N.; Saldner, H.O. Molin, N.-E. combined phase stepping and sinusoidal phase modulation applied to desensitized TV holography. *J. Meas. Sci. Technol.* 1996,7,712-715.

[103] Krishna Mohan, N.; Saldner, H.O. Molin, N.-E. Recent applications of TV holography and shearography. *SPIE* 1996, 2861,248-256.

[104] Joenathan, C.; Khorana. A simple and modified ESPI system. *Optik.* 1991, 88,169-171.

[105] Joenathan, C. ;Torroba, R. Modified electronic speckle pattern interferometer employing an off-axis reference beam. *Appl. Opt.* 1991,30,1169-1171.

[106] Santhanakrishnan,T. Krishna Mohan, N. Sirohi, R.S. New simple ESPI configuration for deformation studies on large structures based on diffuse reference beam. *Proc SPIE* 1996, 2861,253-263.

[107] Flemming, T.; Hertwig, M. Usinger, R. Speckle interferometry for highly localized displacement fields. *Meas. Sci. Technol.* 1993,4,820-825.

[108] Saldner, H.O.; Molin, N.-E. Molin. Krishna Mohan, N. Desensitized TV holography applied to static and vibrating objects for large deformation measurement. *Proc SPIE* 1996.2860,342-349.

[109] Hertwig, M.; Flemming, T. Floureux, T. Aebischen, H.A. Speckle interferometric damage investigation of fiber-reinforced composities. *Opt. Lasers Eng.* 1996,24,485-504.

[110] Holder, L.; Okamoto, T Asakura, T. A. Digitial speckle correlation interferometer using an image fibre. *Meas. Sci. Technol.* 1993,4,746-753.

[111] Joenathan, C. ; Orcutt, C. Fiber optic electronic speckle pattern interferometry operating with a 1550 laser diode. *Optik.* 1993,100,57-60.

[112] Paoletti, D. ; Schirripa Spagnolo, G. Facchini, M. Zanetta, P. Artwork diagnostics with fiber-optic digital speckle pattern interferometry. *Appl. Opt.* 1993,32,6236-6241.

[113] Gülker,G.; Hinsch,K. Hölscher, C. Kramer A. Neunaber, H. ESPI system for in situ deformation monitoring on buildings. *Opt. Eng.* 1990,29,816-820.

[114] Joenathan, C.; Sohmer, A. Burkle, L. Increased sensitivity to in-plane displacements in electronic speckle pattern interferometry. *Appl. Opt.* 1995,34,2880-2885.

[115] Sirohi, R.S.; Burke, H. Helmers, H. Hinsch, K.D. Spatial phase shifting for pure in-plane displacement and displacement-derivative measurements in electronic speckle pattern interferometry (ESPI). *Appl. Opt.* 1997,36,5787-5791.

[116] Krishna Mohan, N. ; Andersson, A. Sjödahl, M. Molin, N.-E. Optical configurations in TV holography for deformation and shape measurement. *Lasers Eng.* 2000,10,147-159.

[117] Krishna Mohan, N. ; Andersson, A. Sjödahl, M. Molin, N.-E. Fringe formation in a dual beam illumination-observation TV holography: An analysis *Proc SPIE* 2000,4101A,204-214.

[118] Krishna Mohan, N. ; Svanbvo, A. Sjödahl, M. Molin, N.-E. Dual beam symmetric illumination –observation TV holography system for measurements. *Opt. Eng.* 2001,40,2780-2787.

[119] Shellabear, M.C.; Tyrer, J.R. Application of ESPI to three dimensional vibration measurement. *Opt. Laser Eng.* 1991,15,43-56.

[120] Bhat, G.K.; An electro-optic holographic technique for the measurement of the components of the strain tensor on three- dimensional object surfaces. *Opt. Lasers Eng.* 1997,26,43-58.

[121] Bowe, B.; Martin, S. Toal, T. Langhoff, A Wheelan, M. Dual in-plane electronic speckle pattern interferometry system with electro-optical switching and phase shifting. *Appl. Opt.* 1999,38,666-673.

[122] Krishna Mohan, N. ; Andersson, A. Sjödahl, M Molin, N.-E. Optical configurations for TV holography measurement of in-plane and out-of-plane deformations. *Appl. Opt.* 2000,39,573-577.

[123] Hung, Y.Y.; Electronic shearography versus ESPI for non destructive testing. *Proc SPIE* 1991, 1554B,692-700.

[124] Steinchen, W.; Yang, L.S. Kupfer,G. Mäckel, P. Non-destructive testing of aerospace composite materials using digital shearography. *Proc Instn Mech Engrs.* 1998,21-30..

[125] Hung, Y.Y.; Shearography a novel and practical approach to non-destructive testing. *J. Non des. Evalu.* 1989,8,55-68.

[126] Krishna Mohan, N. ; Saldner, H.O. Molin, N.-E. Electronic shearography applied to static and vibrating objects. *Opt. Commun.* 1994,108,197-202.

[127] Steinchen, W.; Yang, L.S. Kupfer,G. Mäckel, P. Vössing, F. Strain analysis by means of digital shearography: potential, limitations and demonstrations. *J. Strain Analy.* 1998,3,171-1998.

[128] Yang, L.; Ettemeyer, A.; Strain measurement by 3D-Electronic Speckle pattern interferometry: Potential, Limitations and Applications. *Opt. Eng.* 2003,42,1257-1266.

[129] Hung, Y.Y.; Shang, H.M. Yang, L. Unified approach for holography and shearography in surface deformation measurement and nondestructive testing. *Opt. Eng.* 2003,42,1197-1207.

[130] Ganesan, A.R.; Sharma, D.K. Kothiyal, M.P. Universal digital speckle shearing interferometer. *Appl. Opt.* 1988,27,4731-4734.

[131] Rastogi, P.K.; Measurement of in-plane strains using electronic speckle and electronic speckle-shearing pattern interferometry. *J. Mod. Opt.* 1996, 43,403-407.

[132] Patroski, K; Olszak, A. Digital in-plane electronic speckle pattern shearing interferometry. *Opt. Eng.* 1997,36,2010-2015.

[133] Aebischer, H.A.; Waldner, S. Strain distributions made visible with image-shearing speckle pattern interferometry. *Opt. Lasers Eng.* 1997, 26, 407-420.

[134] Joenathan, C.; Torroba, R. Simple electronic speckle-shearing-pattern interferometer. *Opt. Lett.* 1990,15,1159-1161.

[135] Hung, Y.Y.; Wang, J.Q. Dual-beam phase shift shearography for measurement of in-plane strains. *Opt. Lasers Eng.* 1996,24,403-413.

[136] Nakadate, S.; Phase shifting speckle shearing polarization interferometer using a Birefringent wedge. *Opt. Lasers Eng.* 1997,6,31-350.

[137] Krishna Mohan, N.;Masalkar, P.J. Sirohi, R.S. Electronic speckle pattern interferometry with holo-optical element. *Proc SPIE* 1992,1821, 234-242.

[138] Joenathan, C.; Bürkle, L. Electronic speckle pattern shearing interferometer using holographic gratings. *Opt. Eng.* 1997,36,2473-2477.

[139] Cloud, G.L. Ding, X. Raju, B.B. Real-time digital speckle interferometry to measure crack length. *Exp. Tech.* 2002,26,19-21.

[140] Basanta Bhaduri,; Debnath, S.K. Krishna Mohan, N. Kothiyal, M.P. Analysis of speckle shear fringes for NDT application. ICEM 2005 *Exp. Mech.*12,1-2.

[141] Waldner, S.; Removing the image-doubling in shearography by reconstruction of the displacement fields. *Opt. Commun.* 1996,127,117-126.

[142] Waldner, S.; Removing the image-doubling in shearography – theory and application. *Proc SPIE* 1996,2944,247-255.

[143] Waldner, S.; Brem, S. Compact shearography system for the measurement of 3D deformation. *Proc SPIE* 1999,3745,141-148.

[144] Wegner, W.; Ettemeyer, A. Fast stress and strain analysis with Microstar. Dr. Ettemeyer Application *Report* No. 05-98, Neu-Ulm, Germany.

[145] Schubach, H.R.; Ettemeyer, A. Investigations on Aluminium Alloys with a 3D-ESPI-System. Dr. Ettemeyer Application *Report* No. 01-97, Neu-Ulm, Germany.

[146] Ettermeyer, A. Laser optical strain sensor - Techniques and Applications, *Proc SPIE* 1996,2860, 213-216.

[147] Kastle,R.; Hack, E. Sennhauser, U. Multiwavelength shearography for quantitative measurements of two-dimensional strain distribution. *Appl. Opt.* 1999,38,96-100.

[148] Groves, R.; James, S.W. and Tatam, R.P. Polarisation-multiplexed and phase-stepped fibre optic shearography using laser wavelength modulation. *Proc SPIE* 1999,3744,149-157.

[149] Pezzoni,R.; Krupka, R. Laser shearography for non-destructive testing of large area composite helicopter structures. *Insight* 2001,43,244-248.

[150] Krishna Mohan, N.; Saldner, H.O. Molin, N.-E. Electronic speckle pattern interferometry for simultaneous measurement of out-of-plane displacement and slope. *Opt. Lett.* 1993,18,1861-1863.

[151] Rastogi, P.K. ; Measurement of curvature and twist of a deformed object by electronic speckle shearing pattern interferometry. *Opt. Lett.* 1996,21,905-907.

[152] Fomitchov, P.A.; Krishnaswamy,S. A compact dual-purpose camera for shearography and electronic speckle-pattern interferometry. *Meas. Sci. Technol.* 1997,8,581-583.

[153] Basanta Bhaduri,; Krishna Mohan, N. Kothiyal, M.P. A dual function ESPI system for out-of-plane displacement and shearography. *Opt Lasers Eng.* 2006,44,637-644.

[154] Saldner, H.O.; Molin, N.-E. Krishna Mohan, N. Simultaneous measurement of out-of-plane displacement and slope by electronic holo-shearography. *SEM Conf. Exp. Mech.* 1994, 337-341.

[155] Huang.J.R.; Ford, H.D. Tatam, R.P. Slope measurement by two-wavelength electronic shearography. *Opt. Lasers Eng.* 1997,27, 321-333.

[156] Rastogi, P.K.; An electronic pattern speckle shearing interferometer for the measurement of surface slope variations of three-dimensional objects. *Opt. Lasers Eng.* 1997,26,93-100.

[157] Creath, K.; *Temporal phase shifting methods.* in Robinson, D.W..; Reid.G.T. (Eds.). *Interferogram Analysis.* Bristol and Philadelphia. IOP Publishing Ltd. 1993. 94-140.

[158] Huntley, J.M.; Kaufmann, G.H. Kerr, D. Phase-shifted dynamic speckle pattern interferometry at 1 kHz. *Appl. Opt.* 1999,38,6556-6563.

[159] Svanbro, A.; Huntley, J.M. Davila. Optimal re-referencing rate for in-plane dynamic speckle interferometry. *Appl. Opt.* 2003, 42, 251-258.

[160] Furlong, C.; Pryputniewick. Optoelectronic characterization of shape and deformation of MEMS accelerometers used in transportation application. *Opt. Eng.* 2003.42,1223-1231.

[161] Hung, Y.Y.; Chen, F. *Special section on optical metrology, developments and applications in the transportation industry. Opt. Eng.* 2003,42,1186-1369.

[162] Lokberg, O.J.; Seeberg, B.E. Vestli, K. Microscopic video speckle interferometry. *Opt. Lasers Eng.* 1997,26,313-330.

[163] Kujawinska, M.; Gorecki, C. New challenges and approaches to interferometric MEMS and MOEMS testing. *Proc SPIE* 2002,4900,809-823.

[164] Aswendt, P.; Micromotion analysis by deep-UV speckle interferometry. *Proc. SPIE* 2003,5145,17-22.

[165] Patorski, K.; Sienicki, Z. Pawlowski, M. Styk, A.R. Jozwicka, A. Studies of the properties of the temporal phase-shifting method applied to silicone microelement vibration investigations using the time-average method. *Proc SPIE,* 2004.5458,208-219.

[166] Patorski, K.;Jozwicka, A. Kalinowski, A. Interferometric measurements of the vibration of a silicon microbeam/cantilever unit used in atomic force microscopy (AFM). *Proc SPIE* 2003,5229,275-278.

[167] Murukeshan, V.M.; Sujata, N. Seah, L.K. Ong, L.S. *Proc SPIE* 2005,5651,321-328.

[168] Dyrseth, A.A.; Measurement of plant movement in young and mature plants using electronic speckle pattern interferometry. *Appl. Opt.* 1996,35,3695-3701.

In: New Developments in Lasers and Electrc-Optics Research ISBN 1-59454-771-8
Editor: William T. Arkin, pp. 223-240 

*Chapter 7*

# RECENT PROGRESSES OF MULTI-WAVELENGTH SEMICONDUCTOR LASER DIODES FOR OPTICAL INFORMATION STORAGE SYSTEMS

***Hyunchul Ko[1] and M.W. Cho[2]***

[1] Department of Electrical and Computer Engineering, University of South Alabama, 308 Univ. Blvd. Mobile AL 36695, USA

[2] Institute for Materials Research, Tohoku University, 2-1-1 Katahira, Aoba-ku, Sendai, 980-8577, Japan

## ABSTRACT

Due to the rapid technical advances of semiconductor laser diodes as a light source, the ultra-high-density optical information storage systems have been accelerated to commercialize in various formats, such as CD (Compact Disk), CD-ROM (Read-Only Memory), CD-R (-Recordable), CD-RW (-ReWriteable), DVD (Digital Versatile/Video Disc), DVD-R, and DVD-RW, for many applications, i.e., audio, music, image, movie, and computer data storage. In this complex situation, the compatibility of diverse storage media discs in a playback/writing system is one of the most important issues for system development. This problem has caused complications in the optical components as well as the non-compact system size. A semiconductor laser is an integral part among many optical components because the wavelength and output power of laser light source is a key factor of the media disc specifications.

In order to realize signal compact optical pick-up module that can respond to various media discs, a multiple wavelength laser light source is indispensable. There have been many attempts to develop multi-wavelength laser diodes. The typical approaches are to use two individually packaged laser diodes or hybrid-integrated two laser chips in the same package. However, those techniques require sophisticated assembly processing resulting in low reliability and high cost.

In this chapter, we reviewed the recent progress of multi-wavelength laser diodes and how it can effectively realize dual or multiple wavelength laser diodes in the future. We proposed, and successfully demonstrated, new concept device structures for monolithically-integrated two laser diodes with different wavelengths on a GaAs wafer in parallel and vertical direction.

## 1. INTRODUCTION

In the transitional period of an optical media disc, various kind of optical pick-up module has been developed for various formats, such as CD, CD-ROM, CD-R, CD-RW, DVD, DVD-R, and DVD-RW, for many applications, i.e., audio, music, image, movie, and computer data storage. Figure 1 shows the development trend for optical media disc since 1980's. During recent 5 years, the transition to short wavelength and high output power has drastically accelerated the progress of CD and DVD playback systems. The track pitch distance and minimum pit/land depth on a disc depend on the wavelength of LDs, and also are critical factors to decide recording capacity. To increase of recording capacity, double sided structure disc was also developed. The recording capacity of optical storage media could be extended to ~ 8.5 Gbit in the double sided structure of one layer media disc by the excellent performance of the semiconductor laser diodes as a light source of a playback system.

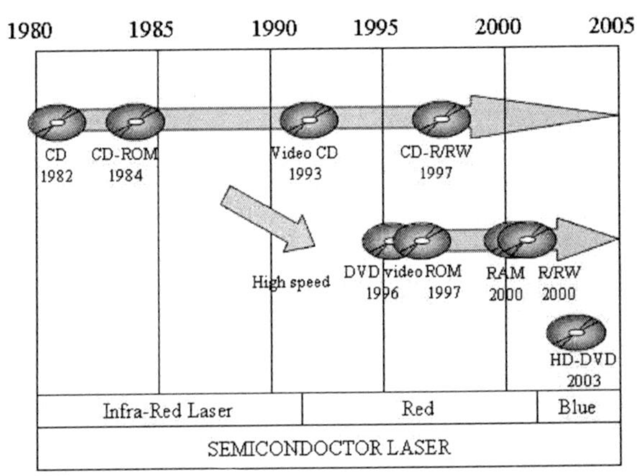


Figure 1. The development trend for optical media discs since 1980's.

Fig 2 shows the disc specifications of CD-R and DVD-R. A 650 or 635 nm laser diodes (LDs) and 780 nm LDs are used as light source for CD-R and DVD-R, respectively. In addition to wavelength, the access speed is mainly dependent on the output power of laser diode in the same wavelength. Recently, HD-DVD (High Density-Digital Versatile Disk) called "Blue-ray disc" as a new generation of storage media, which can record maximum up to 27 Gbit video data on a single sided phase change disc of 12 cm in diameter, has been realized by adopting a 405nm blue-violet GaN-based laser diode [1]. The performance improvement of LDs based on compound semiconductors has leaded a dramatic progress of optical media disc playback systems. All kinds of storage media discs, DVD-ROM, CD-R and CD-RW are immediately required to different optical pick-up heads. Unfortunately, one kind of light source cannot cover all optical media disc. For example, a 780 nm laser source must be added to a DVD pickup-head specially for reading CD-R's data because the reflectivity of CD-R media is lower than 10% for 650 nm light. In order to operate both

DVD-R and CD-RW in one playback system, each optical pick-up head or multi-wavelength optical source corresponding to each disc has to exist within one playback system. Recently, a high-density optical disk system which has a capacity of more than 20 Gbit has been realized using a blue LD (450 nm of wavelength) as the light source of the optical pick up. In such system, a red LD (650 nm of wavelength) will still be need to playback the conventional DVD, DVD-R, DVD-RW, and DVD-RAM discs with the same system.

| | CD-R | DVD-R |
|---|---|---|
| Disc Diameter | 80 mm, 120mm | 80 mm, 120mm |
| Disc Thickness | 1.2 mm | 1.2 mm |
| Laser Wavelength | 780 nm | 635, 650 nm |
| Numerical Aperture | 0.45 | 0.55, 0.60 |
| Track Pitch | 1.6 um | 0.74 um |
| Shortest Pit /Land length | 0.83 | 0.4 um |
| Data capacity | 650 - 700 MB | 1.50 GB and larger |

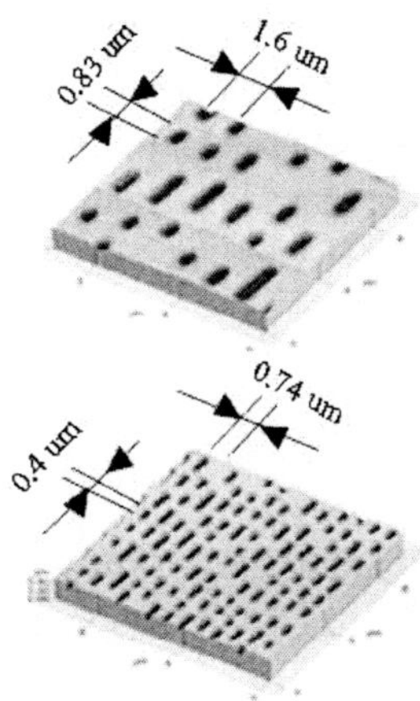


Figure 2. The comparison pf the disc specifications of CD-R and DVD-R.

In this complex situation, the compatibility of diverse storage media discs in a playback/writing system is one of the most important issues for system development. This problem has caused complications in the optical components as well as the non-compact system size. A semiconductor laser is an integral part among many optical components because the wavelength and output power of laser light source is a key factor of the media disc specifications.

In order to realize single compact optical pick-up module that can respond to various media discs, a multiple wavelength laser light source is indispensable. There have been many attempts to develop multi-wavelength laser diodes. The typical approaches are to use two individually packaged laser diodes or hybrid-integrated two laser chips in the same package. However, those techniques require sophisticated assembly processing resulting in low reliability and high cost. The other approach is a monolithically-integrated two-wavelength laser diode on a GaAs wafer in parallel and vertical direction by multi-step epitaxial growth and sophisticated process. Therefore, a novel multiple-wavelength light emitter is strongly required to solve the fundamental problems and to improve the device performance.

In this chapter, we reviewed the recent progress of multi-wavelength laser diodes and how it can effectively realize dual or multiple wavelength laser diodes in the future. We proposed, and successfully demonstrated, new concept device structures for monolithically-integrated two laser diodes with different wavelengths on a GaAs wafer in parallel and vertical direction.

## 2. Hybrid-Integrated Two-Wavelength Laser Diodes

### 2-1. Conventional Type Lasers

A conventional hybrid type two-wavelength laser diode (TWLD) where two distinct laser chips were mounted on a single stem were equipped with two optical paths to read out DVD and CD signals [2]. Two well-developed semiconductor laser diodes with different emitting wavelength are used in a conventional hybrid type. Although the hybrid type TWLD has the advantage of adopting optimized lasers and giving a wide selectivity of laser wavelength and output power, it has difficult in precise definition of the two laser emission spots that could increase the laser packaging cost. Consequently, the opto-mechanical pickup design becomes complicated and the volume of a head is enlarged.

There are two kinds of package type in the conventional hybrid TWLD as shown in fig. 3. Two laser chips are vertically or horizontally bonded on one stem. For the conventional hybrid TWLD, the most crucial issue is "How we can reduce the distance between two emitting spots". In the case of horizontal type, two chips were parallel arranged side by side on the same submount, however with a certain separation between the two emitting spots as shown in fig. 3 (a). In the case of vertical type, two chips were placed on the upper and lower step of the stem as shown in fig. 3 (b). The distance between the two emitting spots was primarily determined by the width of the chip and stem design. According to the stem design, the chip arrangement can be different from that mentioned above. Fig. 4 shows a scanning electron microscopy image of a horizontal type TWLD packaged with 650 and 780 nm LD. The emitting position of the two elements are 250 ~ 300 μm apart, which still requires separate optical alignment.

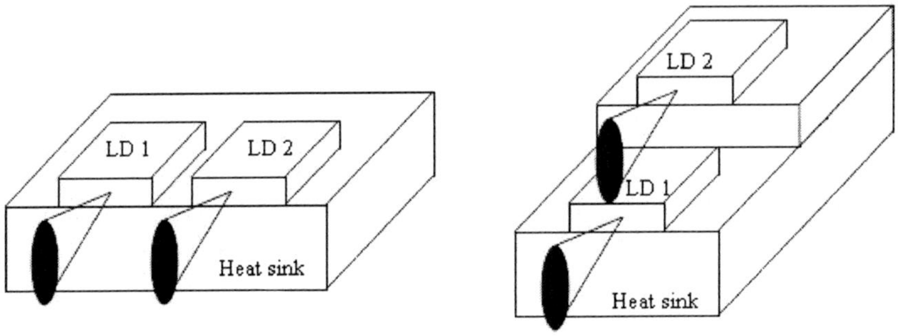


(a) (b)

Figure 3. Two kinds of package type in the conventional hybrid TWLD. Two chips are (a) horizontally and (b) vertically bonded on the same stem.

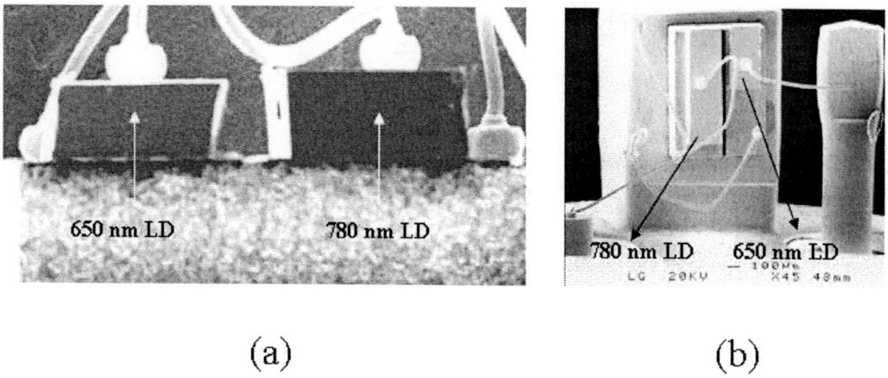


Figure 4. A SEM image of a horizontal type TWLD packaged with 650 and 7801 nm LD. (a) front side view and (b) top side view

## 2-2 Holographic Pick-Up Module Type Lasers

In order to realize compact pick module, a holographic pick up-head module, which can package a light source, photo detectors, beam splitter, and servo generating optical elements, was suggested. Figure 5 shows typical two laser chips packaging for holographic pick-head module [3]. The laser chips are bonded on the upper and lower steps of the stem, respectively. The diachronic micro-mirror has two reflecting surfaces separated by 0.16 mm designed to compensate for the offset between the optical axes of the two beams. It is essentially a 45-degree tilted plate with thickness 0.16 mm bonded to a micro-mirror. The height of the submount is related to the separation of the two reflecting surfaces. The first surface has a wavelength select coating that reflects 650 nm laser light and transmits 780 nm light, while the second surface totally reflects 780 nm light. In the design of the stem and mirror, astigmatism that is introduced when the divergent laser beam passes through twice the 45 degree tilted plate should be considered to meet the strict optical requirement. It results in the advantages of cost reduction in packaging the component and the compactness of the unit. However, the construction of a module requires a mechanical alignment precision on the order of microns as well as allowance for wavelength shift of the laser diodes over several nanometers. To satisfy these strict alignment requirements, a robust focusing and tracking method have been being studied.

## 2-3 Chip-to-Chip Bonding Type Lasers

Though the conventional hybridized LD was attempted for compactness, it was needed to align separate optical paths corresponding to each LD because the distance between the two emitting spots is above a few hundred μm. In order to make the two emitting spots closer, we have vertically stacked two chips on the same stem. Fig. 6 shows the schematic diagram of a vertically hybridized chip-to-chip bonding package on a stem, and patterned solder to bond two chips. Three electrodes were formed, and *p*-electrode was common for two LDs.

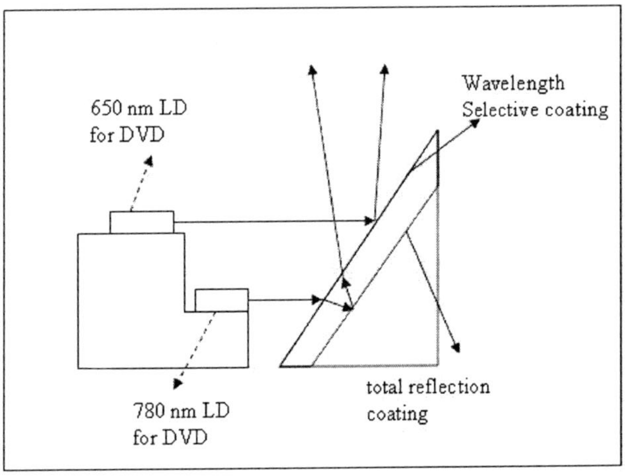


Figure 5. Holographic pick-head module packaging with two laser chips. The laser chips are bonded on the upper and lower steps of the stem, respectively.

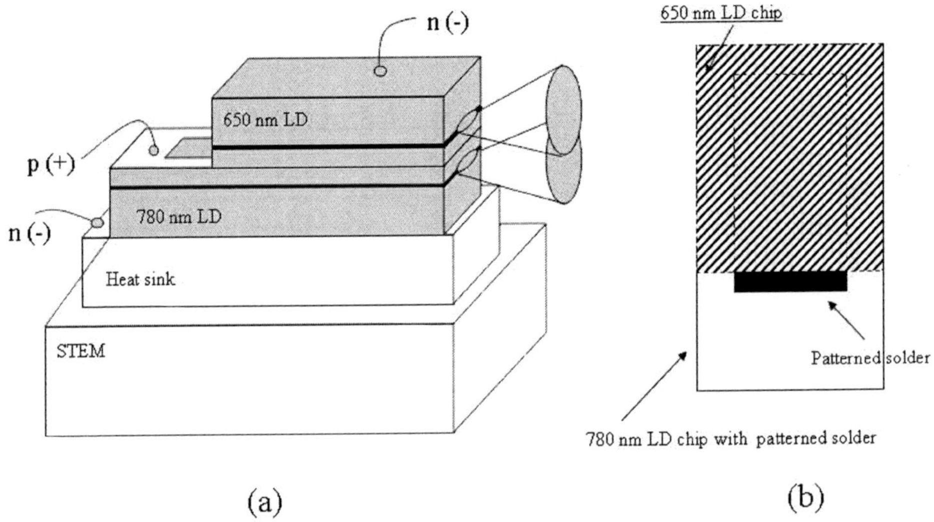


Figure 6. (a) The schematic drawing of a vertically hybridized chip-to-chip bonding on a stem. (b) Top side view of two LDs patterned with solder.

In this experimental, a high power 780 nm LD was used for high speed CD-RW, and a 650 nm LD was standard for DVD-R, respectively. It is the first attempt to make a CD-RW and DVD-R playback system with chip-to-chip bonding type. Fig. 7 shows the schematic diagram of the 650 and 780 nm LD structures, respectively. For 780 nm LD, the optimized graded index separate confinement heterostructure-multiple quantum well (GRINSCH-MQW) structure was used. The active layer was consisted of $Al_{0.3}GaAs/Al_{0.1}GaAs$ MQW, and the optical confinement layer changed the Al composition of AlGaAs layer from 32 % to 55 % within 516 Å. Epitaxial growth was carried out by low pressure MOCVD on an n-type (100)-5° off toward <011> n-GaAs substrate. A window mirror structure was adapted in the facet region so as to increase the catastrophic optical damage level. For 650 nm LD, AlGaInP-GRINSCH-MQW structure was used. The structure is a buried ridge stripe that is widely used in laser diodes applied to light sources in an optical disc system. An InGaP/AlGaInP MQWs and the AlGaInP layers were grown on n-type on a (100)-15° off toward <011> n-GaAs substrate as active and optical confinement layers, respectively.

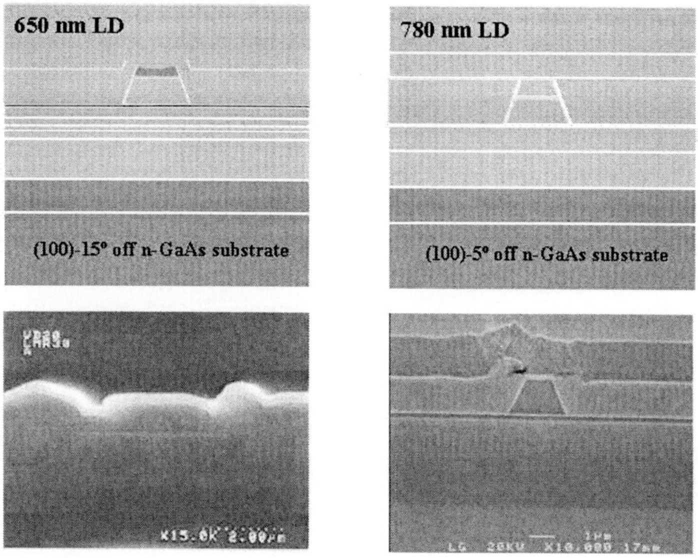


(a) (b)

Figure 7. The schematic diagram and SEM images of (a) 650 and (b) 780 nm LD structures.

**Table 1. Thermal conductivity of various materials.**

| | GaAs | GaN | $Al_2O_3$ | AlN | Si |
|---|---|---|---|---|---|
| Thermal conductivity (W/cm K) | 0.56 | 1.3 | 2.51 | 2.85 | 1.51 |

First, a high power 780 nm laser chip was junction-up bonded on a submount using a hard solder at high temperature. The chip size is 250 × 500 $\mu m^2$ and 250 × 900 $\mu m^2$ for 650 and 780 nm LDs, respectively. Then, a 650 nm LD is junction-down bonded on a 780 nm LD with the patterned soft solder at low temperature. Before packaging, hard solder metal was deposited on p-metal of 780 LD with 200 × 500 $\mu m^2$. Solder selection is the most important to bond two LDs without re-melted phenomenon of the hard solder. Three electrodes were formed with p-common. The distance between emitting spots can be minimized below a few μm apart because only solder and p-metal existed in interface between the two LDs. One optical path is distinguishable merit as compared with other TWLD LDs, and is an ultimate solution for TWLD.

Figure 8 shows the output power versus current (*L-I*) characteristics and far field angles for the both LDs under continuous wave (cw) operation at room temperature after packaging.

Two LDs were uncoated for both facets. The threshold current is 48 mA, maximum power ($P_0$) above 100 mW with kink-free operation for 780 LD. The threshold current is 43 mA, and maximum power ($P_0$) 10 mW for 650 LD. The beam divergence perpendicular to the junction at 5 mw and 100 mW was 27.7° and 14.7°, and that parallel to the junction was 7.0° and 8.7°, respectively. The cavity length is 250 × 500 $\mu m^2$ and 500 × 900 $\mu m^2$, respectively.

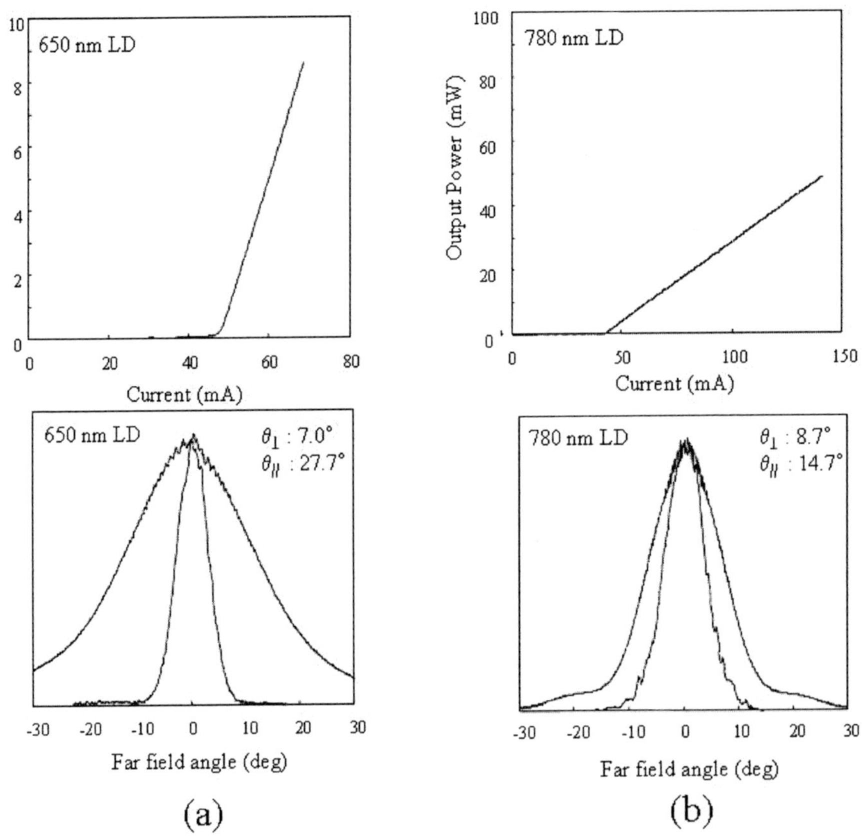


Figure 8. The output power versus current (*L-I*) characteristics and far field patterns for the (a) 650 nm LD and (b) 780 nm LD under continuous wave (cw) operation at room temperature after chip-to-chip bonding on a stem.

Basically, the 780 nm LD on the submount has not any problems for thermal dissipation due to junction-up bonding. But, in the case of the 650 nm LD, the thermal dissipation is a critical problem because it puts on the 780 LD of the thick ~ 80 μm. Generally, the thermal accumulated during operation of LD is one of the critical damages to device degradation. Although thermal conductivity of GaAs material is lower than that of the other heat-sink materials such as Si and AlN, GaAs LD can be expected as a role of heat sink as operating in the low power for DVD-R. The thermal of 650 LD during operation can flow through the 780 LD into heat sink because of metal solder with the high thermal conductivity in the interface between 780 LD and 650 LD. The heat sink material, the package type, the heat capacity of device, the metal system for p-electrode and the solder material much critically influence the thermal dissipation. Table 1 shows the thermal conductivity of various materials. In order to improve heat dissipation, we have attempted to insert the conductive type submount between two LDs as shown in fig. 9. Then, the generated heat can be effectively dissipated through the side edge of submount. Modifying the thickness of the submount can easily reduce the distance between the two emitting spots.

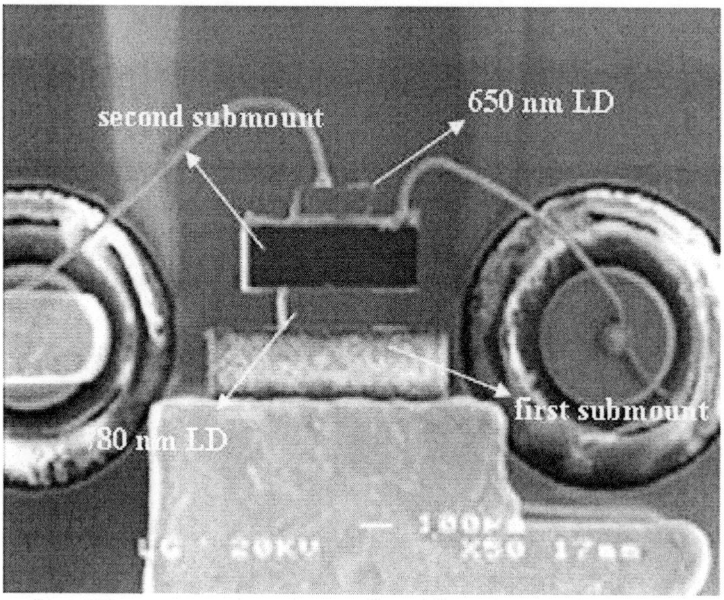


Figure 9. SEM image of chip-to-chip bonding packaged by the conductive type submount inserted between two LDs.

If the concept of chip-to-chip bonding technique applies to a blue GaN laser, the various combinations such as HD-DVD and DVD, HD-DVD and CD can be obtained. Generally, a GaN device was grown on sapphire substrate, and the thermal conductivity of GaN and $Al_2O_3$ substrate is higher than that of GaAs and the other submount materials. It is expected that GaN LD on $Al_2O_3$ totally acts on as a superior heat sink.

# 3. Monolithic-Integrated Two Wavelength Laser Diodes

The first monolithic type LD is originated from a laser diode array for high power operation above a few watts. However, it is not a light source for playback system of optical media disc but for applications like laser marker in the industrial or laser operation in the medical. Recently, a monolithic type two-wavelength laser diode on GaAs substrate is suggested for further reductions in cost and size of the optical disk playback system. TWLD combining two laser diodes is monolithically integrated on a single chip by photolithography processes and re-growth. The device structure and lasing characteristics of monolithic-type dual wavelength LD developed to date was summarized in the table 2.

**Table 2. Device structure and lasing characteristics of Monolithic TWLD.**

| Device structure of Monolithic TWLD | cavity length ($\mu m^2$) | stripe width ($\mu m$) | facet coating | Emitting position ($\mu m$) | lasing characteristics | | | Ref. |
|---|---|---|---|---|---|---|---|---|
| | | | | | $I_{th}$ (mA) | $\theta_{\perp}/\theta_{//}$ (deg.) | Po (mW) | |
| 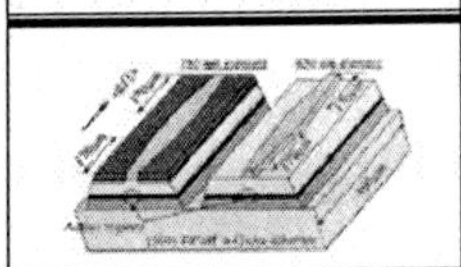<br>(a) VILD | 400 $\mu m$ (650 LD)<br>400 $\mu m$ (780 LD) | 6.5 (center) 3.5 (edge) (650 LD)<br>8.0 (center) 5.0 (edge) (780 LD) | 32 %/75 % (650 LD)<br>28 %/75 % (780 LD) | 120 | ~50 (650 LD)<br>~55 (780 LD) | 35.7/10.0 (650 LD)<br>34.9/14.4 (780 LD) | low power (650 LD)<br>low power (780 LD) | [4] |
| 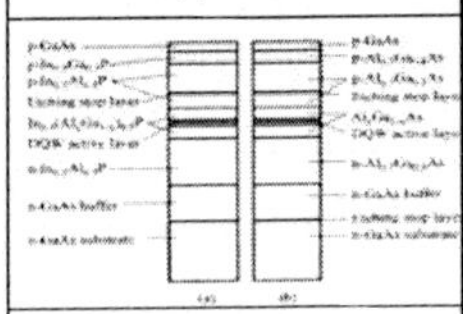<br>(b) TWINLD | 300 $\mu m$ (650 LD)<br>300 $\mu m$ (780 LD) | 3.5 (650 LD)<br>2 (780 LD) | uncoated | 150 | ~12 (650 LD)<br>~14 (780 LD) | 37/6 (650 LD)<br>37/17 (780 LD) | low power (650 LD)<br>35 (780 LD) | [5] |
| 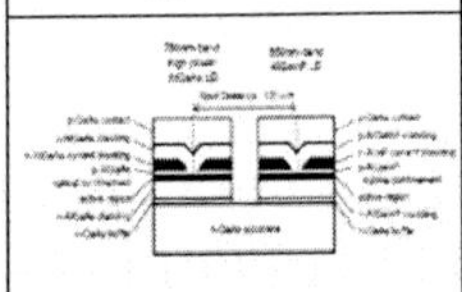<br>(c) RISA LD | 800 $\mu m$ (650 LD)<br>800 $\mu m$ (780 LD) | - | 19 %/86 % (650 LD)<br>10 %/96 % (780 LD) | 120 | ~45 (650 LD)<br>~25 (780 LD) | - (650 LD)<br>17/8.2 (780 LD) | 10 (650 LD)<br>210 (780 LD) | [6,7] |
| ZnCdSe/ZnSe/ZnMgBeSe SCH-SQW<br>InGaP/InGaAlP SCH-MQW<br>GaAs 15°-off sub.<br>(d) Green & red LD | 300 $\mu m$ (650 LD)<br>600 $\mu m$ (780 LD) | - | | ~ 3 $\mu m$ | 115 kW/cm$^2$ (650 LD)<br>84kW/cm$^2$ (780 LD) | - | | [8] |

## 3-1 Visible and Infrared Bands Integrated Type Lasers

Recently, Sony's research group reported a reliable monolithic-integrated two-wavelength visible and infrared laser diode structure (VILD) using two-step MOCVD growth [4]. The 650 nm element's epitaxial islands, and the 780 nm elements' epitaxial islands parallel to the <01-1> direction, are positioned side by side on a GaAs substrate as shown in table 2 (a). Both elements have a gain guiding structure fabricated by stripe pattern. A separated-confinement-heterostructure multi-quantum-well (SCH-MQW) structure is used for each active region to minimize the threshold current. The wave-guide stripe of the 650 nm element is a mesa-shaped tapered stripe, which forms a gain-guiding cavity. For the 780 nm element, a tapered-stripe gain-guiding structure is used, which is prepared using ion implantation. The length of the tapered stripe is 170 μm in both LDs. The stripe width is 6.5 μm at the center and 3.0 μm at the facet edge for the 650 nm LD, and 8.0 μm at the center and 5.0 μm at the facet edge for the 780 nm LD. The emitting positions of the two LDs are 120 μm apart. The laser chip is 120 mm thick and 300 mm wide. The cavity length is 400 μm. The reflectivity of the coated front and rear facets of for the 780 nm LD are 28 % and 75 % respectively, for the 650 nm LD 32 % and 75 % respectively. The waveguide stripes were formed using one-step photolithography, so the stripes for the 650 nm element and the 780 nm element were precisely positioned. Finally, after conventional processes, the laser chip is mounted on a Si-submount, which is mounted on a Cu heat sink in a conventional 5.6 mm $\phi$ package with four electrodes.

For the 650 and 780 nm LD, the operating currents at 5 mW were 57.0 and 61.5 mA, and kink-free operation was achieved up to 12 and 15 mW, respectively. The characteristics temperatures were estimated to be about 150 and 170 K, respectively. The beam divergence perpendicular to the junction at 5 mW was 35.7° and 34.9°, and parallel to the junction was 10.0° and 14.4°, respectively. Both elements showed stable fundamental transverse mode operation. The astigmatism was about 25 μm for both elements, which was a little smaller than that of a conventional gain-guiding laser with a straight stripe. The lasing spectrum of the 780 nm LD showed multimode emission with narrow linewidth longitudinal modes, which is typical for the tapered-stripe gain guiding laser diode that is widely used in the CD playback system. The lasing spectrum of the 650 nm LD was also multimode, and sometimes showed linewidth broadening of the longitudinal modes. The mean time to failure of the 650 and 780 nm elements was over 10,000 hr and over 100,000 hr from aging test under cw operation at 70 °C for the 650 and 780 nm elements at 6 and 7 mW, respectively. This performance allows VILD to support the 6X-speed DVD and 48X-speed CD playback system..

Although the reliable two-wavelength laser diode was realized using VILD structure, it has some restriction for various applications such as CD-RW and DVD-RW media that require high output power laser characteristics. In order to realize high power operation with stable fundamental transverse mode, it is critical to suppress spatial hole-burning in the mirror facet by reducing the operating current. Typically, the high power laser diode has adopted the index-guiding structure and facet coating process suitable to each wavelength for lowering threshold current and increasing the COD (catastrophic optical damage) level. The more complicated lithography and re-growth process were required for high power operation to achieve recording CD-R and DVD-RAM discs.

Recently, T. Lu et al. has also reported on TWINLD (Two-wavelength integrated laser diode), which has been realized by monolithically combining two different laser material structures in a single chip utilizing the MOCVD re-growth technique [5]. The TWINLD emits a red wavelength $\lambda$=650 nm and an infrared wavelength $\lambda$=780 nm for operation of DVD and CD, respectively. This is nearly the same structure as the VILD structure except for using aluminum-free DQW (double quantum well) in the active layer and a $SiO_2$ mask in the re-growth process as shown in table 2 (b). The TWINLD structure was remarkable for using an aluminum-free active area structure that could allow high-power operation for 780 nm LD. The 780 nm laser structure has an active region consisting of two –0.1% tensile strained $In_{0.08}Ga_{0.92}As_{0.83}P_{0.17}$ wells and one $Al_{0.3}Ga_{0.7}As$ barrier, which is adopted as p- and n-cladding layer with the same thickness as the cladding layers of the 650 nm laser. The first one is grown under the buffer layer in order to obtain a flat etched surface and to accurately control the depth of the re-growth area. And the second one is aimed for the p-type cladding layer to ensure the precision of etching depth when making a ridge wave-guide structure.

A 2 μm wide ridge for the 780 nm laser and a 3.5 μm wide ridge for the 650 nm laser were formed with a 150 μm separation between the two emission spots. From typical laser characteristics, the threshold current is 12 mA at 20 °C and the one end slope efficiency is 0.6 W/A for both facets uncoated for 650 nm laser. Similarly, for 780 nm lasers, the threshold current is 14 mA at 20 °C and the one end slope efficiency is 0.4 W/A for both facets uncoated for 780 nm lasers. The single ended output power under cw operation can reach up to 35 mW for both facets uncoated. Far field angles of 650 and 780 nm LDs are 6°/37° ($\theta_{\|}/\theta_{\perp}$) and 17°/37° for the direction parallel and perpendicular to the junction plane, respectively. Although TWINLD structure shows much lower threshold current as compared with VILD structure, it is very closely related with ridge width and cavity length. Considering threshold current density ($J_{th}$) and aspect ratio, the device structure should be optimized. Finally, these factors give critical effect to device reliability.

Matsushita group reported a real refractive index guided self-aligned (RISA) structure to realize high power operation with low threshold current density [6, 7]. RISA structure is a promising optical waveguide structure for the reduction of the operating current because of negligible absorption loss in the current blocking and the built-in refractive index difference. Self-sustained pulsation (SSP) is advantageous for the reduction of additional components such as a quarter-wavelength phase plate or a high frequency superposition circuit in the optical head. In this means, SSP LDs are attractive devices as they can suppress optical feedback noise, due to the fact that mode competition can be suppressed by multi-longitudinal mode oscillation. Using RISA structure, they succeeded in developing SSP 650 nm band LD structure maintained at temperature up to 75°C at an average output power of 7 mW due to reduced current density. In order to obtain higher temperature SSP, a SA (saturable absorbing) layer, which was used for maintaining SSP operation under high power and high temperature, is incorporated in the 650 nm band LD.

The 780 nm AlGaAs and 650 nm AlGaInP lasers are monolithically integrated on a GaAs substrate as shown in table 2 (c). The RISA waveguide structure is adopted for both LDs. The 780 LD has an AlGaAs bulk active layer and an AlGaAs current blocking layer transparent in the lasing wavelength. The 650 nm LD has GaInP/AlGaInP compressively strained multiple quantum wells as the active region, and an AlInP current blocking layer, which are also transparent in the lasing wavelength. The LDs are fabricated by four-step MOCVD growth. The distance between the emitting points of 780- and 650-nm band LDs is set to be 120 μm, and cavity length is 800 μm. The front and rear facets are coated with reflectivity of 10 and 96% for the wavelength of 780 nm, respectively. For the wavelength of 650 nm, the reflectivity of each facet is 19 and 86 %, respectively. As the material with high refractive index for the high reflectivity coating on the rear facet, $TiO_2$ films are selected in order to avoid optical loss and subsequent heat generation in the coating films.

For 780 nm LD, the maximum output power over 200 mW is attained and the operating current is as low as 240 mA at 200 mW under cw operation. Such a high output power and low operating current at a high power level are realized due to reduced internal loss in RISA structure. No kinks are observed up to the COD level, which indicates that spatial hole-burning is effectively suppressed. For 780 nm band LD, beam divergence angles perpendicular and parallel to the junction plane are 17° and 8.2°, respectively, resulting in a smallest aspect ratio of 2.1 as compared with the other TWLDs. It is shown that the RISA structure is highly effective for improving the SSP characteristics in monolithically integrated dual wavelength LDs. Moreover, it means that high power operation based on index-guiding type on monolithic laser chip is applicable to CD-RW media.

## 3-2 Green and Red Bands Integrated Type Lasers

A monolithic type dual wavelength developed up to date has inherent problems such as wavelength selectivity limited to GaAs-based material and the distance between the emitting points between two LDs due to horizontally integrated arrangement. In order to overcome the fundamental problems, we have attempted for realizing vertically integrated ZnSe-based II-VI material grown on InGaP-based III-V structure as shown in table 2 [d] [8].

The merits of this structure are as follows: 1) lattice matching epitaxial growth could be achieved both III-V and II-VI materials on a GaAs substrate, 2) the band gap of these material systems covers wide range from blue to infrared, therefore various combination is possible, 3) simple packaging will be possible as a result of monolithic integration because emitting light will be placed in a few μm distances, which will greatly simplify the optics design, 4) in the point view of epitaxial growth, thermal damage of III-V laser structure due to second MBE growth can be ignored, because II-VI growth temperature of MBE is considerably lower than III-V of MOCVD.

However, InGaP epitaxial layers grown on (001) GaAs substrate by MOCVD has an ordering problem. It is well known that (001) GaAs substrate misoriented toward [110] prevent ordering in the InGaP/GaAs system [9, 10]. On the other hand, there are some disadvantages for MBE growth of ZnSe-base II-VI on misoriented GaAs substrate, since the surface of the misoriented toward [110] GaAs substrate is covered with Ga-step. It is hard to grow ZnSe epitaxial layer on this substrate by MBE due to increased possibility of reaction with Ga and Se.

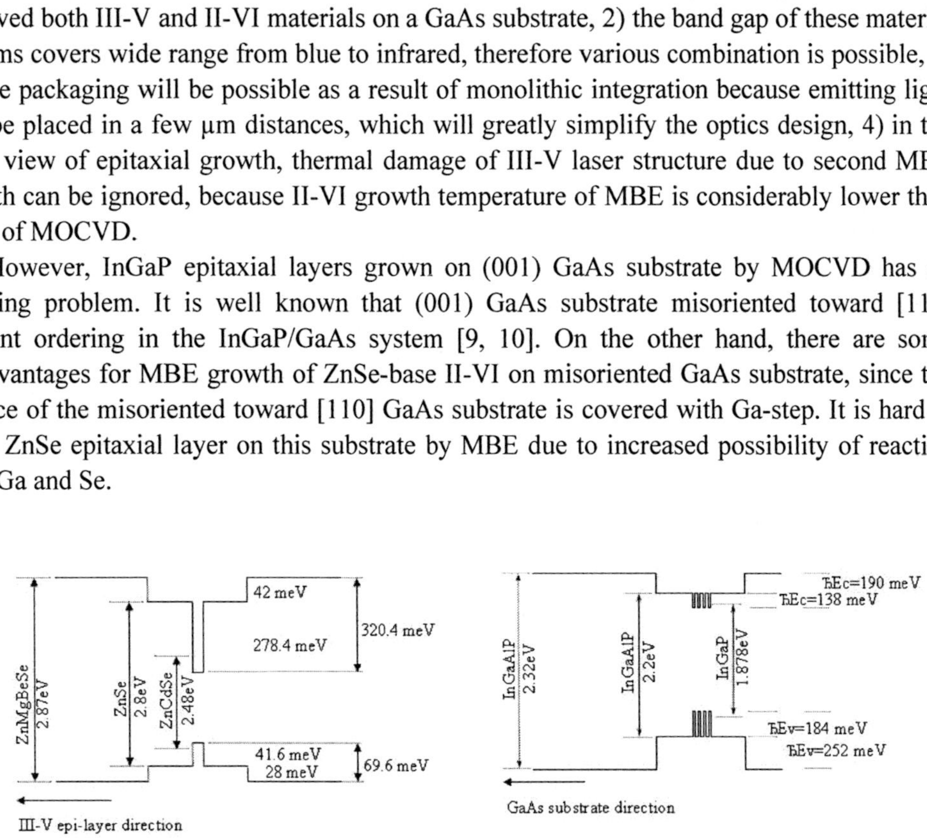


Figure 10. (a) band alignment for II-VI blue-green section (b) band alignment for III-V red section

Since the ZnSe-based light emitting devices have been fabricated by MBE on GaAs (001) substrates, one of the important growth issues would be the growth of high-quality ZnSe-related materials on tilted GaAs (001) substrates. GaAs buffer layers, the pretreatment of Zn exposure on the GaAs in the initial growth stage, and low-temperature-grown ZnSe (LT-ZnSe) buffer have been adopted. In particular, LT-ZnSe buffer layer and the pretreatment of Zn exposure are introduced to obtain a smooth surface and to facilitate high temperature growth of ZnSe layers.[13)] We succeeded in improving ZnSe crystallinity by inserting low temperature (LT)-ZnSe buffer between the high temperature (HT)-ZnSe epitaxial layer and the GaAs buffer on GaAs substrate. LT-ZnSe buffer was grown to reduce the formation of Ga-Se bonding, a well-known source of defect generation [11], due to inter-diffusion through the hetero-interface in the initial stage of growth. The stacking fault density of HT-ZnSe layer is considerably reduced to $2x10^6/cm^2$ by inserting a thin LT-ZnSe buffer layer in the ZnSe/GaAs interface.

In order to show the feasibility of one-chip-two-wavelength light emitting device structure with vertical line-up of MBE grown ZnSe-based II-VI structures on MOCVD grown InGaP-based III-V structures, the SCH-SQW ZnCdSe/ZnSe/ZnMgBeSe structure is grown by MBE on the SCH-MQW InGaP/InGaAlP device structure which is grown on a GaAs (001) substrate tilted by 15° toward [110]. Two-step epitaxy growth is adopted for the one-chip-multiple-wavelength LDs. The MOCVD growth condition of the InGaP/InGaAlP structure was reported previously in detail [12, 13]. Based on such performances in ZnSe layers grown on (001) GaAs substrate tilted by 15° toward [110] direction, high-quality ZnCdSe/ZnSe/ZnMgBeSe structures were fabricated. Based on the previous reports of the band offsets in a ZnCdSe/ZnSe [14] and ZnSe/ZnMgBeSe QW structure [15], the band line up for the ZnCdSe/ZnSe/ZnMgBeSe QW is derived as shown in fig. 10 (a). The band offsets for the conduction and valence bands at the ZnCdSe/ZnSe interface are roughly estimated to be 278.4 meV and 41.6 meV, while those at the ZnSe/ZnMgBeSe are 42 meV and 28 meV. Although the valence band offset of 41.6 meV at the ZnCdSe/ZnSe interface is small to achieve enough carrier confinement at room temperature, the valence band offset of 28 meV at the ZnSe/ZnMgBeSe combined with the ZnCdSe/ZnSe band offset can help carrier confinement. The conduction band offset at the ZnCdSe/ZnSe interface is 278.4 meV, while 42 meV at the ZnSe/ZnMgBeSe. On the other hand, the band line up for the InGaP/InGaAlP QW structure is shown in fig. 10 (b) [16]. The band offsets for the conduction and valence bands of the InGaP/InGaAlP are 190 meV and 252 meV.

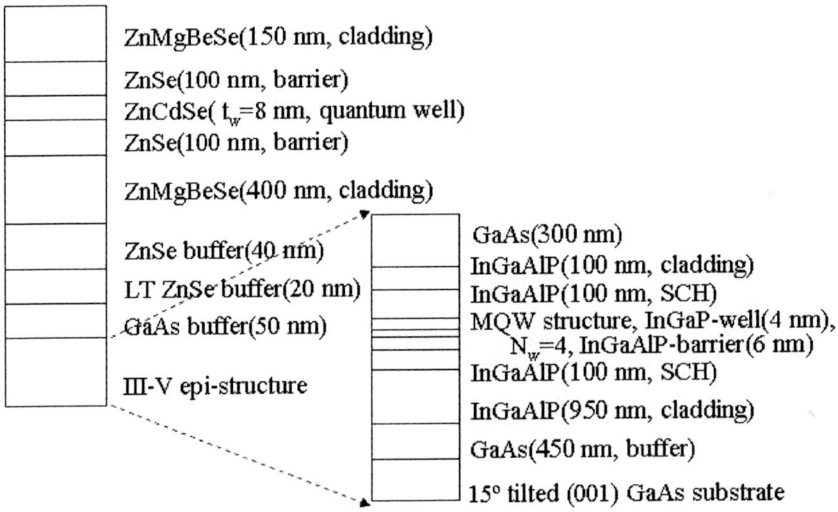


Figure 11. The schematic drawing of the QW structure of the multiple wavelengths light emitter for optical pumping experiments.

The degradation mechanisms of II-VI laser diodes based on sulfur containing II-VI materials have been studied in detail over the past years [17]. The relative abundance of lattice defects like dislocations, point defects, defect complexes in II-VI compounds, which lead to device degradation and limited lifetime, has been attributed to the relatively large ionicity of these materials. Beryllium chalcogenides are somewhat special II-VI materials with respect to lattice rigidity, which is due to their pronounced covalent bonding, in contrast to the much more ionic bonding counterparts like, e.g., ZnSe, and CdTe. Beyond the fundamental interest in these new materials, the increased bond strength can be expected to have an impact on the defect generation and propagation, and therefore the lifetime, of light emitting devices based on II-VI compounds.

We performed optical-pumping experiments to confirm the feasibility of our proposed structure as a new candidate for the multiple-wavelength LDs. Specimens were mechanically polished to a thickness of about 80 μm and cleaved for optical pumping experiments with various cavity lengths. A frequency-doubled, tripled Nd:YAG laser (532 nm, 355 nm) was used as an excitation source for optical pumping experiments, III-V structure is optically pumped with 532 nm and II-VI structure is pumped with 355 nm excitation light source, respectively. The excitation light was irradiated normal to the sample surface and the emission was detected from one of the cleaved facets. Figure 11 shows the schematic diagrams of the QW structure of the multiple wavelengths light emitter for optical pumping experiments. ZnSe layers are inserted in between the ZnCdSe QW layer and ZnMgBeSe cladding layers to minimize interface fluctuations. The QW structures of the multiple wavelength light emitter consist of 8 nm thick ZnCdSe QW layer, 100 nm thick ZnSe act as barrier and SCH layer, 150 nm thick upper ZnMgBeSe and 400 nm thick lower ZnMgBeSe cladding layer in the blue-green region, while 4 nm thick InGaP QWs, 6 nm thick InGaAlP barriers, 100 nm thick InGaAlP SCH layers and 100 nm thick upper InGaAlP and 950 nm thick lower InGaAlP cladding layers in the red emitting III-V region.

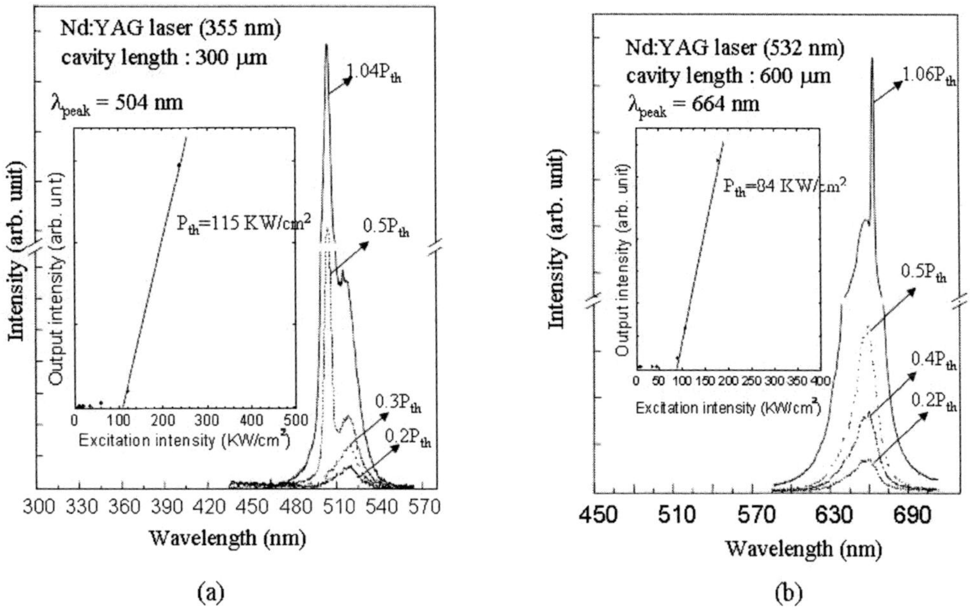


Figure 12. (a) The edge emission spectra measured from the cleaved facet of the II-VI region with a cavity length of 300 μm. (b) The emission spectra from the III-V region with a cavity length of 600 μm

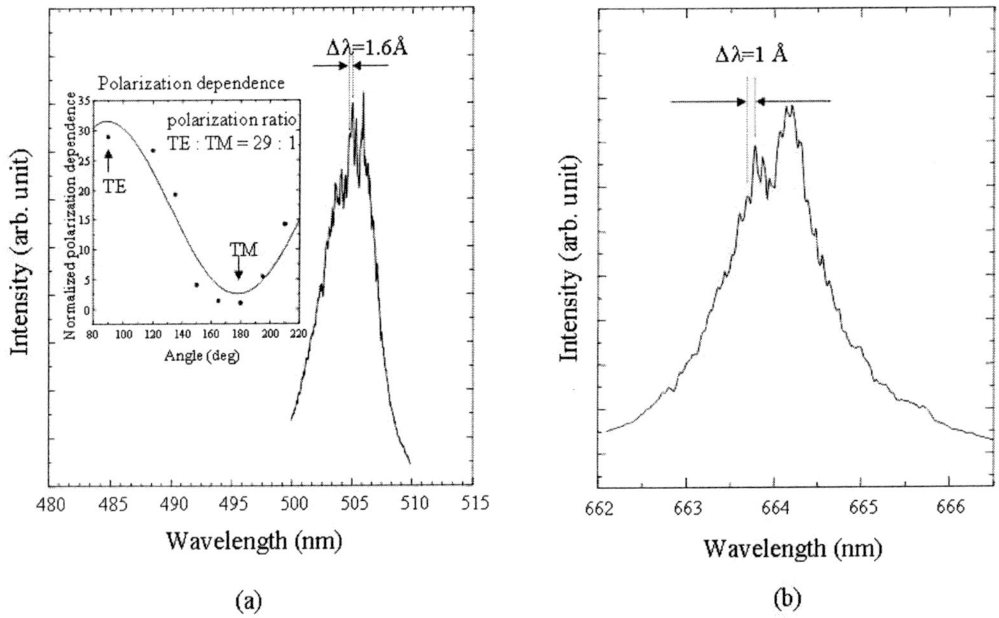


Figure 13. (a) Longitudinal maodes and polarization dependence for the II-VI section. (b) Mode spacing for the III-V section.

Figure 12 (a) shows the evolution of edge emission spectra measured from the cleaved facet of the II-VI region with a cavity length of 300μm. Lasing occurred at the wavelength of 504 nm above threshold at the room temperature, as can be seen from the clear threshold behavior. The threshold power for lasing is estimated at 115 KW/cm$^2$. The emission spectra from the III-V region with a cavity length of 600 μm are shown in fig. 12 (b). The lasing wavelengths and threshold power are 664 nm, and 84 KW/cm$^2$, respectively. The combination of 504 nm and 664 nm wavelength could emit white color light, color temperature in the International Commission on Illumination (CIE) chromaticity coordinates is 3000 K. Longitudinal modes and polarization dependence for the II-VI section are observed as shown in fig. 13 (a). The mode spacing is estimated to be 1.6 Å. The cavity length is calculated as 300 μm using the relation $\Delta\lambda = \lambda^2/2nL$, where $\Delta\lambda$, L, n and λ are the mode spacing, cavity length, refractive index, and wavelength, respectively. We used the refractive index of 2.69 for ZnCdSe, which is approximated by the value of ZnSe. It does not make any difference due to small Cd composition of about 23%. This difference is within the resolution of our high excitation PL system. This value agrees with the measured cavity length of 300 μm by optical microscopy. As an additional evidence for lasing, we have measured the polarization properties of the emission above threshold. The emitted light shows strong transverse-electric (TE) polarization mode against transverse-magnetic (TM) mode with a polarization ratio of 29:1 (TE:TM). The angle of 90° means the emitted TE mode light and analyzer has the same direction. And 180° represent the emitted TM mode light and analyzer has the same direction. It strongly indicates that the electron to heavy hole transition is dominant in the quantum well, which corresponds to the compressive strain regime in the ZnCdSe quantum well. In fig. 13 (b), mode spacing for the III-V section are shown, the mode spacing is estimated to be 1.0 Å at the cavity length of 600 μm. We used the refractive index of 3.625 for InGaP.

This new approach has clearly shown the feasibility of one-chip integration of multiple-wavelength LDs with vertically integrated structure of II-VI and III-V compounds. The new integration technology of ZnSe-based blue-green light emitter grown by MBE on InGaP/InGaAlP red light emitter grown by MOCVD will be useful for a variety of application fields.

## 4. Conclusion

The Hybrid integrated two-wavelength laser diode is a desirable approach for compact optical pick-ups, since the simple package process and wide selectivity of wavelength and output power is to be possible. But, the distance between two emitting spots is still problem. To reduce the spot distance below a few μm, we have suggested vertically stacking chip-to-chip bonding technique, and it could make the spot distance very close below a few μm. But, the thermal dissipation of the upper laser has emerged as a new problem.

A monolithically integrated type two-wavelength laser diode has required high epitaxial growth technique and complicated process. Also, it was limited to the wavelength selectivity based on GaAs material, and very difficult to realize the high power operation of red laser diode on monolithic TWLD. Recent commercialized high-speed RW 52× CD and 16× DVD system require high power operation above ~240 mW. New index-guide laser structure with device performance equivalent to single high power LD chip have been required for stable high power operation of monolithic TWLD.

With appearance of "Blue ray disc" based on GaN material, it is expected that the whole market of the optical playback system may be moved to DVD or HD-DVD in the near future. In the present epitaxial growth technique, it is very difficult to monolithically combine two or three LDs based on GaAs and GaN materials. To hybridize blue and red or infrared lasers, new concepts like new package technique or wafer fusion will be needed. In this means, the chip-to-chip bonding technique is the most realizable technique in combining GaN-based blue laser and GaAs-based red or infrared laser, and then both the distance between two emitting spots and thermal dissipation can be solved due to GaN on $Al_2O_3$ with high conductivity and the small gap between two laser chips.

## References


[1] Ichimura, I.; Maeda, F.; Osato, K.; Yamamoto, K.; Kasami Y. *Jpn. J. Appl. Phys.* P1 2000, 39, 937-940.

[2] Lee, C.W.; Chung, C.S.; Yoo, J.H.; Lee, Y.H., Kim, T.K., Son, Y.K.; Kim; S.J.; Seong, P.Y.; Kim, K.S. *Jpn. J. Appl. Phys.* P1 1998, 37, 2197-2200.

[3] Shih, H.; Yang, T.; Freeman, M.; Wang, J.; Yau, H.; Huang, D. *Jpn. J. Appl. Phys.* P1 1999, 38, 1750-1754.

[4] Nemoto, K.; Kamei, T.; Abe, H.; Imanishi, D.; Narui, H.; Hirata, S. *Appl. Phys. Lett.* 2001, 78, 2270-2272.

[5] Lu, T.; Fu, R.; Shieh, H.M.; Huang, K.J.; Wang, S.C. *Appl. Phys. Lett.* 2001, 78, 853-856.

[6] Onishi, T.; Imafuji, O.; Fukuhisa, T.; Mochida, A.; Kobayashi, Y.; Yuri, M.; Itoh, K.; Shimizu, H. *IEEE Phot. Tech. Lett.* 2001, 13, 550-552.

[7] Onishi, T.; Imafuji, O.; Fukuhisa, T.; Mochida, A.; Kobayashi, Y.; Yuri, M.; Itoh, K.; Shimizu, H. *Jpn. J. Appl. Phys.* P1 2001, 40, 6401-6405.

[8] Song, J.S.; Cho, M.W.; Oh, D.C.; Makino, H.; Hanada, T.; Yao, T.; Zhang, B. P.; Segawa, Y.; Chang, J.H.; Song, H.S.; Cho, I.S.; Kim, H.W.; Jung, J.J. *Appl. Phys. Lett.* 2003, 82, 4095-4097.

[9] Gomyo, A.; Kawata, S.; Suzuki, T.; Iijima, S.; Hino, I. *Jpn. J. Appl. Phys.* 1989, 28, L1728-L1730.

[10] Gomyo, A.; Suzuki, T.; Kobayashi, K.; Kawata, S.; Hino, I. *Appl. Phys. Lett.* 1987, 50, 673-675.

[11] Kuo, L. H., Kimura, K., Ohtake, A., Miwa, S., Yasuda, T., Yao, T. 1997, *J. Vac. Sci. Tech.* B, 15, 1241-1253.

[12] Choi, W.J.; Kim, J.S.; Ko, H.C.; Chung, K.W.; Yoo, T.K. *J. Appl. Phys.* 1995, 77, 3111-3114.

[13] Choi, W.J.; Chang, J.H.; Choi, W.T.; Kim, S.H.; Leem, S.J.; Yoo, T.K. *IEEE J. Selected Topics in Quantum Electronics.* 1995, 1, 717-720.

[14] Guenaud, C.; Deleporte, E.; Filoramo, A.; Lelong, Ph.; Delalande, C.; Morhain, C.; Tournie, E.; Faurie, J. P. *J. Cryst. Growth* 1998, 184/185, 839-843.

[15] Chang, J.H.; Cho, M.W.; Godo, K.; Makino, H.; Yao, T. *Appl. Phys. Lett.* 1999, 75, 894-896.

[16] Tanaka, H., Kawamura, Y.,; Nojima, S.; Wakita, K.; Asahi, H. J. *Appl. Phys.* 1987, 61, 1713-1719.

[17] Gua, S.; de Puydt, J.M.; Haase, M.A.; Qui, J.; Cheng, H *Appl. Phys. Lett.* 1993, 63, 3107-3109.

In: New Developments in Lasers and Electrc-Optics Research ISBN 1-59454-771-8
Editor: William T. Arkin, pp. 241-265 

*Chapter 8*

# SHAPE DETECTION BY MEANS OF A LASER LINE AND APPROXIMATION NEURAL NETWORKS

***J. Apolinar Muñoz-Rodríguez and Ramon Rodríguez-Vera***
Centro de Investigaciones en Optica, A. C.
Leon, Gto, 37150 Mexico, Tel: (477) 441 42 00
e-mail: munoza@foton.cio.mx

## ABSTRACT

We present a review of our computer algorithms developed for the shape detection technique. The study of this chapter involves the laser line projection technique, image processing and approximation neural networks. To extract topographic information of the shape of an object, it is scanned by means of a laser line. From the scanning procedure, a set of images is captured by a CCD camera. By processing these images, the object shape is recovered. The shape information is extracted from an image by detecting the behaviour of the position of the laser line in the image plane. To determine the mathematical model of the relationship between the laser line behaviour and the object shape, approximation neural networks are applied. Approximation networks such as Radial Basis Function neural networks, General Regression neural networks and Bezier neural networks are used to create a model of the behaviour of the laser line. The architecture of these approximation networks is built using data of images of a laser line projected on objects, whose dimensions are known. The data correspond to the object dimensions of the region where the laser line is projected. In this manner, the approximation network is constructed. The approach of the approximation networks in this technique is the calibration procedure of shape detection by computer algorithms. In this form, the accuracy of the shape measurements is improved. It is because errors of parameters measured on the set-up are not introduced to the system. The results of this examination are presented by computer simulation and in experimental form.

## 1. Introduction

The use of optical methods to extract three-dimensional shapes plays an increasingly important role in several applications. One important advantage of optical sensors is that no contact is required with the object being measured. Another approach of these methods is that they are significantly faster than contact methods. Typical industrial applications of the optical methods are production quality control, digitization of three-dimensional shapes, and computer vision matters [1-3]. In Optical Metrology and Robotics Vision, several techniques of structured light have been developed for shape detection [4-5]. The use of structured light generally makes the system more reliable and the acquired data are easier to interpret. A particular technique is the laser line projection [6-12]. When a laser line is projected onto an object surface at an angle, it is moved to another position in the image plane due to surface variation. The main aim of this technique is the detection of the behaviour of the position of the laser line. From the position of the laser line and the geometry of the optical set-up, the object profile is deduced by triangulation. In some cases, the position of the laser line is measured by using a peak detection method [6]. For this method, the pixel of highest intensity is assigned to the centre of the peak. The centre is obtained using a small window and applying a threshold crossing on the same area. Another method consists in measuring the thickness of a laser line as a function of the surface height [7]. This thickness is measured directly on a screen by using an eyepiece or a micrometer. Measurements of the laser line position using a coordinate-measuring machine (CMM) have been also used [8]. This technique uses a gauge blocks as calibrators. The spatial coordinates of the position of the laser line are determined detecting the laser line on the gauge block. Another peak method uses an electronic sensor for the detection of the position of the laser line by means of a numerical integration [9]. In time delay and integration methods, a camera is used to record line by line to obtain a fringe pattern [10-12]. The object shape is obtained by calculating the phase from the fringe pattern. In all of these methods, the calibration system includes the distances of the geometry of the components of the optical set-up to determine the dimensions of object shape. This procedure involves measurements on the set-up geometry. After that, these data are given to the computer to obtain the object shape dimensions. To perform this procedure by computer, approximation neural networks are proposed in this chapter. These networks can describe the geometric relationship between the laser line position and the object surface. The proposed approximation networks for this technique are radial basis function neural networks (RBFNN), General Regression neural networks (GRNN) and Bezier neural networks (BNN). The data used to construct these networks are obtained measuring in the image plane the position of the laser line projected on pattern objects, whose dimensions are known. In this manner, a mathematical model is obtained. This model determines the surface dimension for a particular position of the laser line. To know the best accuracy of the result obtained by the networks, an *rms* (root mean squared) value is calculated using data of the networks and data of the contact method. The set-up used in this study is very simple. In this arrangement, object is moved on an axis and scanned by a laser line. In the scanning procedure, images of laser line are captured by a CCD camera. These images are processed to detect the laser line position by measuring the maximum intensity. This position is processed by a neural network to determine the dimensions of an object. The produced information by each laser line is stored in an array memory to obtain the complete object shape. A computer controls this set-up, therefore, the performance of these technique is achieved in a good manner. The content of this chapter is organized as follow. In section 2, the technique of laser line projection is described. In section 3, the image processing of the laser line is presented. In section 4, the approximation neural networks are presented to perform the calibration. In section 5, experimental results of shape detection are presented. Finally, in section 6 conclusion and future trends are draw.

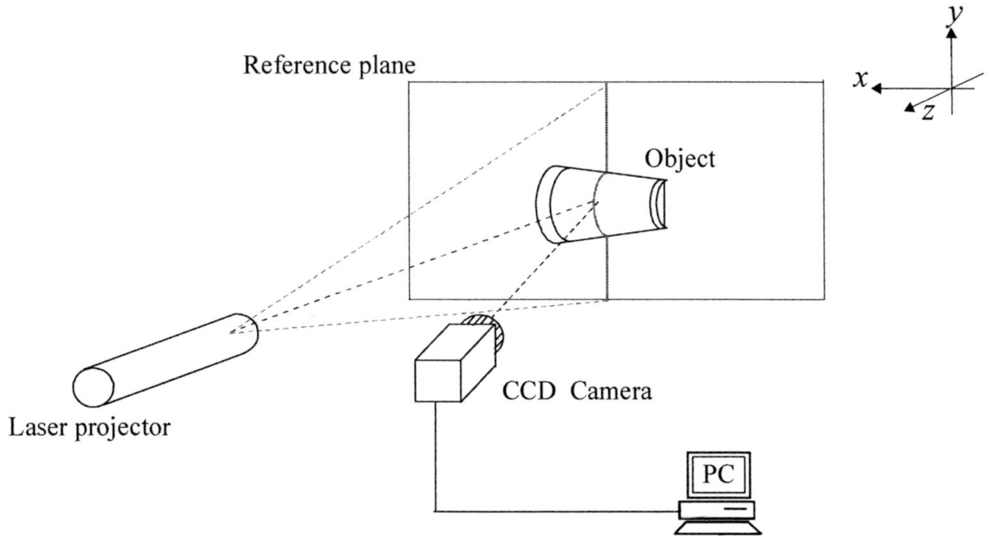


Figure 1. Optical set-up for scanning by means of a laser line.

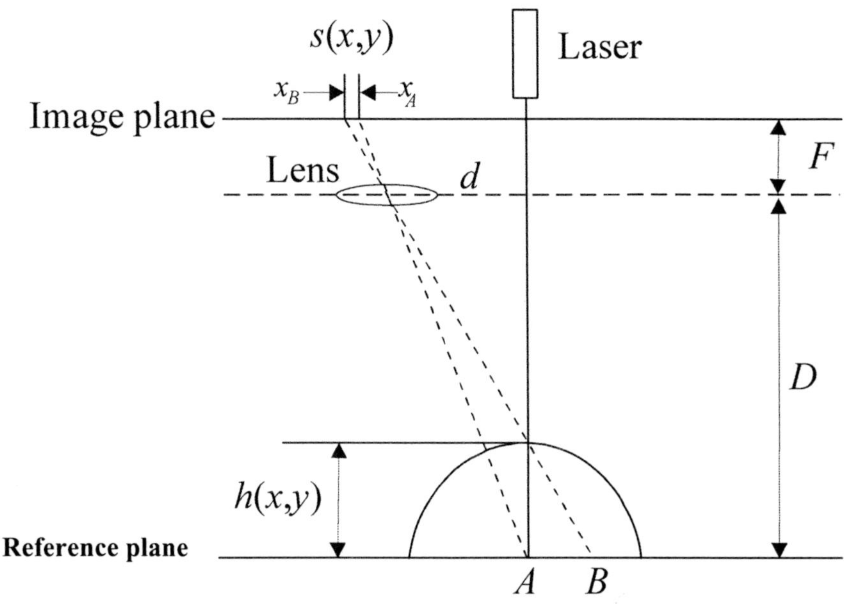


Figure. 2. Geometry of the optical set-up.

## 2. Laser Line Projection

Scanning techniques combining laser line projection are simple and efficient. The three-dimensional shape of a surface can be achieved using profilometry based on scanning laser line projection. Many diverse industrial inspection tasks can be solved using scanning laser profiler. The applications for laser line profiler are dimensional-control, reverse engineering, and rapid prototyping. The basic scheme of the optical set-up of scanning by means of a laser line is shown in Fig.1. For this set-up, there are three kinds of geometric configurations. In the first configuration, both the camera and a laser projector are aligned at an angle. For the second configuration, the laser projector is placed perpendicularly at the object surface, and the camera is aligned at an oblique angle. In the last, configuration the camera is collocated perpendicularly at the object surface and the projector is aligned at an oblique angle. The scanning laser line method used in this study can be used in these three kinds of configurations. The configuration used in this description, the laser projector is placed perpendicularly at the object surface and the camera is aligned at an oblique angle. In this arrangement the object is fixed on an electromechanical device. This platform uses a stepping motor to move the object in steps with an accuracy of a fraction of millimetre. To perform the scanning procedure, a vertical laser line is projected onto the object by means of a laser diode. In each step of the scanning, the laser line changes of position in $x$-axis according to the object topography. Every laser line is captured and digitized by a CCD camera and a frame grabber, respectively. By detecting the position of the laser line, the dimension of the object shape is determined by a geometric relationship. To describe the relationship between the position of the laser line and the object dimension, the geometry of the optical set-up is shown in Fig. 2. In this geometry, on the reference plane are located the $x$-axis and $y$-axis, and the object dimension is indicated by $h(x, y)$. The point $A$ is the position of the laser line projected onto the reference plane. The point $B$ is a projection of the laser line projected onto the object surface. The distance from the reference plane to the CCD camera is indicated by $D$. The distance from the CCD camera to the laser line projector is $d$. The distance from the lenses to the CCD array is $F$. When a laser line is projected on the target, the position of the laser line on the image plane is moved from $x_A$ to $x_B$. This displacement can be represented as:

$$s(x,y) = x_A - x_B. \tag{1}$$

The distance $AB$ is proportional to $s(x,y)$[13]. Therefore, the dimension of the surface height can be calculated with the distance $AB$, $D$ and $d$. By using triangulation, the height dimension is obtained as

$$h(x,y) = \frac{(AB)D}{AB + d}. \tag{2}$$

To compute the Eq.(2) is necessary to measuring the distances $D$, and $d$. Other way for obtaining the dimension $h(x, y)$ is by means of the displacement $s(x,y)$. It is because the variation of displacement $s(x,y)$ is directly proportional to dimension $h(x, y)$. Using several objects with known dimensions, a mathematical model $h(x, y)$ based on $s(x,y)$ can be obtained. To construct the mathematical model $h(x, y)$, approximation neural networks are used. In this manner, a network describes the object dimension $h(x, y)$ for a particular displacement $s(x,y)$. Therefore, the object dimensions are obtained by processing the displacement by means of the approximation network. To detect the displacement, laser line position is measured in every row of the image. Processing the displacement of an image of laser line by the network, a transverse section of the object is obtained. Then, the information of each laser line is stored in an array memory to reconstruct the complete object shape. The approach of the neural networks in this technique is for performing all steps of the process by computer algorithms. Applying networks in this technique, makes the process of shape detection easier than the process given by Eq.(2). In section 3, the measurement of laser line displacement is described.

## 3. Laser Line Image Processing

Gathering depth data from images of structured light has been a procedure for many different environments and uses. Many of these systems are utilized because quickness of laser line scanning. The simplicity of the image processing provides a quickly digitizing of an object surface with good accuracy and resolution. Using laser line scanning, image processing and geometric triangulation, the surface coordinates of objects are determined. When a laser line is projected onto the object, it is displaced in the image plane due to variation of the surface height. By detecting the position of the laser line in the image, the object surface can be deduced. Therefore, the displacement of the laser line in the image should be detected. To carry it out, the position of the laser line is measured by detecting the maximum intensity in every row of the image. To elucidate this procedure, a laser line image is shown in Fig. 3. From this image, the pixels of a row are used to detect laser line position. The projected intensity by a laser diode is a Gaussian distribution in lateral direction ($x$-axis) [14]. Intensity values of every row of the line image are represented as: $(x_0, z_0)$, $(x_1, z_1)$, $(x_2, z_2)$, $(x_3, z_3)$,.....,$(x_n, z_n)$, where $x_i$ is the pixel position and $z_i$ is the pixel intensity. To detect the maximum intensity from this set of pixel, methods such as *a) Least squared, b) Gaussian approximation and c) Bezier curves* can be used [15]. These methods are described as follow.

Figure 3. Image of a laser line.

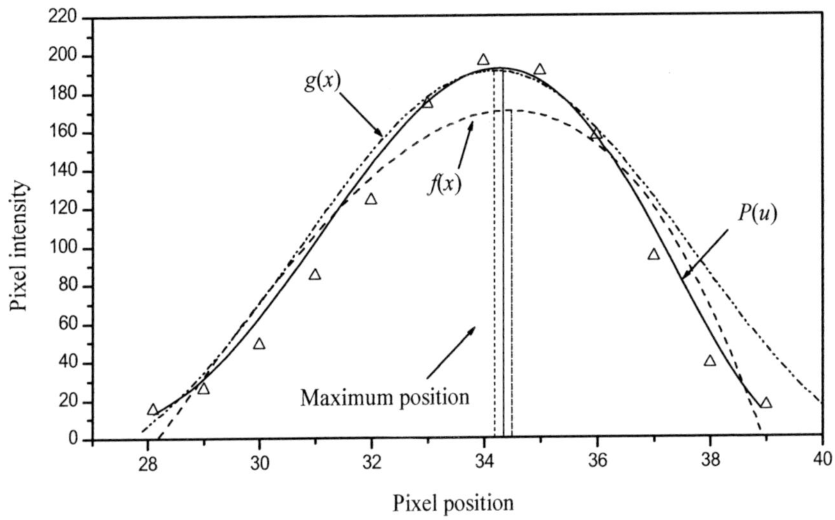


Figure 4. A set of pixels fitted to a continuous function.

## a) Least Squared

The projected intensity by a laser can be represented by a polynomial function. Least squared method is very useful to obtain a smooth continuous curve of a set of data with a minimum error [16]. A set of pixels is shown in Fig. 4. In this figure the symbol "Δ" indicates pixel intensity. These pixels correspond to a row of the laser line image. To adjust a set of pixels to a cubic function by least squared method a third equation is chosen:

$$f(x) = a + bx + cx^2 + dx^3 \text{ ,} \tag{3}$$

where $f(x)$ is a function that represents the laser line intensity in lateral direction ($x$-axis), $a$, $b$, $c$, and $d$ are the constants for a particular curve. The least squared criterion requires that

$$S = \sum_{i=1}^{n} (z_i - a - bx_i - cx_i^{\ 2} - dx_i^{\ 3})^2 \text{ ,} \qquad (4)]$$

be a minimum. At minimum for $S$ the partial derivatives will be zero:

$$\frac{\partial S}{\partial a} = 0 \text{ ,} \quad \frac{\partial S}{\partial b} = 0 \text{ ,} \quad \frac{\partial S}{\partial c} = 0 \text{ ,} \quad \frac{\partial S}{\partial d} = 0 \text{ .}$$

Expanding the summation of these derivatives, the next equations system is obtained:

$$
\begin{aligned}
&aN + b\sum x_i + c\sum x_i^2 + d\sum x_i^3 = \sum z_i \\
&a\sum x_i + b\sum x_i^2 + c\sum x_i^3 + d\sum x_i^4 = \sum z_i\, x_i \\
&a\sum x_i^2 + b\sum x_i^3 + c\sum x_i^4 + d\sum x_i^5 = \sum z_i\, x_i^2 \\
&a\sum x_i^3 + b\sum x_i^4 + c\sum x_i^5 + d\sum x_i^6 = \sum z_i\, x_i^3
\end{aligned}
\qquad (5)
$$

this system is solved to obtain the constants $a$, $b$, $c$, and $d$. These constants are substituted into Eq.(3) to obtain the curve, which represents a set of pixels. This function is evaluated for $x_1$ to $x_n$ to fit the curve $f(x)$ shown in Fig. 4. To extract the position of the laser line from this continuous curve, the maximum is computed. To calculate the maximum, it is necessary to find the first and the second derivative to determine the local maximum or minimum. In this case the value of the second derivative for this curve is greater than zero ( $f''(x) > 0$ ) in the interval from $x_1$ to $x_n$[17]. Therefore, maximum position is obtained by calculating the $x^*$ value where the first derivative is zero $f'(x) = 0$. To find this $x^*$ value, the Bisection method is used [16]. The strategy to carry this out, is begin with a pair of values that converge at root of $f'(x) = 0$. Beginning with a pair of values $x_i = x_1$ and $x_s = x_n$, and $x^*$ is halfway between $x_i$ and $x_s$. If $f'(x)$ evaluated at $x = x^*$ is positive, then $x_i = x^*$. If $f'(x)$ evaluated at $x = x^*$ is negative, then $x_s = x^*$. Next $x^*$ is taken as the mid point of the last pair of values that converges to the root. This method converges with a few iterations. To evaluate the derivative, the secant method is used [16]. For the curve, which represents the pixels of Fig. 4 the result is $x^* = 34.456$. Therefore, the position of the light line is $x_A = 34.456$ pixels, as also is shown in Fig. 5.

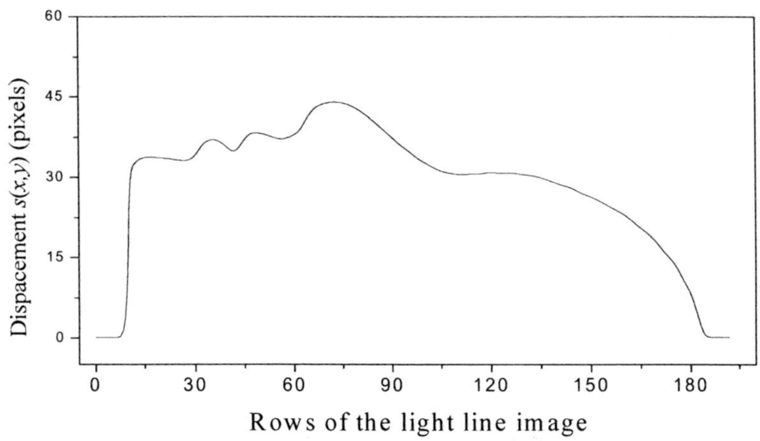


Figure 5. Profile of the displacement corresponding to light line of Fig. 4.

## b) Gaussian Approximation

A way for representing the pixels of a row of a laser line image is a Gaussian function. The equation of a Gaussian function is

$$g(x) = \frac{Ni}{\sigma\sqrt{2\pi}} e^{-\frac{1}{2}\left(\frac{x-\mu}{\sigma}\right)^2} \tag{6}$$

where $Ni$ is the area under the curve, $Ni/(2\pi)^{1/2}$ is a factor of normalization, $\mu$ is the mean of the function, and $\sigma$ is the standard deviation [18]. To determine the position of the laser line, the most important parameter of the Gaussian function is the mean value. This is because $\mu$ represents the center of the Gaussian function. The mean can be calculated using the set of pixels as

$$\mu = \frac{\sum_{i=1}^{n} z_i x_i}{\sum_{i=1}^{n} z_i} , \tag{7}$$

Using the pixels of Fig. 4, a Gaussian function is constructed to determine the position of the intensity maximum. To calculate the mean $\mu$, position $x_i$ and intensity $z_i$ of these pixels are replaced into Eq.(7), for $i = 1,2,3,....,n$. In this case the result is $\mu = 34.182$. Therefore, the position of the laser line is $x_A = 34.182$ pixels. Also, the factor of normalization and the standard deviation are calculated to fit the Gaussian function shown in Fig.4. Usually for a Gaussian approximation, the center of this function is calculated by using the arithmetic mean [18]. The result of this mean is always a mid point of the pixels position in $x$-direction. But the intensity maximum not always is at the middle point of the pixel's position in $x$ direction. Most of the times the laser line or CCD camera is placed at an angle respect to the object surface. Therefore, the maximum is on the left hand or right of the mid point of the pixel set. The result of the arithmetic mean for the set of pixels of Fig.4 is $\mu = 33.5$. This result is slightly different of the maximum obtained by the mean of Eq. (7), which is called weighted arithmetic mean [18]. For the weighted arithmetic mean, a weight is assigned to every point. In this case the weight is the pixel intensity. It means that the pixels with higher intensity are more important than pixels with lower intensity. Therefore the weighted arithmetic mean is used to detect the maximum intensity of a Gaussian function.

## c) Bezier Curves

A polynomial function that represents the intensity of a laser line also can be approximate by Bezier Curves. The $n$th-degree Bezier polynomial function is determined by $n+1$ points [19]. In this case the point number is the pixel number. The parametric form represents the relation of the position $x$ and the intensity $z$ by others two equations, $x = F_1(u)$, $z = F_2(u)$, with $u$ as a parameter [20]. In the Bezier Curves method a continuous function $f(x)$ is represented in parametric form. The set of pixels is represented in parametric form as:

$$P_i = \begin{pmatrix} x(u) \\ z(u) \end{pmatrix}, \qquad 0 \le u \le 1$$

The $n$th-degree Bezier polynomial is determined by $n$+1 points, which is given by

$$P(u) = \sum_{i=0}^{n} \binom{n}{i} (1-u)^{n-1} u^i P_i \quad , 0 \le u \le 1 \tag{8}$$

where

$$\binom{n}{i} = \frac{n!}{i!(n-i)!}.$$

The preceding Eq. (8) really represents others two equations, one for $x$ and the other for $z$:

$$x(u) = \binom{n}{0} (1-u)^n u^0 x_0 + \binom{n}{1} (1-u)^{n-1} u\, x_1 + \ldots\ldots + \binom{n}{n} (1-u)^0 u^n x_n, 0 \le u \le 1. \tag{9}$$

$$z(u) = \binom{n}{0} (1-u)^n u^0 z_0 + \binom{n}{1} (1-u)^{n-1} u\, z_1 + \ldots\ldots + \binom{n}{n} (1-u)^0 u^n z_n, 0 \le u \le 1. \tag{10}$$

Then the position of the pixels: $x_0, x_1, x_2, \ldots\ldots, x_n$, are substituted into Eq. (9) and intensity: $z_0, z_1, z_2, \ldots, z_n$, are substituted into Eq. (10). These two equations $x(u)$ and $z(u)$ are evaluated in the interval $0 \le u \le 1$ to fit the Bezier curve $P(u)$ shown in Fig. 4. To extract the position of the laser line from the curve $P(u)$, its maximum is calculated. In this case the second derivative value close to the function peak is major than zero ( $z''(u) > 0$ ). Therefore, the maximum is obtained by finding $u$ value where the first derivative is zero $z'(u) = 0$. To find this value $u$, the bisection method is also used. Beginning with a pair of values $u_i = 0$ and $u_s = 1$, because $z(u)$ is defined for the interval $0 \le u \le 1$, $u^*$ is halfway between $u_i$ and $u_s$. If $z'(u)$ evaluated at $u = u^*$ is positive, then $u_i = u^*$. If $z'(u)$ evaluated at $u = u^*$ is negative, then $u_s = u^*$. Next, $u^*$ is taken as the mid point of the last pair values that converges to the root. The value $u$ where $z'(u) = 0$ is substituted into Eq.(9) to find maximum position $x^*$ of the function $P(u)$. For the curve $P(u)$, which represents the pixels of Fig.4 the result is $x^* = 34.274$. Therefore, the position of the laser line is $x_A = 34.274$ pixels, as is also shown in Fig.4. The procedure to obtain the laser line position is applied to all rows of the image to extract an object profile by means of the displacement. The object profile of Fig.3 is show in Fig. 5. The information produced by a laser line image is processed by a network to extract a transverse of the object. The approximation networks are described in the section 4.

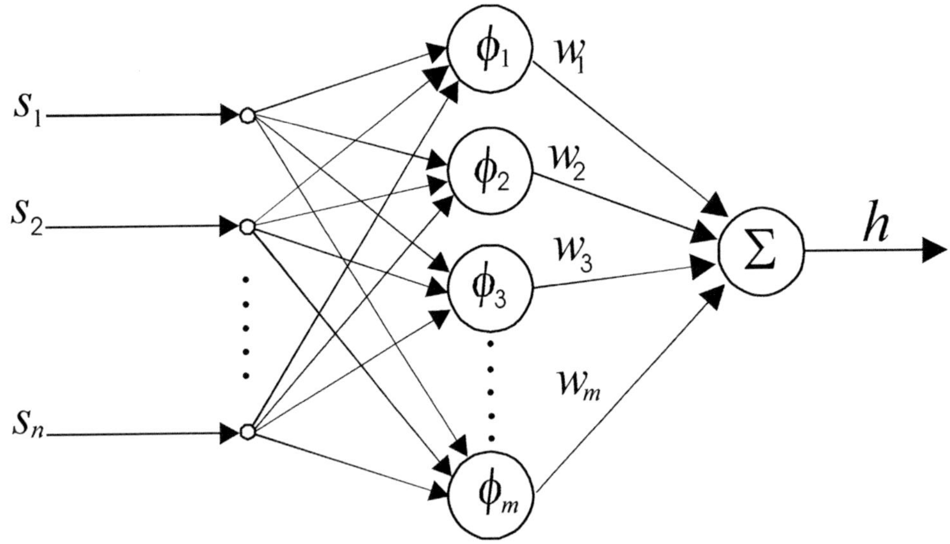


Figure 6. Structure of a RBFNN.

# 4. Approximation Neural Networks Applied to a Laser Line

## 4.1 Structure of a Radial Basis Function Neural Network

Radial basis function neural networks (RBFNN) is an artificial network applied to solve problems of supervised learning, regression, classification, shape identification, object detection and time series prediction [21-25]. The basic structure for an RBFNN is an input vector, a hidden layer and output layer [26]. This structure is shown in Fig. 6. The input vector is only a fan-out to connect the input data with the hidden layer. The hidden layer is constructed using radial basis function as neurons. The output layer is obtained by means of the summation of the neurons of the hidden layer, which are multiplied by a weight. Each layer of the network structure is defined in the next form. The data of input vector $s_1$, $s_2$, $s_3$,…,$s_n$ are extracted measuring the laser line position by image processing, as is described in section 3. The radial basis function of the hidden layer is a Gaussian neuron, which is described by the next equation

$$\phi_i = \exp\left(-\frac{\|s - s_i\|^2}{2\sigma^2}\right), \tag{11}$$

where $s_i$ is the neuron centre and $\sigma$ is the neuron width, which is calculated as

$$\sigma = \sqrt{\frac{1}{q}\sum_{i=1}^{q}(s - s_i)^2} \quad , \tag{12}$$

where $q = 2$, $s$ - $s_i$ is de distance between each couple of nearest neurons. The output layer is computed by the summation of the weighted neurons, which is represented by

$$h(s) = \sum_{i=1}^{m} w_i \phi(\|s - s_i\|) \quad , \tag{13}$$

where $m$ is the number of neurons. To construct the complete network architecture, the parameters of the Eq.(13) should be determined. These parameters include: the neuron shape, neuron centre, neuron width, number of neurons and the weights. The first parameter to be determined is the neuron shape. Some functions can be used to construct the neuron shape, such as: cubic, inverse quadratic, inverse multi-quadratic and Gaussian [24]. To determine the suitable neuron shape, the *MSE* (mean squared error) criterion is used [27]. Based on neuron shape, the *MSE* provides the network error respect to the desired output. The *MSE* criterion is defined by

$$MSE = \frac{1}{n}\sum_{i=1}^{n}(hc_i - h_i)^2 \quad , \tag{14}$$

where $hc_i$ is the output calculated by the RBFNN, $h_i$ is the desired output and $n$ is the number of analyzed data. Using the functions: cubic, inverse quadratic, inverse multi-quadratic and Gaussian, four RBFNN are constructed. Then, a *MSE* is calculated for each RBFNN, to determine which network provides the best output. Based on the *MSE*, the best output response is given by the Gaussian function. Therefore, the Gaussian function Eq.(11) is used to built the neuron shape. In this manner, the Gaussian function Eq.(11) is used to define the neuron shape. The second parameter is the neuron centre $s_i$, which is given by each data of the input vector. The third parameter to be determined is the number of neurons of the hidden layer. Again, the *MSE* criterion is used. This criterion, also, provides the error according to the number of neurons. To define neuron number, the network is constructed with different neuron number. Then, the neuron number that provides the best *MSE* is selected. The parameter sigma, which corresponds to the neuron width is calculated by Eq.(12). Finally, the training process is applied to determine the weights. This training is supervised, because the outputs are known and the network is being forced to produce the correct outputs. The requirements for a supervised learning are a suitable weight adjustment mechanism. To carry it out, the input data $s_1$, $s_2$, $s_3$,...,$s_n$ are connected to the hidden layer. Therefore, for each input $s_i$ an output $h_i$ is produced and an equation system is generated:

$$h_1 = w_1 \exp\left(-\frac{\|s_1 - s_1\|^2}{2\sigma^2}\right) + w_2 \exp\left(-\frac{\|s_1 - s_2\|^2}{2\sigma^2}\right) + \dots\dots\dots + w_m \exp\left(-\frac{\|s_1 - s_m\|^2}{2\sigma^2}\right)$$

$$h_2 = w_1 \exp\left(-\frac{\|s_2 - s_1\|^2}{2\sigma^2}\right) + w_2 \exp\left(-\frac{\|s_2 - s_2\|^2}{2\sigma^2}\right) + \dots\dots\dots + w_m \exp\left(-\frac{\|s_2 - s_m\|^2}{2\sigma^2}\right)$$

$$\vdots \qquad \vdots \qquad \vdots \qquad \vdots \tag{15}$$

$$h_n = w_1 \exp\left(-\frac{\|s_n - s_1\|^2}{2\sigma^2}\right) + w_2 \exp\left(-\frac{\|s_n - s_2\|^2}{2\sigma^2}\right) + \dots\dots\dots + w_m \exp\left(-\frac{\|s_n - s_m\|^2}{2\sigma^2}\right)$$

The linear system of Eq.(15) can be represented as

$$\begin{aligned} h_1 &= w_1\phi_{1,1} + w_2\phi_{1,2} + \dots\dots\dots + w_m\phi_{1,m} \\ h_2 &= w_1\phi_{2,1} + w_2\phi_{2,2} + \dots\dots\dots + w_m\phi_{2,m} \\ &\vdots \\ h_n &= w_1\phi_{n,1} + w_2\phi_{n,2} + \dots\dots\dots + w_m\phi_{n,m} \end{aligned} \tag{16}$$

Eq.(16) can be rewritten in matrix form as $\boldsymbol{\phi}\mathbf{W} = \mathbf{H}$, which is represented as

$$\begin{bmatrix} \phi_{1,1} & \phi_{1,2} & \phi_{1,3} & \cdots & \phi_{1,m} \\ \phi_{2,1} & \phi_{2,2} & \phi_{2,3} & \cdots & \phi_{2,m} \\ \vdots & \vdots & \vdots & & \vdots \\ \phi_{n,1} & \phi_{n,2} & \phi_{n,3} & \cdots & \phi_{n,m} \end{bmatrix} \begin{bmatrix} w_1 \\ w_2 \\ \vdots \\ w_m \end{bmatrix} = \begin{bmatrix} h_1 \\ h_2 \\ \vdots \\ h_n \end{bmatrix}. \tag{17}$$

By solving the linear equation system Eq.(17), the weights $w_i$ are determined. This system can be solved by the Chelosky method [28]. This method uses a Lower and Upper (***LU***) decomposition to transform the matrix system. By using the lower and upper triangular matrix, linear system Eq.(17) $\mathbf{H}=\boldsymbol{\phi}\mathbf{W}$ is transformed into $\boldsymbol{LU}\mathbf{W}=\mathbf{H}$. In this ***LU*** decomposition $\boldsymbol{\phi} = \boldsymbol{LU}$ is determined as

$$\begin{bmatrix} \phi_{1,1} & \phi_{1,2} & \phi_{1,3} & \cdots & \phi_{1,m} \\ \phi_{2,1} & \phi_{2,2} & \phi_{2,3} & \cdots & \phi_{2,m} \\ \vdots & \vdots & \vdots & & \vdots \\ \phi_{n,1} & \phi_{n,2} & \phi_{n,3} & \cdots & \phi_{n,m} \end{bmatrix} = \begin{bmatrix} L_{1,1} & 0 & 0 & \cdots & 0 \\ L_{2,1} & L_{2,2} & 0 & \cdots & 0 \\ \vdots & \vdots & \vdots & & \vdots \\ L_{n,1} & L_{n,2} & L_{n,3} & \cdots & L_{n,m} \end{bmatrix} \begin{bmatrix} U_{1,1} & U_{1,2} & U_{1,3} & \cdots & U_{1,m} \\ 0 & U_{2,2} & U_{2,3} & \cdots & U_{2,m} \\ \vdots & \vdots & \vdots & & \vdots \\ 0 & 0 & 0 & \cdots & U_{n,m} \end{bmatrix}, \tag{18}$$

Then, triangular system is solved as $\boldsymbol{LY}=\mathbf{H}$ and $\boldsymbol{U}\mathbf{W}=\boldsymbol{Y}$. In this manner, ***Y*** is calculated using ***L*** and ***H***. Then, ***W*** is solved with ***Y*** and ***U***. With this step the weight values $w_1$, $w_2$, $w_3$,…,$w_n$ are determined. With these calculated weights, the RBFNN has been completed. In this manner, for a particular given $s_i$ a $h_i$ is calculated by the RBFNN.

## 4.2 Structure of a General Regression Neural Network

The GRNN was developed by D. Specht in 1991 for systems modeling and identification [29]. The GRNN is an architecture that can solve any approximation problem. The basic idea of a GRNN is that a data set can be approximated by a continuous function. Assuming that a function has $x$ inputs from $x_1$ to $x_n$ and one output $y$, the expected mean value of the output for a particular input $x$ can be found using the next equation:

$$\bar{y} = \frac{\int_{-\infty}^{\infty} y\, f(x,y)dy}{\int_{-\infty}^{\infty} f(x,y)dy} \qquad (19)$$

In this equation, $f(x,y)$ is a joint probability density function (PDF), the probability of the output being $y$ and input being $x$. The PDF can be approximated by a sum of Gaussians [29]. This process is done by placing the center of Gaussian over the input data and multiplying the Gausian by the corresponding output. The Eq.(19) can be approximated by the next equation

$$\hat{y} = \frac{\sum_{i=1}^{n} w_i \exp(-d_i^2 / 2\sigma^2)}{\sum_{i=1}^{n} \exp(-d_i^2 / 2\sigma^2)} \qquad (20)$$

From this equation $d_i$ is the distance between each input, $n$ is the number of input/output, $\sigma$ is the width of the Gaussian function, and $w_i$ is a weight that multiplies at Gaussian function. The architecture to construct a GRNN to obtain the Eq.(20) is shown in Fig. 7. In this structure, the output $h$ is the object dimension and $s_i$ is its corresponding laser line displacement. This GRNN is formed by an input vector, a hidden layer, two nodes of summation and output layer. The data of input vector $s_1$, $s_2$, $s_3$,....,$s_n$ are the laser line displacement measured by image processing. The hidden layer is constructed by a set of neurons, whose shape is a Gaussian function centered at the input data. The effective range of the Gaussian function is determined by the values allocated to the center and width of it. The function used to obtain each neuron is determined by the next equation

$$f_i = \exp\left( - \frac{\left\| s - s_i \right\|^2}{2\sigma^2} \right), \qquad (21)$$

where $||s-s_i||$ is the distance $d_i$ used in Eq.(20), $s_i$ and $\sigma$ are center and the width of the Gaussian function respectively. The two nodes contain the summation of the neurons of the hidden layer, which are multiplied by a weight. The output layer is determined with the division of the two summation nodes. In this manner, the output of this architecture is defined by the next equation

$$h = \frac{\sum_{i=1}^{n} w_i f_i}{\sum_{i=1}^{n} f_i} \quad . \tag{22}$$

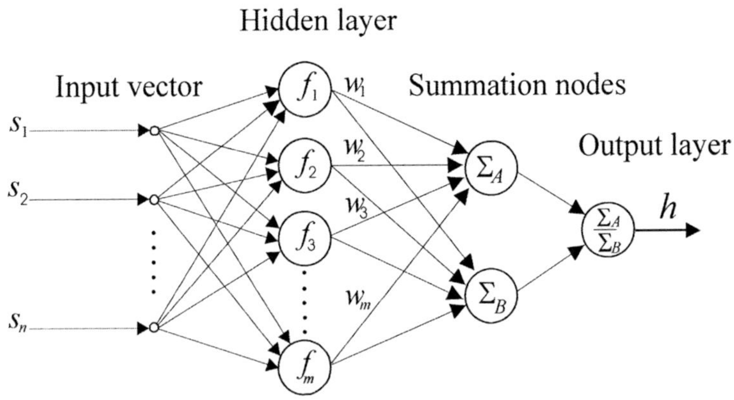


Figure 7. Structure of a GRNN.

To construct the GRNN via Eq.(22), neuron width, and weights should be determined. The width of the Gausian neuron is the smother parameter $\sigma$, which is calculated via Eq.(12). To determine the weights, also, the supervised learning is applied. It is because the outputs are known and the network should be forced to produce the correct outputs. The requirements for this training process are a suitable weight adjustment mechanism. To carry it out, the data of input vector $s_1$, $s_2$, $s_3$, .....,$s_n$ are connected to hidden layer to produce the outputs $h_1$, $h_2$, $h_3$,.....,$h_n$. The output of this GRNN is defined by the division of two summations nodes. Therefore, for each input $s_i$ an output is generated as a linear equation system:

$$h_1 = \frac{w_1}{k_1}\exp\left(-\frac{\|s_1 - s_1\|^2}{2\sigma^2}\right) + \frac{w_2}{k_1}\exp\left(-\frac{\|s_1 - s_2\|^2}{2\sigma^2}\right) +,..........,+ \frac{w_m}{k_1}\exp\left(-\frac{\|s_1 - s_m\|^2}{2\sigma^2}\right)$$

$$h_2 = \frac{w_1}{k_2}\exp\left(-\frac{\|s_2 - s_1\|^2}{2\sigma^2}\right) + \frac{w_2}{k_2}\exp\left(-\frac{\|s_2 - s_2\|^2}{2\sigma^2}\right) +,..........,+ \frac{w_n}{k_2}\exp\left(-\frac{\|s_2 - s_n\|^2}{2\sigma^2}\right) \tag{23}$$

$$\vdots \quad \vdots \qquad \vdots \qquad \vdots$$

$$h_n = \frac{w_1}{k_n}\exp\left(-\frac{\|s_m - s_1\|^2}{2\sigma^2}\right) + \frac{w_2}{k_n}\exp\left(-\frac{\|s_m - s_2\|^2}{2\sigma^2}\right) +,..........,+ \frac{w_m}{k_n}\exp\left(-\frac{\|s_n - s_m\|^2}{2\sigma^2}\right)$$

where

$$k_1 = \exp\left(-\frac{\|s_1 - s_1\|^2}{2\sigma^2}\right) + \exp\left(-\frac{\|s_1 - s_2\|^2}{2\sigma^2}\right) +,\dots\dots\dots,+ \exp\left(-\frac{\|s_1 - s_m\|^2}{2\sigma^2}\right)$$

$$k_2 = \exp\left(-\frac{\|s_2 - s_1\|^2}{2\sigma^2}\right) + \exp\left(-\frac{\|s_2 - s_2\|^2}{2\sigma^2}\right) +,\dots\dots\dots,+ \exp\left(-\frac{\|s_2 - s_m\|^2}{2\sigma^2}\right)$$

$$\vdots \quad \vdots \qquad \vdots \qquad \vdots$$

$$k_n = \exp\left(-\frac{\|s_n - s_1\|^2}{2\sigma^2}\right) + \exp\left(-\frac{\|s_n - s_2\|^2}{2\sigma^2}\right) +,\dots\dots\dots,+ \exp\left(-\frac{\|s_n - s_m\|^2}{2\sigma^2}\right)$$

The linear system of Eq.(23) can be represented as

$$\begin{aligned} h_1 &= w_1 f_{1,1} + w_2 f_{1,2} + \dots\dots\dots + w_n f_{1,m} \\ h_2 &= w_1 f_{2,1} + w_2 f_{2,2} + \dots\dots\dots + w_n f_{2,m} \\ &\vdots \quad \vdots \quad \vdots \qquad \vdots \\ h_n &= w_1 f_{n,1} + w_2 f_{n,2} + \dots\dots\dots + w_m f_{n,m} \end{aligned} \tag{24}$$

This equation can be rewritten in matrix form as the product between the matrix of the input data and the matrix of the corresponding output values: **H**=***F*****W**. The linear system represented as matrix form is described as:

$$\begin{bmatrix} f_{1,1} & f_{1,2} & f_{1,3} & \cdots & f_{1,m} \\ f_{2,1} & f_{2,2} & f_{2,3} & \cdots & f_{2,m} \\ \vdots & \vdots & \vdots & & \vdots \\ f_{n,1} & f_{n,2} & f_{n,3} & \cdots & f_{n,m} \end{bmatrix} \begin{bmatrix} w_1 \\ w_2 \\ \vdots \\ w_m \end{bmatrix} = \begin{bmatrix} h_1 \\ h_2 \\ \vdots \\ h_n \end{bmatrix} \tag{25}$$

By solving linear system Eq.(25), the weights $w_i$ are obtained. This system is solved by Chelosky method. With this step the weights $w_1$, $w_2$, $w_3$,….,$w_m$ are calculated. In this manner, the GRNN has been constructed. The result of this GRNN is a continuous function, which describes the object dimension according to the light line displacement. In this manner, for a particular given $s_i$ an $h_i$ is calculated by the GRNN. To select the neuron shape for this GRNN, functions as cubic, inverse quadratic, inverse multi-quadratic and Gaussian were analyzed. Based on *MSE*, the function used as neuron that provides the best approximation respect to desired output is the Gaussian function. Therefore, the Gaussian function Eq.(21) is used to construct the neurons of the hidden layer of this GRNN. The number of neurons used in the hidden layer for this GRNN is determined again by the *MSE* criterion.

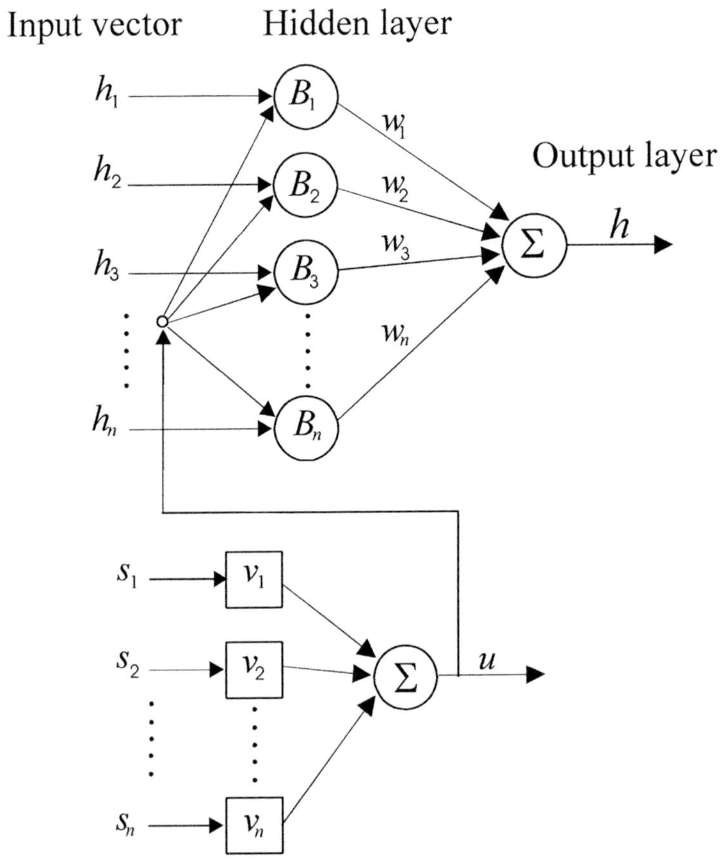


Figure 8. Structure of a BNN.

### 4.3 Structure of a Bezier Neural Network

The Bezier basis function has been used to represent curves and surfaces [30-33]. This is the main aim to propose a network of Bezier basis function. The structure proposed for this Bezier neural network (BNN) consists of an input vector, a parametric input, a hidden layer and an output layer. The structure of this BNN is shown in Fig. 8. The data of input vector $h_1$, $h_2$, $h_3$,...,$h_n$ are dimensions of the known objects. The parametric input is a value in the interval $0 \leq u \leq 1$, which is obtained by a procedure of weighted linear input combination (WLIC) [34]. In this procedure, the input data $s_1$, $s_2$, $s_3$,.....,$s_n$ are laser line displacement, which are used to perform the linear combination with the weights. The hidden layer is constructed by a set of neurons, whose shape is given by the Bezier basis function. The output layer is formed with the summation of the neurons of the hidden layer, which are multiplied by a weight. Each layer of the network structure is obtained in the next form. The input data $s_1$, $s_2$, $s_3$,....,$s_n$ are laser line displacement obtained by image processing, as is described in section 3. Using these displacements, the parametric $u$ value can be determined by means of the VLIC. In this manner, the relationship between the displacement and the parametric $u$ value is represented by

$$u = \sum_{i=0}^{n} v_i s_i \quad , \tag{26}$$

where $v_i$ are the weights, $s_i$ is the laser line displacement. Therefore, for each input $s_i$ an $u$ value is produced and an equation system is generated:

$$\begin{aligned} u_1 &= v_1 s_1 + v_2 s_2 + .........+ v_n s_n \\ u_2 &= v_1 s_1 + v_2 s_2 + .........+ v_n s_n \\ &\vdots \\ u_n &= v_1 s_1 + v_2 s_2 + .........+ v_n s_n \end{aligned} \tag{27}$$

By solving the linear system Eq.(27), the weights $v_i$ are determined. To carry it out, displacement and its corresponding $u$ value are substituted into Eq.(27). As the Bezier curves are defined in the interval $0 \le u \le 1$. Therefore, the value $u_i$ for the Eq.(27) is selected in this interval. In this manner, the weights $v_i$ are computed and Eq.(26) is completed. Therefore for a particular displacement $s_i$, a parametric $u$ valued is obtained by mean of Eq.(26). The other input data $h_1$, $h_2$, $h_3$,...,$h_n$ are obtained measuring objects with a contact method. The neurons of the hidden layer are defined by the Bezier basis function. The next expression describes a Bezier basis function

$$B_i(u) = \binom{n}{i} u^i (1-u)^{n-i} \quad , \tag{28}$$

and $\binom{n}{i} = \frac{n!}{i!(n-i)!}$, denotes the binomial distribution from probability and statistics. The output layer is obtained by the summation of the neurons, which are multiplied by a weight. The output response is the object dimension $h(u)$, which is given by

$$h(u) = \sum_{i=0}^{n} w_i B_i(u) h_i \quad , 0 \le u \le 1, \tag{29}$$

where $w_i$ are the weights, $h_i$ is the known object dimension and $B_i$ is the Bezier basis function represented by Eq.(28). To construct the complete structure of the BNN, the suitable weights $w_i$ should be determined. For the weights $w_1$, $w_2$, $w_3$,.....,$w_n$, the network is being forced to produce the correct outputs $h_1$, $h_2$, $h_3$,.....,$h_n$. To determine the weights $w_i$ an adjustment mechanism is applied. To carry it out, the values $u_i$ and $h_i$ are substituted in Eq.(29). Then, for each one of the values $u_i$, an output $h_i$ is obtained. In this manner, the next equation system is obtained

$$h_0 = w_1 \binom{n}{0} (1 - u)^n u^0 h_0 + w_2 \binom{n}{1} (1 - u)^{n-1} uxh_1 +,\ldots\ldots,+ w_n \binom{n}{n} (1 - u)^0 u^n h_n, 0 \le u \le 1.$$

$$h_1 = w_1 \binom{n}{0} (1 - u)^n u^0 h_0 + w_2 \binom{n}{1} (1 - u)^{n-1} uxh_1 +,\ldots\ldots,+ w_n \binom{n}{n} (1 - u)^0 u^n h_n, 0 \le u \le 1.$$

$$h_2 = w_1 \binom{n}{0} (1 - u)^n u^0 h_0 + w_2 \binom{n}{1} (1 - u)^{n-1} uxh_1 +,\ldots\ldots,+ w_n \binom{n}{n} (1 - u)^0 u^n h_n, 0 \le u \le 1. \quad (30)$$

$$\vdots \quad \vdots \quad \vdots \qquad\qquad \vdots \quad \vdots$$

$$h_n = w_1 \binom{n}{0} (1 - u)^n u^0 h_0 + w_2 \binom{n}{1} (1 - u)^{n-1} uxh_1 +,\ldots\ldots,+ w_n \binom{n}{n} (1 - u)^0 u^n h_n, 0 \le u \le 1.$$

This linear system of Eq.(30) can be represented as

$$\begin{aligned} h_1 &= w_1\beta_{1,1} + w_2\beta_{1,2} + \ldots\ldots\ldots + w_n\beta_{1,n} \\ h_2 &= w_1\beta_{2,1} + w_2\beta_{2,2} + \ldots\ldots\ldots + w_n\beta_{2,n} \\ &\vdots \\ h_n &= w_1\beta_{n,1} + w_2\beta_{n,2} + \ldots\ldots\ldots + w_n\beta_{n,n} \end{aligned} \quad (31)$$

This equation can be rewritten in matrix form as the product between the matrix of the input data and the matrix of the corresponding output values: **H**=$\boldsymbol{\beta}$**W**. The linear system represented by a matrix is described as:

$$\begin{bmatrix} \beta_{1,1} & \beta_{1,2} & \beta_{1,3} & \cdots & \beta_{1,n} \\ \beta_{2,1} & \beta_{2,2} & \beta_{2,3} & \cdots & \beta_{2,n} \\ \vdots & \vdots & \vdots & & \vdots \\ \beta_{n,1} & \beta_{n,2} & \beta_{n,3} & \cdots & \beta_{n,n} \end{bmatrix} \begin{bmatrix} w_1 \\ w_2 \\ \vdots \\ w_n \end{bmatrix} = \begin{bmatrix} h_1 \\ h_2 \\ \vdots \\ h_n \end{bmatrix} \quad (32)$$

By solving the linear system Eq.(32), the weights $w_i$ are obtained. This system is solved by the Chelosky method. This method uses a Lower and Upper (***LU***) decomposition to transform the matrix system. In ***LU*** decomposition a symmetric and positive definided matrix can be efficiently decomposed into a lower and upper triangular matrix. In this ***LU*** decomposition $\boldsymbol{\beta}$ = ***LU***. By using the lower and upper triangular matrix, the linear system Eq.(32) is transformed into ***LU*W**=**H**. Then, the triangular system is solved as ***L*Y**=**H** and ***U*W**=**Y**. In this manner, **Y** is calculated by using ***L*** and ***H***. Then ***W*** is solved using ***Y*** and ***U***. With this step the weights $w_1$, $w_2$, $w_3$,....,$w_n$ are calculated. In this manner, the BNN has been completed. In this manner, for a particular given $s_i$ an $h_i$ is calculated by the BNN. The number of neurons used in the hidden layer for this BNN is determined by the *MSE* criterion.

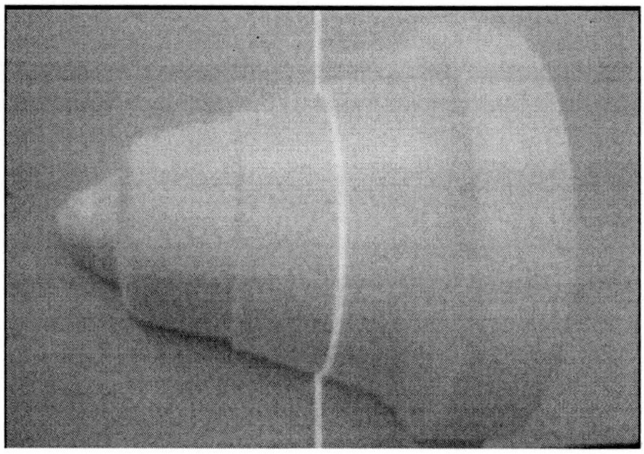

Figure 9. (a) A dummy bell to be profiled.

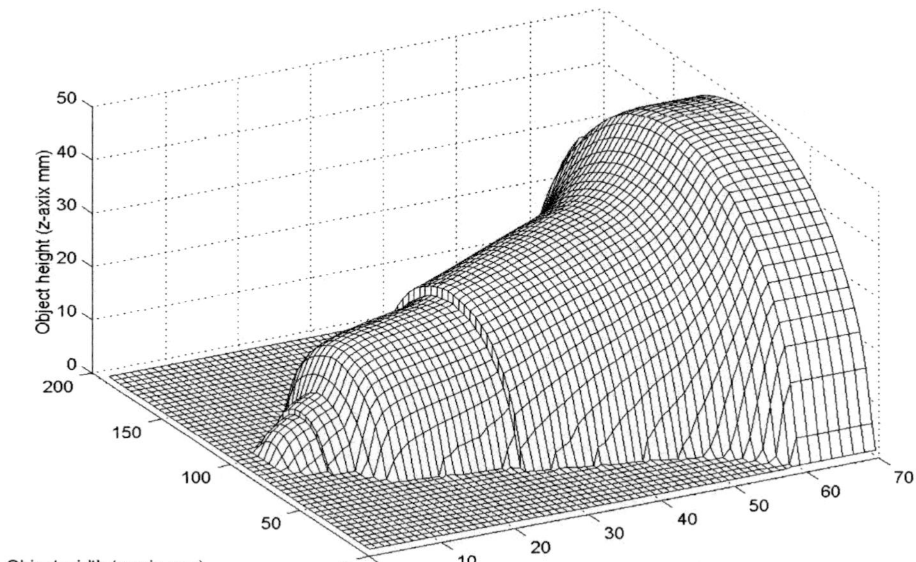


Figure 9. (b) Shape of dummy bell.

## 5. Object Shape Detection Applying Neural Networks

Fig. 1 shows the experimental set-up for shape detection. In this arrangement the object is moved in steps of 1.27 mm along $x$-axis. A vertical laser line is projected on the surface target by a 10 mW laser diode. In the scanning process, the laser line is deformed in the image plane according to the object topography. Each one of these deformed laser lines are captured by a

CCD camera and digitized by a frame grabber. The images have a resolution of 320 x 200 pixels and 256 grey levels. Each captured laser line is processed to detect the displacement $s(x,y)$ in the image plane. To extract the displacement, every image is processed by the Gaussian approximation method, which is described in section 3. The object dimension $h$ is recovered substituting the displacement $s(x,y)$ into approximation network. From each processed image by the network, a transverse section of the object is extracted. All transverse sections are stored in array memory to obtain the complete object shape. The approach of the approximation neural networks in this technique is based on good adjustment to a set data. In this manner, the calibration to obtain object dimension is achieved in good manner. The experimental results of this technique are performed with three objects. The first object to be profiled is the surface of a dummy bell, shown in Fig. 9(a). This object is scanned by means of the laser line to obtain surface information. By applying Gaussian approximation method Eq.(7) and Eq.(1), the displacement $s(x,y)$ is measured. The result of this step is a set of data $s_1, s_2, \ldots, s_n$, corresponding to the laser line displacement in the image plane. Then, these data are processed by the approximation networks to determine the object dimension. It is performed by applying the Eq.(13), Eq.(22) and Eq.(29). In this manner, the object dimension is obtained by each one of the approximation networks. Processing an image of a laser line by means of an approximation network, a transverse section of the object is obtained. The obtained height data from each transverse section are stored in array memory to construct the complete object shape. To know the accuracy of the result obtained by each neural network, a *root mean squared error* (*rms*) [35] is calculated. It is performed using data calculated by the networks and data obtained by coordinates measuring machine (CMM). The *rms* is described as

$$rms = \sqrt{\frac{1}{n}\sum_{i=1}^{n} (ho_i - hc_i)^2} \quad , \tag{33}$$

where $h_{oi}$ is the data obtained by the CMM, $h_{ci}$ is the calculated data by the each network and $n$ is the number of analysed data. The *rms* value obtained for the dummy bell using the RBFNN is $rms = 0.183$ *mm*. The *rms* value obtained using the GRNN is $rms = 0.213$ *mm*. By using the BNN the result is $rms = 0.161$ *mm*. Seventy two lines are processed to obtain the complete shape of the dummy bell. The topography obtained by BNN is shown in Fig. 9(b). In this figure, the scale of the axis is in *mm*. The second object profiled is a pyramid, which is shown in Fig. 10(a). This object is scanned by the laser line. The images captures in this step are processed to obtain the displacement. Then, displacement $s_i$ is processed by the RBFNN, GRNN and BNN to obtain its corresponding height data $h$. In this manner, height data of the transverse sections of the pyramid are obtained. These transverse sections are stored in an array memory to construct the complete pyramid shape. The *rms* value obtained for the pyramid using the RBFNN is $rms = 0.178$ *mm*. The *rms* value obtained using the GRNN is $rms = 0.204$ *mm*. By using the BNN the result is $rms = 0.156$ *mm*. Sixty four lines are processed to obtain the complete pyramid shape. The pyramid shape calculated by BNN is shown in Fig. 10(b). In this figure, the scale of the axis is in *mm*. The third object profiled is a dummy face shown in Fig. 11(a). The shape of this object is obtained by processing one hundred thirty two laser lines captured in the scanning step. The *rms* value is calculated for various characteristic points, which are measured by the CMM. The result of the error obtained for the dumy face using the RBFNN is *rms* = 0.283 *mm*. The *rms* value obtained using the GRNN is *rms* = 0.321 *mm*. By using the BNN the result is *rms* = 0.186 *mm*. The dummy face shape obtained by BNN is shown in Fig.11(b). In this figure, the scale of the axis is in *mm*. According to the result obtained for these three objects, BNN produces the best accuracy respect to the result produced by the MMC. The employed computer in this process is a PC to 333 MHz. Each light line image is processed in 0.031 sec. This time processing is given because the data from the image is extracted whit a few operations via Eq.(7). By using neural networks in this technique, the object shape is determined by image processing only. The set-up parameters as projection angle, distances of the camera and laser line are not necessary. Therefore, the procedure is easier than those techniques that use distances of the components of optical set-up. In this manner, this technique is performed by computer processing and measurements on optical set-up are avoided. Therefore, a good repeatibility is achieved in each measurement.

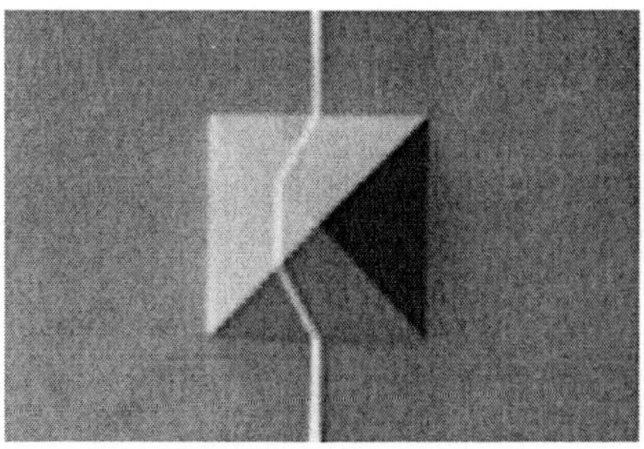

Figure 10. (a) Pyramid used to extract its shape.

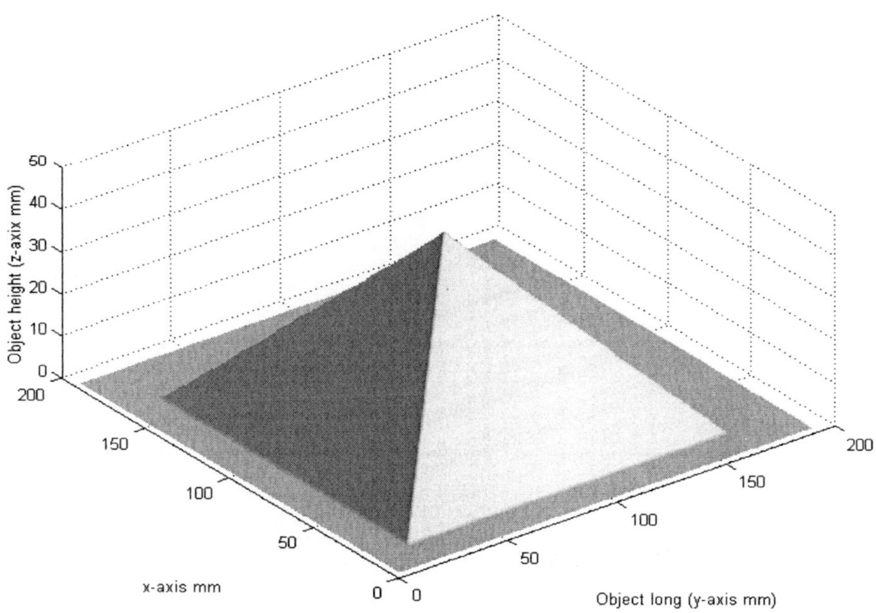


Figure 10. (b) Shape of the pyramid.

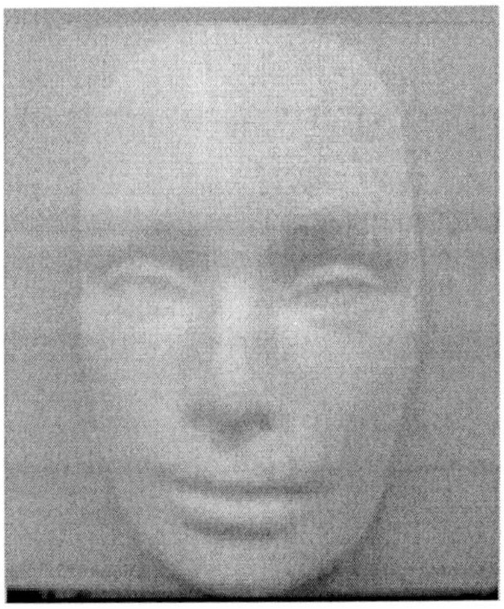

Figure 11. (a) Image of a dummy face.

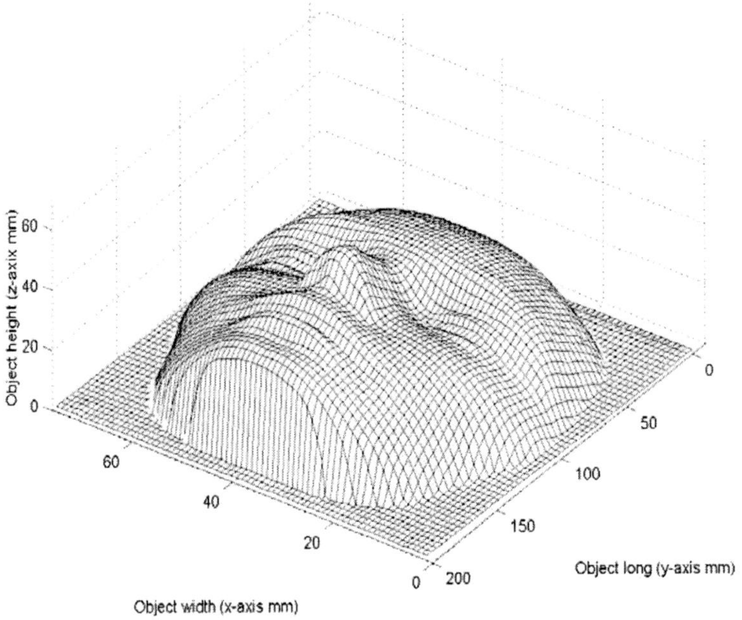


Figure 11. (b) Shape of the dummy face.

# CONCLUSION

A review of the shape detection technique based on a laser line and approximation networks has been presented. The described technique here provides a valuable tool for shape detection in industrial inspection. In this work the ability to obtain the object dimension has been achieved by applying RBFNN, GRNN and BNN. The results of these three methods have been examined in accuracy and the architecture. The described results here, shows that the BNN produces the best accuracy respect to the contact method. Also, the BNN has an easier architecture than RBFNN, and GRNN. The technique described in this study avoids the measurements of the geometry of the experimental set-up, as is common in the laser line projection methods. It improves the accuracy of measurement, because errors are not introduced in the computer system. By using this experimental set-up a good repeatability is achieved in every measurement, therefore this technique is performed in good manner. The future work consist in obtain a geometric model o a surface by means of a neural network.

## ACKNOWLEDGMENTS

J. Apolinar Muñoz Rodríguez would like to thank to CONCYTEG of Guanajuato and CONACYT of Mexico, for the partial support of this research.

## REFERENCES

[1] D. Vaughnn, C. Watkins and D. Anderson, (2001), "Rapid Confocal Sensor: a non contact profilometer for fast 3D submicron inspection and metrology of large formats," *Proc. SPIE* Vol. 4449, 169-177.
[2] G. Sansoni, R. Rodella, M. Carocci and V. Carbone, (2000), "Machine vision: Optical digitization of free-form complex surface using the projection of structured light," *Optics and Photonics News*, 23-29.
[3] G. Sansoni, M. Carocci, and R. Rodella, (1999), "3D vision based on the combination of gray code and phase shift light projection: analysis and compensation of the systematic errors," *Appl. Opt.* Vol. 38 No.31, 6565-6573.
[4] O. Gray, B. Liagre, A. Poupinet and L. Pujol, (2004), "Laser metrology for distance and displacement measurement: application to large optical systems," *Proc. SPIE* Vol. 5494, 418-425.
[5] R. Klette, K. Schluns and A. Koschan, (1997), *Computer Vision: Three-dimensional data from images*, U.S.A. Springer.
[6] X. X. Cheng, X.Y. Su, and L.R. Guo, (1991), "Automated measurement method for 360 profilometry of 3-D diffuse objects," *Appl. Optics* Vol. 30 No.10, 1274-1278.
[7] L. P. Thomas, R. Gratton, B. M. Marino, and J. M. Simon, ( 1994 ), "Measurements of free surface profiles in transient flow by a simple light slicing method, " *Appl. Optics* Vol. 33 No. 13, 2455-2458.
[8] W. Ch. Tai and M. Chang, (1996), "Non contact profilometric measurement of large form parts," *Opt. Eng.* Vol. 35 No. 9, 2730-2735.
[9] M. Baba, T. Konishi and N. Kobayashi, (1998), "A novel fast rangefinder with non-mechanical operation," *Journal of Optics.* Vol. 29 No. 3, 241-249.
[10] S. R. Marokkey, Ch. J. Tay, H. M. Shang and A. K. Asundi, (1997), "Time Delay an integration imaging for inspection and profilometry of moving objects," *Opt. Eng. Vol.* 36 No.9, 2573-2578.
[11] M. R. Sajan, C. J. Tay, H.M. Shang, and A. Asundi, (1998), "Improved spatial phase detection for profilometry using a TDI imager," *Optics Comm.* Vol. 150 No. 1-6, 66-70.

[12] Asundi and W. Zhou, (1999), "Mapping algorithm for 360-deg profilometry with time delay integration imaging," *Opt. Eng.* Vol. 38 No. 2, 339-344.

[13] J. A. Muñoz-Rodriguez, R. Rodriguez-Vera, and M. Servin, (2000), "Direct object shape detection base on skeleton extraction of a light line," *Opt. Eng.* Vol. 39 No. 9, 2463-2471.

[14] W. D. Herzog, M.S. Unlu, B.B. Goldberg, and G.H. Rhodes, (1997) "Beam divergence and waist measurement of laser diodes by near field scanning optical microscopy," *Appl. Phys. Lett.* Vol. 70 No.6, 688-690.

[15] J. A. Muñoz Rodríguez and R. Rodríguez-Vera, (2003), "Evaluation of the light line displacement location for object shape detection ," *Journal of Modern Optics*, Vol. 50 No. 1 pag. 137-154.

[16] C. F. Gerald and P. O. Wheatley, (1992), *Applied numerical analysis*, U.S.A. Addison Wesley.

[17] H. Frederick and G. J. Lieberman, (1982), *Introduction to operations research*, U.S.A. McGraw- Hill.

[18] J. S. Milton and J. C. Arnold, (2003) *Introduction to probability and statistics: principles and applications for engineering and computing sciences*, U.S.A. Mc Graw Hill, Fourth edition.

[19] Peter C. Gasson, (1989), *Geometry of spatial forms*, U.S.A. John Wiley and sons.

[20] M. E. Mortenson, (1997), *Geometric Modeling*, U.S.A.Willey, Second edition.

[21] M. Neifeld and D. Psaltis, (1993), "Optical implementations of radial basis classifiers," *Appl. Optics* Vol. 32 No.8, 1370-1379.

[22] E. W. Foor and A. M. Neifeld, (1995), "Adaptive optical radial basis function neural network for handwritten digit recognition," *Appl. Optics* Vol. 34 No.32, 7545-7555.

[23] T. Sabisch, A. Ferguson, H. Bolouri, (2000) "Identification of complex shapes using a self organizing neural system," *IEEE Transactions on Neural Networks*, Vol. 11 Issue: 4, 921 -934.

[24] Rojas, H. Pomares, J. Gonzalez, J. L.Bernier, E. Ros, F. J. Pelayo, A.Prieto, (2000), "Analysis of the functional block involved in the design of Radial basic Functions," *Neural Processing Letters*, Vol.12 No.1, 1-17.

[25] J.A. Muñoz Rodríguez, R. Rodríguez-Vera and A. Asundi, (2003), "Shape detection of moving objects based on a neural network of a light line". *Optics communication* Vol. 221, 73-86.

[26] P. Picton, (1998), *Neural networks*, Polgrave, U.S.A.

[27] Leung, H.; Dubash, N.; Xie, N, (2002)"Detection of small objects in clutter using a GA-RBF neural network," *IEEE Transactions on Aerospace and Electronic Systems*, Volume: 38 Issue: 1, 98 -118.

[28] W.H. Press, B. P. Flannery, S. A. Teukolsky, W. T. Vetterling, (1993), *Numerical Recipes in C*, U.S.A. Cambridge Press.

[29] Specht, D. F. (1991), "A general regression neural network," *IEEE Trans. Neural Networks*, vol. 2, 568-576.

[30] L. Shao and H. Zhou, (1996), "Curve Fitting with Bezier cubics", *Graphical models and image processing* Vol. 58 No.3, 223-232.

[31] Zhang , (1999), "C- Bezier Curves and Surface", *Graphical models and Image processing* Vol. 61, 2-15.

[32] Mohamed, N. Horoshi and A. Takeshi, (1993) "Bezier and polynomial methods of making curves," *SPIE* Vol. 2644, 290-295.

[33] A. Muñoz Rodríguez, Miguel Rosales Ciseña and R. Rodriguez-Vera, (2005), "3D shape detection based on a Bezier neural network of a light line," *SPIE Proc.* Vol. 5776, 630-639.

[34] Robert J. Schalkoff, (1997), *Artificial Neural Networks*, U.S.A. Mc Graw Hill.

[35] T. Masters, (1993), *Practical Neural Networks Recipes in C++*, U.S.A. Academic Press.

# INDEX

## D

## E



## F



## G

## M



## N



## O

## P



## Q



## R

## T



## U



## V



## W